Hyers–Ulam–Rassias Stability of Functional Equations in Nonlinear Analysis

For further volumes:
http://www.springer.com/series/7393

Springer Optimization and Its Applications

VOLUME 48

Aims and Scope
Optimization has been expanding in all directions at an astonishing rate during the last few decades. New algorithmic and theoretical techniques have been developed, the diffusion into other disciplines has proceeded at a rapid pace, and our knowledge of all aspects of the field has grown even more profound. At the same time, one of the most striking trends in optimization is the constantly increasing emphasis on the interdisciplinary nature of the field. Optimization has been a basic tool in all areas of applied mathematics, engineering, medicine, economics and other sciences.

The series *Springer Optimization and Its Applications* publishes undergraduate and graduate textbooks, monographs and state-of-the-art expository works that focus on algorithms for solving optimization problems and also study applications involving such problems. Some of the topics covered include nonlinear optimization (convex and nonconvex), network flow problems, stochastic optimization, optimal control, discrete optimization, multi-objective programming, description of software packages, approximation techniques and heuristic approaches.

Soon-Mo Jung

Hyers–Ulam–Rassias Stability of Functional Equations in Nonlinear Analysis

 Springer

Soon-Mo Jung
Mathematics Section
College of Science and Technology
Hongik University
339-701 Jochiwon
Korea, Republic of (South Korea)
smjung@hongik.ac.kr

ISSN 1931-6828
ISBN 978-1-4614-2862-6   ISBN 978-1-4419-9637-4 (eBook)
DOI 10.1007/978-1-4419-9637-4
Springer New York Dordrecht Heidelberg London

Mathematics Subject Classification (2010): 39B82, 39B62, 39B72

Springer is part of Springer Science+Business Media (www.springer.com)

To Themistocles M. Rassias
Creator of the Hyers–Ulam–Rassias Stability
of Functional Equations in Nonlinear
Analysis

Preface

An intriguing and famous talk presented by Stanislaw M. Ulam in 1940 triggered the study of stability problems for various functional equations. In his talk, Ulam discussed a number of important unsolved mathematical problems. Among them, a question concerning the stability of homomorphisms seemed too abstract for anyone to reach any conclusion. In the following year, Donald H. Hyers was able to give a partial solution to Ulam's question that was the first significant breakthrough and step toward more solutions in this area. Since then, a large number of papers have been published in connection with various generalizations of Ulam's problem and Hyers's theorem. In particular, Themistocles M. Rassias succeeded in extending the result of Hyers's theorem by weakening the condition for the Cauchy difference. This remarkable result of Rassias led the concern of mathematicians toward the study of stability problems of functional equations.

Unfortunately, no books dealing with a comprehensive illustration of the fast developing field of nonlinear analysis had been published for the mathematicians interested in this field for more than a half century until D. H. Hyers, G. Isac and Th. M. Rassias published their book, *Stability of Functional Equations in Several Variables*, Birkhäuser, 1998.

This very book will complement the books of Hyers, Isac and Rassias and of Czerwik (*Functional Equations and Inequalities in Several Variables*, World Scientific, 2002) by presenting mainly the results applying to the Hyers–Ulam–Rassias stability. Many mathematicians have extensively investigated the subjects on the Hyers–Ulam–Rassias stability. This book covers and offers almost all classical results on the Hyers–Ulam–Rassias stability in an integrated and self-contained fashion.

- In Chapter 2, we discuss the Hyers–Ulam–Rassias stability problems and the related topics of the additive Cauchy equation. In Section 2.1, we explain the behaviors of additive functions. We, then, begin to discuss the Hyers–Ulam and the Hyers–Ulam–Rassias stability problems of additive functional equation in Sections 2.2 and 2.3. The stability on restricted domains and its applications are introduced in Section 2.4. We explain briefly the method of invariant means and the fixed point method in Sections 2.5 and 2.6. In Section 2.7, the composite functional congruences are surveyed. The stability of the Pexider equation will be proved in Section 2.8.

- The Hyers–Ulam–Rassias stability of some generalized additive functional equations is proved in Chapter 3. Moreover, we discuss the Hyers–Ulam stability problem in connection with a question of Th. M. Rassias and J. Tabor.
- Chapter 4 deals with the Hosszú's functional equation. In Section 4.1, we prove that the Hosszú's equation is stable in the sense of C. Borelli. The Hyers–Ulam stability problem is discussed in Section 4.2. In Section 4.3, we present that the generalized Hosszú's equation is stable in the sense of Borelli. In the next section, we prove that the Hosszú's equation is not stable on the unit interval. Moreover, the Hyers–Ulam stability of the Hosszú's equation of Pexider type is proved in Section 4.5.
- We survey the stability problems of the homogeneous functional equation in Chapter 5. In Section 5.1, we prove the Hyers–Ulam–Rassias stability of the homogeneous functional equation between real Banach algebras. Section 5.2 deals with the superstability on restricted domains. The stability problem of the equation between vector spaces will be discussed in Section 5.3. Moreover, we present the Hyers–Ulam–Rassias stability of the homogeneous equation of Pexider type in Section 5.4.
- There are a number of functional equations including all the linear functions as their solutions. In Chapter 6, we introduce a few functional equations among them. We survey the superstability property of the system of functional equations $f(x + y) = f(x) + f(y)$ and $f(cx) = cf(x)$ in Section 6.1. Section 6.2 deals with the stability problem for the functional equation $f(x+cy) = f(x)+cf(y)$. In Section 6.3, we discuss stability problems of other systems which describe linear functions.
- Jensen's functional equation is the most important equation among a number of variations of the additive Cauchy equation. The Hyers–Ulam–Rassias stability problems of Jensen's equation are proved in Section 7.1, and the Hyers–Ulam stability on restricted domains is discussed in Section 7.2. In Section 7.3, we prove the stability of Jensen's equation by using the fixed point method. The superstability and Ger type stability of the Lobačevskiĭ functional equation will be surveyed in Section 7.4.
- Chapter 8 is dedicated to a survey on the stability problems for the quadratic functional equations. We prove the Hyers–Ulam–Rassias stability of the quadratic equation in Section 8.1. The stability problems on restricted domains are discussed in Section 8.2. Moreover, we prove the Hyers–Ulam–Rassias stability by using the fixed point method in Section 8.3. Section 8.4 deals with the Hyers–Ulam stability of another quadratic functional equation. We prove the stability of the quadratic equation of Pexider type in Section 8.5.
- In Chapter 9, we discuss the stability problems for the exponential functional equations. In Section 9.1, the superstability of the exponential Cauchy equation is proved. Section 9.2 deals with the stability of the exponential equation in the sense of R. Ger. Stability problems on restricted domains are discussed in Section 9.3. Another exponential functional equation $f(xy) = f(x)^y$ is introduced in Section 9.4.

- Chapter 10 deals with the stability problems for the multiplicative functional equations. In Section 10.1, we discuss the superstability of the multiplicative Cauchy equation and a functional equation connected with the Reynolds operator. The results on δ-multiplicative functionals on complex Banach algebras are presented in Section 10.2. We describe δ-multiplicative functionals in connection with the AMNM algebras in Section 10.3. Another multiplicative functional equation $f(x^y) = f(x)^y$ is discussed in Section 10.4. In Section 10.5, we prove that a new multiplicative functional equation $f(x + y) = f(x)f(y)$ $f(1/x + 1/y)$ is stable in the sense of Ger.
- In Chapter 11, we introduce a new functional equation $f(x^y) = yf(x)$ with the logarithmic property. Moreover, the functional equation of Heuvers $f(x + y) = f(x) + f(y) + f(1/x + 1/y)$ will be discussed.
- The addition and subtraction rules for trigonometric functions can be represented by using functional equations. Some of these equations are introduced and the stability problems are surveyed in Chapter 12. Sections 12.1 and 12.2 deal with the superstability phenomena of the cosine and the sine equations. In Section 12.3, some trigonometric functional equations with two unknown functions are discussed. In Section 12.4, we deal with the Hyers–Ulam stability of the Butler-Rassias functional equation following M. Th. Rassias's solution.
- Chapter 13 deals with the Hyers–Ulam–Rassias stability of isometries. The historical background for Hyers–Ulam stability of isometries is introduced in Section 13.1. The Hyers–Ulam–Rassias stability of isometries on restricted domains is proved in Section 13.2. Section 13.3 is dedicated to the fixed point method for studying the stability problem of isometries. In Section 13.4, we discuss the Hyers–Ulam–Rassias stability of the Wigner equation on restricted domains.
- Section 14.1 deals with the superstability of the associativity equation. In Section 14.2, the Hyers–Ulam stability of a functional equation defining multiplicative derivations is proved for functions on $(0, 1]$. In Section 14.3, the Hyers–Ulam–Rassias stability of the gamma functional equation and a generalized beta functional equation is proved. The Hyers–Ulam stability of the Fibonacci functional equation will be proved in Section 14.4.

I would like to express my humble gratitude to Professor Themistocles M. Rassias for his valuable suggestions and comments. I am also very much indebted to my wife, Min-Soon Lee, as she has always stood by me and helped me type the manuscript in LaTeX system.

This research was supported by the Basic Science Research Program through the National Research Foundation of Korea (NRF) funded by the Ministry of Education, Science and Technology (No. 2010-0007143).

Finally, I would also like to acknowledge the fine cooperation and assistance that Ms. Elizabeth Loew of Springer Publishing has provided in the publication of this book.

December 2010   *Soon-Mo Jung*

Contents

Chapter 1
Introduction

In the fall of 1940, S. M. Ulam gave a wide-ranging talk before a Mathematical Colloquium at the University of Wisconsin in which he discussed a number of important unsolved problems. Among those was the following question concerning the stability of homomorphisms (cf. [354]):

Let G_1 be a group and let G_2 be a metric group with a metric $d(\cdot, \cdot)$. Given $\varepsilon > 0$, does there exist a $\delta > 0$ such that if a function $h : G_1 \to G_2$ satisfies the inequality $d(h(xy), h(x)h(y)) < \delta$ for all $x, y \in G_1$, then there is a homomorphism $H : G_1 \to G_2$ with $d(h(x), H(x)) < \varepsilon$ for all $x \in G_1$?

If the answer is affirmative, we say that the functional equation for homomorphisms is *stable*.

D. H. Hyers was the first mathematician to present the result concerning the stability of functional equations. He brilliantly answered the *question of Ulam* for the case where G_1 and G_2 are assumed to be Banach spaces (see [135]). This result of Hyers is stated as follows (cf. Theorem 2.3):

Theorem 1.1 (Hyers). *Let $f : E_1 \to E_2$ be a function between Banach spaces such that*

$$\| f(x + y) - f(x) - f(y) \| \le \delta \tag{1.1}$$

for some $\delta > 0$ and for all $x, y \in E_1$. Then the limit

$$A(x) = \lim_{n \to \infty} 2^{-n} f(2^n x) \tag{1.2}$$

exists for each $x \in E_1$, and $A : E_1 \to E_2$ is the unique additive function such that

$$\| f(x) - A(x) \| \le \delta$$

for every $x \in E_1$. Moreover, if $f(tx)$ is continuous in t for each fixed $x \in E_1$, then the function A is linear.

Taking this famous result into consideration, the additive Cauchy equation $f(x + y) = f(x) + f(y)$ is said to have the *Hyers–Ulam stability* on (E_1, E_2) if for every function $f : E_1 \to E_2$ satisfying the inequality (1.1) for some $\delta \ge 0$ and for all $x, y \in E_1$, there exists an additive function $A : E_1 \to E_2$ such that $f - A$ is bounded on E_1.

S.-M. Jung, *Hyers–Ulam–Rassias Stability of Functional Equations in Nonlinear Analysis*, Springer Optimization and Its Applications 48, DOI 10.1007/978-1-4419-9637-4_1, © Springer Science+Business Media, LLC 2011

The method in (1.2) provided by Hyers which produces the additive function A will be called a *direct method*. This method is the most important and powerful tool to study the stability of various functional equations.

Ten years after the publication of Hyers's theorem, D. G. Bourgin extended the *theorem of Hyers* and stated it in his paper [28] without proof. Unfortunately, it seems that this result of Bourgin failed to receive attention from mathematicians at that time. No one has made use of this result for a long time.

In 1978, Th. M. Rassias addressed the Hyers's stability theorem and attempted to weaken the condition for the bound of the norm of *Cauchy difference*

$$f(x + y) - f(x) - f(y)$$

and proved a considerably generalized result of Hyers by making use of a direct method (cf. Theorem 2.5):

Theorem 1.2 (Rassias). *Let* $f : E_1 \to E_2$ *be a function between Banach spaces. If* f *satisfies the functional inequality*

$$\|f(x + y) - f(x) - f(y)\| \leq \theta \big(\|x\|^p + \|y\|^p \big)$$

for some $\theta \geq 0$, p *with* $0 \leq p < 1$ *and for all* $x, y \in E_1$, *then there exists a unique additive function* $A : E_1 \to E_2$ *such that*

$$\|f(x) - A(x)\| \leq \frac{2\theta}{2 - 2^p} \|x\|^p \tag{1.3}$$

for each $x \in E_1$. *If, in addition,* $f(tx)$ *is continuous in* t *for each fixed* $x \in E_1$, *then the function* A *is linear.*

This exciting result of Rassias attracted a number of mathematicians who began to be stimulated to investigate the stability problems of functional equations.

By regarding a large influence of S. M. Ulam, D. H. Hyers, and Th. M. Rassias on the study of stability problems of functional equations, the stability phenomenon proved by Th. M. Rassias is called the *Hyers–Ulam–Rassias stability*. In this book, the Hyers–Ulam stability will be regarded as a special case of the Hyers–Ulam–Rassias stability.

For the last thirty years many results concerning the Hyers–Ulam–Rassias stability of various functional equations have been obtained, and a number of definitions of stability have been introduced. Hence, it is necessary to introduce the exact definition of the Hyers–Ulam–Rassias stability which is applicable to all functional equations appearing in this book.

Let E_1 and E_2 be some appropriate spaces. For some $p, q \in \mathbb{N}$ and for any $i \in \{1, \ldots, p\}$ let

$$g_i : E_1^q \to E_1 \quad \text{and} \quad G : E_2^p \times E_1^q \to E_2$$

be functions. Assume that φ, $\Phi : E_1^q \to [0, \infty)$ are functions satisfying some given conditions. If for every function $f : E_1 \to E_2$ satisfying the inequality

$$\left\| G\big(f(g_1(x_1, \ldots, x_q)), \ldots, f(g_p(x_1, \ldots, x_q)), x_1, \ldots, x_q\big) \right\|$$
$$\leq \varphi(x_1, \ldots, x_q) \tag{1.4}$$

for all $x_1, \ldots, x_q \in E_1$ there exists a function $H : E_1 \to E_2$ such that

$$G\big(H(g_1(x_1, \ldots, x_q)), \ldots, H(g_p(x_1, \ldots, x_q)), x_1, \ldots, x_q\big) = 0$$

for all $x_1, \ldots, x_q \in E_1$ and

$$\|f(x) - H(x)\| \leq \Phi(x, \ldots, x) \tag{1.5}$$

for any $x \in E_1$, then we say that the functional equation

$$G\big(f(g_1(x_1, \ldots, x_q)), \ldots, f(g_p(x_1, \ldots, x_q)), x_1, \ldots, x_q\big) = 0 \tag{1.6}$$

has *Hyers–Ulam–Rassias stability* on (E_1, E_2) or we say that it is *stable in the sense of Hyers, Ulam, and Rassias.*

If $\varphi(x_1, \ldots, x_q)$ in (1.4) and $\Phi(x, \ldots, x)$ in (1.5) are replaced by δ and $K\delta$ ($K > 0$), respectively, then we say that the corresponding phenomenon of the functional equation (1.6) is the *Hyers–Ulam stability* on (E_1, E_2).

If each solution $f : E_1 \to E_2$ of the inequality (1.4) is either a solution of the equation (1.6) or satisfies some strong regular conditions, then we say that the equation (1.6) is *superstable* on (E_1, E_2). J. Baker, J. Lawrence, and F. Zorzitto discovered the *superstability* phenomenon for the first time. Indeed, they assumed that V is a vector space over rational numbers $\mathbb{Q}$ and proved that if a function $f : V \to \mathbb{R}$ satisfies $|f(x + y) - f(x)f(y)| \leq \delta$ for a given $\delta > 0$ and for all $x, y \in V$, then either $f(x)$ remains bounded or f is an exponential function (ref. [17]). If there is no confusion, we can omit (E_1, E_2) in the terminologies.

R. Ger pointed out that the superstability phenomenon of the exponential equation is caused by the fact that the natural group structure in the range space is disregarded, and he suggested a new type of stability for the exponential equation (ref. [122]):

$$\left| \frac{f(x + y)}{f(x)f(y)} - 1 \right| \leq \delta. \tag{1.7}$$

If for each function $f : (G, +) \to E \setminus \{0\}$ satisfying the inequality (1.7) for some $\delta > 0$ and for all $x, y \in G$, there exists an exponential function $M : G \to E \setminus \{0\}$ such that

$$\|f(x)/M(x) - 1\| \leq \Phi(\delta) \quad \text{and} \quad \|M(x)/f(x) - 1\| \leq \Psi(\delta)$$

for all $x \in G$, where $\Phi(\delta)$ and $\Psi(\delta)$ depend on δ only, then the exponential functional equation is said to be *stable in the sense of Ger*.

We will use the following notations. $\mathbb{C}, \mathbb{N}, \mathbb{Q}, \mathbb{R}$, and $\mathbb{Z}$ denote the set of complex numbers, of positive integers, of rational numbers, of real numbers, and of integers, respectively. Furthermore, let $\mathbb{N}_0 = \mathbb{N} \cup \{0\}$.

Now, let us review the contents of this book briefly.

In Chapter 2, we discuss the most famous functional equation, namely, the *additive Cauchy functional equation*

$$f(x + y) = f(x) + f(y)$$

of which properties have been widely applied to almost every field of science and engineering.

In Section 2.1, the behaviors of the continuous or discontinuous *additive functions* will be briefly described.

In Section 2.2, the very historically important theorem of Hyers will again be presented.

Section 2.3 will entirely be devoted to the Hyers–Ulam–Rassias stability of the *additive functional equation* for functions between Banach spaces. The *theorem of Rassias* (see above or [285]) which is also very important from a historical point of view will be described in Theorem 2.5. Furthermore, a counterexample of Z. Gajda [112], stating that the theorem of Rassias is no more valid if $p = 1$, will be introduced in Theorem 2.6.

In relation to Rassias's theorem, Th. M. Rassias asked in his paper [290] whether the inequality (1.3) provides the best possible estimate of the difference $\| f(x) - A(x) \|$ between an *"approximately additive function"* f and the *additive function* A constructed by making use of a direct method. This question was affirmatively answered by Th. M. Rassias and J. Tabor in the case of $p = 1/2$ (ref. [313]). In the case of $p > 0$ ($p \neq 1$), the answer is also affirmative as was demonstrated by J. Brzdęk (ref. [30]). The result of Brzdęk will be formulated in Theorem 2.10.

In the next part of Section 2.3, more generalized results on the Hyers–Ulam–Rassias stability of the additive Cauchy equation will further be introduced (see, e.g., [113, 118, 119, 142–144, 156, 280–282, 302, 311]). These results will be applied to the study of some important problems in nonlinear analysis; for example, the existence of fixed points on cones for nonlinear functions, the study of eigenvalues for a couple of nonlinear operators, and the study of bifurcations to the infinity, with respect to a convex cone, of solutions of the *Hammerstein equation* (see [144]).

In Section 2.4, the *Hyers–Ulam stability* of the additive Cauchy equation *on a restricted domain* will be proved. F. Skof was the first person to address the *stability on a bounded domain*. Indeed, Skof obtained the following result for $N = 1$, and Z. Kominek extended it for any $N \in \mathbb{N}$ (cf. [224, 330] or Lemma 2.28):

Let E be a Banach space. Given $c > 0$, let a function $f : [0, c)^N \to E$ satisfy the functional inequality

$$\| f(x + y) - f(x) - f(y) \| \le \delta$$

for all $x, y \in [0, c)^N$ *with* $x + y \in [0, c)^N$. *Then there exists an additive function* $A : \mathbb{R}^N \to E$ *such that*

$$\|f(x) - A(x)\| \leq (4N - 1)\delta$$

for $x \in [0, c)^N$.

A theorem of L. Losonczi [238] concerning the stability on a restricted domain will also be introduced in Theorem 2.31. This result will be reserved for further use in the study of the Hyers–Ulam stability of *Hosszú's equation* (see Theorem 4.5).

In contrast to the previous one, F. Skof [331] also proved the Hyers–Ulam stability of the additive Cauchy equation on an unbounded and restricted domain. She applied this result to the study of an interesting *asymptotic behavior of additive functions* (Theorems 2.32 and 2.34 and Corollary 2.35):

The function $f : \mathbb{R} \to \mathbb{R}$ *is additive if and only if* $f(x + y) - f(x) - f(y) \to 0$ *as* $|x| + |y| \to \infty$.

The stability result of the additive Cauchy equation on a restricted domain presented by D. H. Hyers, G. Isac, and Th. M. Rassias will be discussed in Theorem 2.36, and its applications to *p-asymptotical derivatives* will also be presented (ref. [136]).

Section 2.5 will address the *method of invariant means*. Until now, it has been assumed that the domains of functions involved are vector spaces. In this section, the theorem of Hyers will be generalized by extending the domain spaces of involved functions to abelian semigroups. The results of G. L. Forti and Z. Gajda will be described (cf. [105, 111, 316, 342, 343]).

Recently, L. Cădariu and V. Radu proved the Hyers–Ulam–Rassias stability of the additive Cauchy equation by using the fixed point method (see [57, 279]). Many mathematicians try to prove the Hyers–Ulam–Rassias stability of various functional equations by applying the fixed point method. This method appears to be powerful and successful. In Section 2.6, a theorem of Cădariu and Radu will be introduced.

In Section 2.7, a theorem of R. Ger and P. Šemrl concerning the *composite functional congruence* will be demonstrated (Corollary 2.48 or [123]):

Let $(G, +)$ *be a cancellative abelian semigroup, and let* $\varepsilon \in (0, 1/4)$. *If a function* $f : G \to \mathbb{R}$ *satisfies the congruence*

$$f(x + y) - f(x) - f(y) \in \mathbb{Z} + (-\varepsilon, \varepsilon)$$

for all $x, y \in G$, *then there exists a function* $p : G \to \mathbb{R}$ *such that*

$$p(x + y) - p(x) - p(y) \in \mathbb{Z} \quad \text{and} \quad |f(x) - p(x)| \leq \varepsilon$$

for any $x, y \in G$.

This result will be applied to the proof of stability of the exponential functional equation in the sense of Ger (see Theorem 9.7).

The Hyers–Ulam–Rassias stability of the Pexider equation, $f(x + y) = g(x) + h(y)$, will be surveyed in the last section of Chapter 2. A theorem proved by K.-W. Jun, D.-S. Shin, and B.-D. Kim is introduced (see Theorem 2.49 or [155]):

Let G and E be an abelian group and a Banach space, respectively. Let φ : $G^2 \to [0, \infty)$ be a function satisfying

$$\Phi(x) = \sum_{i=1}^{\infty} 2^{-i} \left(\varphi(2^{i-1}x, 0) + \varphi(0, 2^{i-1}x) + \varphi(2^{i-1}x, 2^{i-1}x) \right) < \infty$$

and

$$\lim_{n\to\infty} 2^{-n} \varphi(2^n x, 2^n y) = 0$$

for all $x, y \in G$. If functions $f, g, h : G \to E$ satisfy the inequality

$$\|f(x + y) - g(x) - h(y)\| \le \varphi(x, y)$$

for all $x, y \in G$, then there exists a unique additive function $A : G \to E$ such that

$$\|f(x) - A(x)\| \le \|g(0)\| + \|h(0)\| + \Phi(x),$$

$$\|g(x) - A(x)\| \le \|g(0)\| + 2\|h(0)\| + \varphi(x, 0) + \Phi(x),$$

$$\|h(x) - A(x)\| \le 2\|g(0)\| + \|h(0)\| + \varphi(0, x) + \Phi(x)$$

for all $x \in G$.

In Chapter 3, the Hyers–Ulam–Rassias stability of some *generalized additive Cauchy equations* will be discussed. In the paper [312], Th. M. Rassias and J. Tabor asked whether the functional equation

$$f(ax + by + c) = Af(x) + Bf(y) + C$$

with $abAB \ne 0$ is stable in the sense of Hyers, Ulam, and Rassias.

Section 3.1 will deal with the results of C. Badea [8] concerning the Hyers–Ulam–Rassias stability of the functional equation

$$f(ax + by) = af(x) + bf(y).$$

These results of Badea provide a partial answer to the *question of Rassias and Tabor*.

In Section 3.2, in the paper [162] S.-M. Jung introduced another generalized additive Cauchy equation

$$f\left(x_0 + \sum_{j=1}^{m} a_j x_j\right) = \sum_{i=1}^{m} b_i f\left(\sum_{j=1}^{m} a_{ij} x_j\right). \tag{1.8}$$

In the case of $x_0 = 0$ in (1.8), the Hyers–Ulam–Rassias stability of (1.8) will be proved in Theorem 3.6. This result will be applied to a partial answer to the question of Rassias and Tabor.

In Section 3.3, the Hyers–Ulam stability phenomenon of a new functional equation

$$f(x + y)^2 = (f(x) + f(y))^2$$

introduced by S.-M. Jung [157] will be presented.

In Chapter 4, a special form of generalized additive Cauchy equations, the presumed *Hosszú's functional equation*

$$f(x + y - xy) = f(x) + f(y) - f(xy),$$

will be discussed.

C. Borelli was the first person to deal with the stability problem of this equation. He proved in his paper [23] that the *Hosszú's equation* is stable in the sense of Hyers, Ulam, and Borelli (cf. Theorem 4.4). The result of Borelli will be introduced in Section 4.1.

The result of Borelli was a predecessor to the proof of the Hyers–Ulam stability of the Hosszú's equation. In fact, L. Losonczi proved the Hyers–Ulam stability of the Hosszú's functional equation (Theorem 4.5 or [238]):

Let E be a Banach space and suppose that a function $f : \mathbb{R} \to E$ satisfies the inequality

$$|f(x + y - xy) - f(x) - f(y) + f(xy)| \leq \delta$$

for some $\delta \geq 0$ and for all $x, y \in \mathbb{R}$. Then there exists a unique additive function $A : \mathbb{R} \to E$ and a unique constant $b \in E$ such that

$$\|f(x) - A(x) - b\| \leq 20\delta$$

for any $x \in \mathbb{R}$.

In Section 4.3, a generalized form of the Hosszú's functional equation

$$f(x + y + qxy) = f(x) + f(y) + qf(xy),$$

where q is a fixed rational number, will be discussed. In Theorem 4.10, the *Hyers–Ulam–Borelli stability* of the functional equation mentioned above will be proved under the assumption that $q \notin \{-1/2, 0, 1/2\}$ (cf. [191]).

We will see that the Hosszú's equation is not stable on the unit interval. J. Tabor proved this surprising theorem in the paper [348] and will be addressed in Section 4.4.

In Section 4.5, we will survey the Hyers–Ulam stability of the Hosszú's functional equation of Pexider type.

In Chapter 5, the Hyers–Ulam–Rassias stability of the *homogeneous functional equation*

$$f(yx) = y^k f(x)$$

of degree k will be proved.

Section 5.1 will be devoted to the Hyers–Ulam–Rassias stability property of the homogeneous equation for the functions between Banach algebras. Notably, J. Tabor and J. Tabor proved that the homogeneous equation for the functions between Banach algebras is superstable if some weak conditions are satisfied (Theorem 5.3 or [161]).

In Section 5.2, a few results of S.-M. Jung [166] will be introduced, i.e., the superstability property of the homogeneous equation will be applied to the study of the superstability phenomenon of the same equation on a restricted domain, and this result for a restricted domain will be applied to the proof of an asymptotic behavior of the homogeneous functions.

In Section 5.3, the stability results of the homogeneous equation for the functions between vector spaces will be presented. The superstability results of J. Tabor [347] will be formulated in Theorem 5.9. It is interesting to observe that the bound condition for the norm of the difference $f(cx) - cf(x)$ determines the type of stability as we may see in Theorems 5.9 and 5.11. Indeed, S. Czerwik [88] proved the Hyers–Ulam–Rassias stability of the homogeneous equation (Theorem 5.11), while J. Tabor [347] observed the superstability phenomenon for the same equation under different conditions from those of Czerwik.

In the last section of Chapter 5, the Hyers–Ulam–Rassias stability of the homogeneous functional equation of Pexider type

$$f(\alpha x) = \psi(\alpha)g(x)$$

will be discussed, where f and g are unknown functions and ψ is a given function (Theorem 5.15 or [91]).

There are a number of (systems of) functional equations which include all the *linear functions* as their solutions.

In Chapter 6, only a few (systems of) functional equations among them will be introduced. In Section 6.1, the superstability property of the "intuitive" system

$$\begin{cases} f(x + y) = f(x) + f(y), \\ f(cx) = cf(x) \end{cases} \tag{1.9}$$

obtained by J. Tabor will be presented (ref. [347]).

J. Schwaiger introduced the functional equation

$$f(x + cy) = f(x) + cf(y)$$

which is equivalent to the system (1.9) if the related domain and range are vector spaces (ref. [325]). He proved the stability of the given equation, and these related results will be found in Section 6.2.

As indicated above, Rassias's theorem cannot be extended for $p = 1$ (cf. Theorem 2.6). Such a counterexample has stimulated many mathematicians in an attempt to surpass such awkwardness. For example, B. E. Johnson worked for this purpose. In Section 6.3, a theorem of Johnson will be formulated in Theorem 6.7

(ref. [147]). P. Šemrl simplified the functional inequalities appearing in the theorem of Johnson and proved the stability result (cf. Theorem 6.9 or [326]):

Let a continuous function $f : \mathbb{R} \to \mathbb{R}$ satisfy the functional inequalities

$$|f(x_1 + \cdots + x_n) - f(x_1) - \cdots - f(x_n)| \le \delta(|x_1| + \cdots + |x_n|)$$

for some $\delta > 0$ and for all $n \in \mathbb{N}$, $x_1, \ldots, x_n \in \mathbb{R}$. Then there exists a linear function $L : \mathbb{R} \to \mathbb{R}$ such that

$$|f(x) - L(x)| \le \delta|x|$$

for all $x \in \mathbb{R}$.

The simplest and most elegant variation of the additive Cauchy equation is the *Jensen's functional equation*

$$2f\left(\frac{x+y}{2}\right) = f(x) + f(y)$$

whose Hyers–Ulam–Rassias stability properties will be proved in Chapter 7.

In Section 7.1, the Hyers–Ulam–Rassias stability result of S.-M. Jung will be presented (see Theorem 7.1 or [163]):

Let E_1 and E_2 be a real normed space and a real Banach space, respectively. Assume that δ, $\theta \ge 0$ are fixed, and let $p > 0$ be given with $p \ne 1$. Suppose a function $f : E_1 \to E_2$ satisfies the functional inequality

$$\left\| 2f\left(\frac{x+y}{2}\right) - f(x) - f(y) \right\| \le \delta + \theta(\|x\|^p + \|y\|^p) \qquad (1.10)$$

for all $x, y \in E_1$. Furthermore, assume $f(0) = 0$ and $\delta = 0$ in (1.10) for the case of $p > 1$. Then there exists a unique additive function $A : E_1 \to E_2$ such that

$$\|f(x) - A(x)\| \le \begin{cases} \delta + \|f(0)\| + (2^{1-p} - 1)^{-1}\theta\|x\|^p & \text{(for } 0 < p < 1), \\ 2^{p-1}(2^{p-1} - 1)^{-1}\theta\|x\|^p & \text{(for } p > 1) \end{cases}$$

for all $x \in E_1$.

Similarly, as in the case of the additive Cauchy equation, the Jensen's functional equation is not stable if $p = 1$ and $\delta = 0$ are assumed in the inequality (1.10). Jung proved that an example which was constructed by Th. M. Rassias and P. Šemrl [310] as a counterexample for the case of $p = 1$ in the theorem of Rassias also serves as a counterexample for the Jensen's equation (see Theorem 7.3 or [163]).

In Section 7.2, the stability result of Z. Kominek for a bounded domain will be introduced in Theorem 7.5 (ref. [224]). Another stability result of the Jensen's equation on an unbounded and restricted domain was obtained by S.-M. Jung, which will be formulated in Theorem 7.7. By using this result, Jung was able to prove an asymptotic property of the additive functions which may be regarded as a modification of Skof's result mentioned above (see Corollary 7.8 or [163]).

A new method for proving the stability will be introduced in Section 7.3. By applying the fixed point method (Theorem 2.43), L. Cădariu and V. Radu [55] proved the Hyers–Ulam–Rassias stability of the Jensen's functional equation (see Theorem 7.9):

Let E_1 be a (real or complex) vector space and let E_2 be a Banach space. Assume that a function $f : E_1 \to E_2$ satisfies $f(0) = 0$ and the inequality

$$\left\| 2f\left(\frac{x+y}{2}\right) - f(x) - f(y) \right\| \le \varphi(x, y)$$

for all $x, y \in E_1$, where $\varphi : E_1^2 \to [0, \infty)$ is a given function. Moreover, assume that there exists a constant $0 < L < 1$ such that

$$\varphi(x, 0) \le L q_i \varphi\left(q_i^{-1} x, 0\right)$$

for any $x \in E_1$, where $q_0 = 2$ and $q_1 = 1/2$. If φ satisfies the condition

$$\lim_{n \to \infty} q_i^{-n} \varphi(q_i^n x, q_i^n y) = 0$$

for all $x, y \in E_1$, then there exists a unique additive function $A : E_1 \to E_2$ such that

$$\| f(x) - A(x) \| \le \frac{L^{1-i}}{1 - L} \varphi(x, 0)$$

for any $x \in E_1$.

A theorem of P. Găvruta will be introduced in Section 7.4 which concerns the superstability of Lobačevskiĭ's functional equation

$$f\left(\frac{x+y}{2}\right)^2 = f(x) f(y)$$

(see Theorem 7.10 or [114]). Moreover, the Ger type stability of the Lobačevskiĭ's equation is also introduced (see [170]).

In Chapter 8, the Hyers–Ulam–Rassias stability of the *quadratic functional equations* will be discussed. The "original" quadratic functional equation

$$f(x + y) + f(x - y) = 2f(x) + 2f(y)$$

will be presented in Section 8.1.

F. Skof was the first person who proved the Hyers–Ulam stability of the quadratic equation for the functions $f : E_1 \to E_2$, where E_1 is a normed space and E_2 is a Banach space (ref. [331]). P. W. Cholewa demonstrated that the theorem of Skof is also valid if E_1 is replaced by an abelian group (ref. [70]). S. Czerwik finally proved the Hyers–Ulam–Rassias stability of the quadratic equation (see Theorems 8.3 and 8.4 or [87]).

Similarly, as in the cases of the additive Cauchy equation and the Jensen's equation, Czerwik presented a counterexample concerning the special case when $p = 2$ is assumed in Theorem 8.3 or 8.4.

In Section 8.2, the important results concerning the Hyers–Ulam stability of the quadratic functional equation on restricted domains will be introduced.

The Hyers–Ulam–Rassias stability of the quadratic functional equation will be again proved in Section 8.3 by using the fixed point method. In Theorems 8.12 and 8.14, some results of S.-M. Jung, T.-S. Kim, and K.-S. Lee will be presented.

Section 8.4 is devoted to a quadratic functional equation different from the "original" quadratic functional equation:

$$f(x + y + z) + f(x) + f(y) + f(z) = f(x + y) + f(y + z) + f(z + x).$$

In Theorems 8.16 and 8.17, the Hyers–Ulam stability of the above equation will be proved under some suitable conditions.

The stability problem of the quadratic functional equation of Pexider type is treated in Section 8.5. In Theorem 8.18 and Corollary 8.19, we will introduce some results presented by S.-M. Jung (see [173]).

Chapter 9 will discuss the stability problems of the *exponential functional equations*. The exponent law of the exponential functions is intuitively symbolized by the exponential functional equation

$$f(x + y) = f(x)f(y).$$

This equation reveals a different stability behavior from those of other functional equations. Indeed, J. Baker, J. Lawrence, and F. Zorzitto proved the superstability of the exponential equation (ref. [17]):

If a real-valued function defined on a real vector space satisfies the functional inequality

$$|f(x + y) - f(x)f(y)| \leq \delta$$

for some $\delta > 0$ and for all x and y, then f is either bounded or an exponential function.

This theorem was the first result concerning the superstability phenomenon of functional equations. Later, J. Baker, L. Székelyhidi, and S.-M. Jung generalized this result (ref. [16, 164, 341]). These results will be described in Section 9.1.

Section 9.2 will address the subject on the stability in the sense of Ger. As mentioned above, R. Ger first demonstrated that the superstability phenomenon of the exponential equation is caused by the fact that the natural group structure in the range space is disregarded and he suggested the stability problem in the form (1.7) (ref. [122]).

R. Ger and P. Šemrl proved the stability (in the sense of Ger) of the exponential equation (cf. Theorem 9.7 or [123]):

Let $(G, +)$ be a cancellative abelian semigroup, and let $\delta \in [0, 1)$ be given. If a function $f : G \to \mathbb{C} \setminus \{0\}$ satisfies the inequality (1.7) for all $x, y \in G$, then there exists a unique exponential function $M : G \to \mathbb{C} \setminus \{0\}$ such that

$$\max \left\{ |f(x)/M(x) - 1|, \ |M(x)/f(x) - 1| \right\}$$

$$\leq \left(1 + (1 - \delta)^{-2} - 2((1 + \delta)/(1 - \delta))^{1/2} \right)^{1/2}$$

for any $x \in G$.

Section 9.3 will deal with the stability problems of the exponential equation on a restricted domain. S.-M. Jung presented the superstability phenomenon of the exponential equation for functions on an unbounded and restricted domain under a weak condition and applied this result to the proof of an asymptotic property of the exponential functions (see Theorem 9.8, Corollary 9.9 or [180]). Furthermore, he also proved a theorem which may be regarded as a version of the theorem of Ger for a restricted domain (Theorem 9.10).

In Section 9.4, we will introduce a new type of functional equation

$$f(xy) = f(x)^y$$

with the exponential property because every exponential function $f(x) = a^x$ ($a > 0$; $x \in \mathbb{C}$) satisfies this equation. The results of Jung [165] on the stability in the sense of Ger of this equation will be presented.

In Chapter 10, stability problems of the *multiplicative functional equations* will be discussed. First, we will consider the "original" multiplicative functional equation

$$f(xy) = f(x)f(y).$$

Let E be a commutative complex Banach algebra. A linear functional ϕ on E is called δ-*multiplicative* if

$$|\phi(xy) - \phi(x)\phi(y)| \leq \delta \|x\| \|y\|$$

for all $x, y \in E$. Each nonzero continuous linear functional on E is called a *character* of E if it is multiplicative. Let us denote by $\hat{E}$ the set of characters of E. For every continuous linear functional ϕ, let us define

$$d(\phi) = \inf \left\{ \|\phi - \psi\| \mid \psi \in \hat{E} \cup \{0\} \right\}.$$

We state that E is an *algebra in which approximately multiplicative functionals are near multiplicative functionals*, or E is *AMNM* for short, if for any $\varepsilon > 0$ there exists $\delta > 0$ such that $d(\phi) < \varepsilon$ whenever ϕ is a δ-multiplicative linear functional.

A theory of AMNM algebras contributed by B. E. Johnson [146] will be demonstrated in Section 10.3.

A new type of multiplicative functional equation, namely, $f(x^y) = f(x)^y$, will be discussed in Section 10.4. In the paper [171] (cf. [300]) S.-M. Jung dealt with the stability problems of this equation in the sense of Ger, and these results will be presented.

In the last section of Chapter 10, a new multiplicative functional equation,

$$f(x + y) = f(x)f(y)f(x^{-1} + y^{-1}),$$

will be discussed. In particular, S.-M. Jung [175] proved that this functional equation is stable in the sense of Ger (see Theorem 10.23).

It is not difficult to demonstrate the Hyers–Ulam stability of the *logarithmic functional equation* $f(xy) = f(x) + f(y)$ for the functions from $(0, \infty)$ into a Banach space E. More precisely, if a function $f : (0, \infty) \to E$ satisfies the inequality $\|f(xy) - f(x) - f(y)\| \le \delta$ for some $\delta > 0$ and for all $x, y > 0$, then there exists a unique logarithmic function $L : (0, \infty) \to E$ such that $\|f(x) - L(x)\| \le \delta$ for any $x > 0$. Therefore, we will introduce, in Chapter 11, a new functional equation

$$f(x^y) = yf(x).$$

The above functional equation may be regarded as another logarithmic functional equation because each logarithmic function $f(x) = \log x$ $(x > 0)$ is a solution of this equation. S.-M. Jung obtained the superstability results of this equation in various settings in his paper [158], which will be presented in Theorems 11.2, 11.3, and 11.8.

Moreover, the functional equation of Heuvers, $f(x + y) = f(x) + f(y) + f(x^{-1} + y^{-1})$, will be introduced. The Hyers–Ulam stability of the equation of Heuvers will be proved in Theorem 11.10.

The famous addition or subtraction rules for trigonometric functions may be represented by using functional equations. Some of such equations will be introduced and stability problems for them will be surveyed in Chapter 12.

In Section 12.1, we will discuss the superstability phenomenon of the *cosine functional equation*

$$f(x + y) + f(x - y) = 2f(x)f(y) \tag{1.11}$$

which stands for an addition theorem of cosine function. This equation is sometimes called the d'Alembert equation.

J. Baker proved the superstability for this equation for the first time, and later P. Găvruta presented a short proof for the theorem (Theorem 12.2, [16] or [114]):

Let $\delta > 0$ be given and let $(G, +)$ be an abelian group. If a function $f : G \to \mathbb{C}$ satisfies the functional inequality

$$|f(x + y) + f(x - y) - 2f(x)f(y)| \le \delta$$

for all $x, y \in G$, then either $|f(x)| \le (1 + \sqrt{1 + 2\delta})/2$ for any $x \in G$ or f satisfies the cosine functional equation (1.11) for all $x, y \in G$.

Similarly, the superstability of the *sine functional equation*

$$f(x + y)f(x - y) = f(x)^2 - f(y)^2 \tag{1.12}$$

will be presented in Section 12.2. This equation may remind us of one of the trigonometric formulas:

$$\sin(x + y) \sin(x - y) = \sin^2 x - \sin^2 y.$$

P. W. Cholewa obtained the superstability phenomenon of the sine functional equation (1.12) (ref. Theorem 12.7 or [69]):

Let $(G, +)$ be an abelian group in which division by 2 is uniquely performable. If an unbounded function $f : G \to \mathbb{C}$ satisfies the inequality

$$\left| f(x + y) f(x - y) - f(x)^2 + f(y)^2 \right| \le \delta$$

for some $\delta > 0$ and for all $x, y \in G$, then f is a solution of the sine functional equation (1.12).

L. Székelyhidi introduced the following functional equations

$$f(xy) = f(x)f(y) - g(x)g(y) \tag{1.13}$$

and

$$f(xy) = f(x)g(y) + f(y)g(x) \tag{1.14}$$

for complex-valued functions defined on a semigroup $(G, \cdot)$ (ref. [345]). It is not difficult to demonstrate that the equations (1.13) and (1.14) represent addition theorems for cosine and sine.

In Section 12.3, the stability results of L. Székelyhidi in connection with the trigonometric functional equations (1.13) and (1.14) will be surveyed (Theorems 12.10, 12.13 or [345]). It is very interesting that these functional equations for complex-valued functions defined on an amenable group are not superstable, but they are stable in the sense of Hyers and Ulam, whereas the equations (1.11) and (1.12) are superstable.

In Section 12.4, we will deal with the Butler–Rassias functional equation

$$f(x + y) = f(x)f(y) + c \sin x \sin y.$$

M. Th. Rassias solved the equation in the class of real functions (Theorem 12.14):

Let $c < -1$ be a constant. The solution $f : \mathbb{R} \to \mathbb{R}$ of the Butler–Rassias functional equation is $f(x) = a \sin x + \cos x$ or $f(x) = -a \sin x + \cos x$, where $a = \sqrt{|c| - 1}$.

S.-M. Jung proved the Hyers–Ulam stability of the Butler–Rassias functional equation (see Theorem 12.17).

It is worthwhile to note that R. Ger considered the functional equations (1.13) and (1.14) simultaneously and proved that the system is not superstable, but that it is stable in the sense of Hyers, Ulam, and Rassias (ref. Theorem 12.18 or [121]).

An *isometry* is a *distance-preserving map* between metric spaces. For normed spaces E_1 and E_2, a function $f : E_1 \to E_2$ is called a *δ-isometry* if f changes distances at most δ, i.e.,

$$\big| \|f(x) - f(y)\| - \|x - y\| \big| \le \delta$$

for all $x, y \in E_1$.

In Section 13.1, we will introduce the Hyers–Ulam stability results of the isometries. By making use of a direct method, D. H. Hyers and S. M. Ulam proved that the surjective isometries of a complete Euclidean space are stable in the sense of Hyers and Ulam. (Theorem 13.4 or [139]).

Let E be a complete abstract Euclidean space. Assume that $f : E \to E$ is a surjective δ-isometry and $f(0) = 0$. Then there exists a surjective isometry $I : E \to E$ such that

$$\|f(x) - I(x)\| \le 10\delta$$

for all $x \in E$.

D. G. Bourgin [25], R. D. Bourgin [27], P. M. Gruber [128], and J. Gevirtz [124] continued the study of stability problems for isometries. We close Section 13.1 by introducing a theorem of M. Omladič and P. Šemrl in which they presented a sharp stability result for δ-isometries (see Theorem 13.8 or [259]):

Let E_1 and E_2 be real Banach spaces. If $f : E_1 \to E_2$ is a surjective δ-isometry and $f(0) = 0$, then there exists a unique surjective linear isometry $I : E_1 \to E_2$ such that

$$\|f(x) - I(x)\| \le 2\delta$$

for each $x \in E_1$.

The Hyers–Ulam–Rassias stability of isometries on a restricted domain will be surveyed in Section 13.2. Let us define

$$K_0(\delta) = K_1(\delta) = \delta, \quad K_2(\delta) = 3\sqrt{3\delta}, \quad K_i(\delta) = 27\delta^{2^{1-i}}$$

inductively for all integers $i \ge 3$ and for some $\delta > 0$. First, we introduce a theorem of J. W. Fickett (see Theorem 13.10 or [103]). Fickett constructed the related isometry by quite a different method from the (direct) method of Hyers and Ulam:

Let S be a bounded subset of $\mathbb{R}^n$ and let $f : S \to \mathbb{R}^n$ be a δ-isometry, where $0 \le K_n(\delta/\operatorname{diam} S) \le 1/3$. Then there exists an isometry $I : S \to \mathbb{R}^n$ such that

$$\|f(s) - I(s)\| \le K_{n+1}(\delta/\operatorname{diam} S) \cdot \operatorname{diam} S$$

for all $s \in S$.

Moreover, some results of S.-M. Jung concerning the Hyers–Ulam stability of isometries on a restricted domain will be presented in Theorems 13.11 and 13.13 (ref. [178]).

Section 13.3 will be devoted to the fixed point method for studying the stability problem of isometries. Recently, by applying the fixed point method, S.-M. Jung could present a short and simple proof for the Hyers–Ulam–Rassias stability of isometries of which domain is a normed space and range is a Banach space in which the parallelogram law holds true (see Theorems 13.15 and 13.17).

Let E_1 and E_2 be real or complex Hilbert spaces with the inner products and the associated norms denoted by $\langle \cdot, \cdot \rangle$ and $\| \cdot \|$, respectively. A function $f : E_1 \to E_2$ is a solution of the *orthogonality equation* $\langle f(x), f(y) \rangle = \langle x, y \rangle$ if and only if it is a linear isometry. Functions $f, g : E_1 \to E_2$ are called *phase-equivalent* if and only if there exists a function $\sigma : E_1 \to \mathcal{S}$ such that $g(x) = \sigma(x) f(x)$ for each $x \in E_1$, where we set $\mathcal{S} = \{ z \in \mathbb{K} \mid |z| = 1 \}$. A functional equation

$$| \langle f(x), f(y) \rangle | = | \langle x, y \rangle |$$

is called the *Wigner equation* (or the *generalized orthogonality equation*).

In the last section of Chapter 13, we will present a theorem of J. Chmieliński and S.-M. Jung concerning the Hyers–Ulam–Rassias stability of the Wigner equation on a restricted domain (see Theorem 13.20 or [66]):

If a function $f : E_1 \to E_2$ satisfies the inequality

$$\big| | \langle f(x), f(y) \rangle | - | \langle x, y \rangle | \big| \leq \varphi(x, y),$$

for all $x, y \in D$, with the function $\varphi : E_1^2 \to [0, \infty)$ satisfying the property

$$\lim_{m+n \to \infty} c^{m+n} \varphi(c^{-m} x, c^{-n} y) = 0,$$

for all $x, y \in D$, where

$$D = \begin{cases} \{ x \in E_1 \mid \|x\| \geq d \} \ (for \ 0 < c < 1), \\ \{ x \in E_1 \mid \|x\| \leq d \} \ (for \ c > 1) \end{cases}$$

for given constants $c > 0$ $(c \neq 1)$ and $d \geq 0$. Then there exists a function $I : E_1 \to E_2$ satisfying the Wigner equation and such that

$$\| f(x) - I(x) \| \leq \sqrt{\varphi(x, x)}$$

for all $x \in D$. The function I is unique up to a phase-equivalent function.

One of the simplest functional equations is the *associativity equation*

$$F\big(x, F(y, z)\big) = F\big(F(x, y), z\big)$$

representing the famous *associativity axiom* $x \cdot (y \cdot z) = (x \cdot y) \cdot z$ which plays an important role in definitions of algebraic structures. The superstability phenomenon of the associativity equation will be demonstrated in Section 14.1 (Theorem 14.1 or [4]).

In Section 14.2, the stability problems of an important functional equation

$$f(xy) = xf(y) + f(x)y$$

representing *multiplicative derivations* in algebras will be discussed. P. Šemrl [327] obtained the first result concerning the superstability of this equation for functions between operator algebras. In this section, a Hyers–Ulam stability result presented by J. Tabor as an answer to a question of G. Maksa will be formulated in Theorem 14.2 (ref. [239, 260, 349]).

The *gamma function* Γ is very useful to develop other functions which have physical applications. In Section 14.3, the Hyers–Ulam–Rassias stability of the *gamma functional equation*

$$f(x + 1) = xf(x)$$

and a generalized beta functional equation

$$f(x + p, y + q) = \psi(x, y)f(x, y)$$

will be proved (see Theorems 14.3 and 14.4 or [148, 160]).

The *Fibonacci sequence* is one of the most well-known number sequences. If F_n stands for the nth *Fibonacci number* for any $n \in \mathbb{N}$ and $F_0 = 0$, then the Fibonacci numbers satisfy the equation

$$F_n = F_{n-1} + F_{n-2} \quad \text{for all} \ \ n \geq 2$$

(ref. [225]). From this formula, we derive a functional equation

$$f(x) = f(x - 1) + f(x - 2),$$

which is called the *Fibonacci functional equation*. The Hyers–Ulam stability of the generalized Fibonacci functional equation

$$f(x) = pf(x - 1) - qf(x - 2),$$

where p and q are fixed real numbers with $q \neq 0$ and $p^2 - 4q \neq 0$, will be proved in the last section.

Besides, there are a large number of valuable results concerning the stability problems of various functional equations which cannot be presented in this book for lack of space (see, e.g., [198, 243, 299, 305]).

Chapter 2
Additive Cauchy Equation

The functional equation $f(x + y) = f(x) + f(y)$ is the most famous among the functional equations. Already in 1821, A. L. Cauchy solved it in the class of continuous real-valued functions. It is often called the *additive Cauchy functional equation* in honor of A. L. Cauchy. The properties of this functional equation are frequently applied to the development of theories of other functional equations. Moreover, the properties of the additive Cauchy equation are powerful tools in almost every field of natural and social sciences. In Section 2.1, the behaviors of solutions of the additive functional equation are described. The Hyers–Ulam stability problem of this equation is discussed in Section 2.2, and theorems concerning the Hyers–Ulam–Rassias stability of the equation are proved in Section 2.3. The stability on a restricted domain and its applications are introduced in Section 2.4. The method of invariant means and the fixed point method will be explained briefly in Sections 2.5 and 2.6. In Section 2.7, the composite functional congruences will be surveyed. The stability results for the Pexider equation will be treated in the last section.

2.1 Behavior of Additive Functions

The history of the study of functional equations is long. Already in 1821, A. L. Cauchy [60] noted that every continuous solution of the *additive Cauchy functional equation*

$$f(x + y) = f(x) + f(y), \qquad (2.1)$$

for all $x, y \in \mathbb{R}$, is linear. Every solution of the additive Cauchy equation (2.1) is called an *additive function*.

First, we will solve the additive Cauchy equation (2.1) under some weaker conditions than that of A. L. Cauchy (ref. [59]).

Theorem 2.1. *If an additive function* $f : \mathbb{R} \to \mathbb{R}$ *satisfies one of the following conditions, then there exists a real constant* c *such that* $f(x) = cx$ *for all* $x \in \mathbb{R}$:

(i) f *is continuous at a point;*
(ii) f *is monotonic on an interval of positive length;*

S.-M. Jung, *Hyers–Ulam–Rassias Stability of Functional Equations in Nonlinear Analysis*, Springer Optimization and Its Applications 48, DOI 10.1007/978-1-4419-9637-4_2, © Springer Science+Business Media, LLC 2011

(iii) f is bounded from above or below on an interval of positive length;
(iv) f is integrable;
(v) f is Lebesgue measurable;
(vi) f is a Borel function.

Proof. We prove the theorem under the condition (i) only. By induction on n we first prove

$$f(nx) = nf(x) \qquad\qquad (a)$$

for all $x \in \mathbb{R}$ and $n \in \mathbb{N}$. Let x in $\mathbb{R}$ be arbitrary. Obviously, (a) is true for $n = 1$. Assume now that (a) is true for some n. Then, by (a), we get

$$f((n + 1)x) = f(nx) + f(x) = nf(x) + f(x) = (n + 1)f(x).$$

If we substitute x/n for x in (a), we obtain

$$f(x/n) = (1/n)f(x). \qquad\qquad (b)$$

Following from (a) and (b) yields

$$f(qx) = qf(x) \qquad\qquad (c)$$

for any $x \in \mathbb{R}$ and for all $q \in \mathbb{Q}$.

Finally, by letting $x = 1$ in (c) and considering the condition (i), we have $f(x) = cx$ for any $x \in \mathbb{R}$, where $c = f(1)$.   □

As indicated in the previous theorem, if a solution of the additive Cauchy equation (2.1) additionally satisfies one of the very weak conditions $(i) - (v)$, then it has the linearity.

Every additive function which is not linear, however, displays a very strange behavior presented in the following theorem (ref. [2]):

Theorem 2.2. *The graph of every additive function $f : \mathbb{R} \to \mathbb{R}$ which is not of the form $f(x) = cx$, for all $x \in \mathbb{R}$, is dense in $\mathbb{R}^2$.*

Proof. The graph of f is the set

$$G = \{(x, y) \in \mathbb{R}^2 \mid y = f(x)\}.$$

Choose a real number $x_1 \neq 0$. Since f is not of the form $f(x) = cx$ for any real constant c, there exists a real number $x_2 \neq 0$ such that

$$f(x_1)/x_1 \neq f(x_2)/x_2.$$

Namely,

$$\begin{vmatrix} x_1 & f(x_1) \\ x_2 & f(x_2) \end{vmatrix} \neq 0.$$

This means that the vectors $\vec{p}_1 = (x_1, f(x_1))$ and $\vec{p}_2 = (x_2, f(x_2))$ are linearly independent and thus span the whole plane $\mathbb{R}^2$. Let $\vec{p}$ be an arbitrary plane vector. Then there exist rational numbers q_1 and q_2 such that $|\vec{p} - (q_1 \vec{p}_1 + q_2 \vec{p}_2)| \leq \varepsilon$ for any $\varepsilon > 0$, since $\mathbb{Q}^2$ is dense in $\mathbb{R}^2$. Now,

$$
\begin{aligned}
q_1 \vec{p}_1 + q_2 \vec{p}_2 &= q_1(x_1, f(x_1)) + q_2(x_2, f(x_2)) \\
&= (q_1 x_1 + q_2 x_2, q_1 f(x_1) + q_2 f(x_2)) \\
&= (q_1 x_1 + q_2 x_2, f(q_1 x_1 + q_2 x_2)).
\end{aligned}
$$

The last inequality follows from (c) in the proof of Theorem 2.1. Hence,

$$
G_{12} = \{(x, y) \in \mathbb{R}^2 \mid x = q_1 x_1 + q_2 x_2, \ y = f(x), \ q_1, q_2 \in \mathbb{Q}\}
$$

is dense in $\mathbb{R}^2$. From the fact $G_{12} \subset G$ we conclude that G is dense in $\mathbb{R}^2$ which completes the proof of our theorem. $\square$

We now give some results concerning the additive complex-valued functions defined on the complex plane:

If an additive function $f : \mathbb{C} \to \mathbb{C}$ is continuous, then there exist complex constants c_1 and c_2 with $f(z) = c_1 z + c_2 \bar{z}$ for all $z \in \mathbb{C}$, where $\bar{z}$ denotes the complex conjugate of z.

Unlike the case of real-valued additive functions on the reals, the complex-valued continuous additive functions on the complex plane are not linear. However, every complex-valued additive function is linear if it is analytic or differentiable.

2.2 Hyers–Ulam Stability

As stated in the introduction, S. M. Ulam [354] raised the following question concerning the stability of homomorphisms:

Let G_1 and G_2 be a group and a metric group with a metric $d(\cdot, \cdot)$, respectively. Given $\varepsilon > 0$, does there exist a $\delta > 0$ such that if a function $h : G_1 \to G_2$ satisfies the inequality $d(h(xy), h(x)h(y)) < \delta$ for all $x, y \in G_1$, then there exists a homomorphism $H : G_1 \to G_2$ with $d(h(x), H(x)) < \varepsilon$ for all $x \in G_1$?

D. H. Hyers presented the first result concerning the stability of functional equations. Indeed, he obtained a celebrated theorem while he was trying to answer the question of Ulam (ref. [135]).

Theorem 2.3 (Hyers). *Let $f : E_1 \to E_2$ be a function between Banach spaces such that*

$$
\| f(x + y) - f(x) - f(y) \| \leq \delta \tag{2.2}
$$

for some $\delta > 0$ and for all $x, y \in E_1$. Then the limit

$$A(x) = \lim_{n \to \infty} 2^{-n} f(2^n x) \qquad (2.3)$$

exists for each $x \in E_1$ and $A : E_1 \to E_2$ is the unique additive function such that

$$\|f(x) - A(x)\| \le \delta \qquad (2.4)$$

for any $x \in E_1$. Moreover, if $f(tx)$ is continuous in t for each fixed $x \in E_1$, then A is linear.

Proof. For any $x \in E_1$ the inequality $\|f(2x) - 2f(x)\| \le \delta$ is obvious from (2.2). Replacing x by $x/2$ in this inequality and dividing by 2 we get

$$\|(1/2)f(x) - f(x/2)\| \le (1/2)\delta$$

for any $x \in E_1$. Now, make the induction assumption

$$\left\|2^{-n} f(x) - f\left(2^{-n} x\right)\right\| \le \left(1 - 2^{-n}\right)\delta. \qquad (a)$$

It then follows from the last two inequalities that

$$\left\|(1/2)f(2^{-n}x) - f\left(2^{-n-1}x\right)\right\| \le (1/2)\delta$$

and

$$\left\|2^{-n-1} f(x) - (1/2)f(2^{-n}x)\right\| \le (1/2)(1 - 2^{-n})\delta.$$

Hence,

$$\left\|2^{-n-1} f(x) - f\left(2^{-n-1}x\right)\right\| \le \left(1 - 2^{-n-1}\right)\delta.$$

Therefore, the inequality (a) is true for all $x \in E_1$ and $n \in \mathbb{N}$.

Put $q_n(x) = 2^{-n} f(2^n x)$, where $n \in \mathbb{N}$ and $x \in E_1$. Then

$$q_m(x) - q_n(x) = 2^{-m} f(2^m x) - 2^{-n} f(2^n x)$$
$$= 2^{-m}\left(f(2^{m-n}2^n x) - 2^{m-n} f(2^n x)\right).$$

Therefore, if $m < n$, we can apply the inequality (a) to the last equality and we get

$$\|q_m(x) - q_n(x)\| \le (2^{-m} - 2^{-n})\delta$$

for all $x \in E_1$. Hence, the Hyers–Ulam sequence $\{q_n(x)\}$ is a Cauchy sequence for each x, and since E_2 is complete, there exists a limit function

$$A(x) = \lim_{n \to \infty} q_n(x).$$

Let x and y be any two points of E_1. It follows from (2.2) that

$$\|f(2^n x + 2^n y) - f(2^n x) - f(2^n y)\| \leq \delta$$

for any $n \in \mathbb{N}$. Dividing by 2^n and letting $n \to \infty$ we see that A is an additive function. If we replace x by $2^n x$ in (a) and take the limit, we have the inequality (2.4).

Suppose that $A' : E_1 \to E_2$ was another additive function satisfying (2.4) in place of A, and such that $A(y) \neq A'(y)$ for some $y \in E_1$. For any integer $n > 2\delta/\|A(y) - A'(y)\|$ we see that the inequality $\|A(ny) - A'(ny)\| > 2\delta$ holds. On the other hand, this inequality contradicts the inequalities

$$\|A(ny) - f(ny)\| \leq \delta \quad \text{and} \quad \|A'(ny) - f(ny)\| \leq \delta.$$

Hence, A is the unique additive function satisfying the inequality (2.4).

Assume that f is continuous at y. If A is not continuous at a point $x \in E_1$, then there exist an integer k and a sequence $\{x_n\}$ in E_1 converging to zero such that $\|A(x_n)\| > 1/k$ for any $n \in \mathbb{N}$. Let m be an integer greater than $3k\delta$. Then

$$\|A(mx_n + y) - A(y)\| = \|A(mx_n)\| > 3\delta.$$

On the other hand,

$$\begin{aligned}
\|A(mx_n + y) - A(y)\| &\leq \|A(mx_n + y) - f(mx_n + y)\| \\
&\quad + \|f(mx_n + y) - f(y)\| + \|f(y) - A(y)\| \\
&\leq 3\delta
\end{aligned}$$

for sufficiently large n, since $f(mx_n + y) \to f(y)$ as $n \to \infty$. This contradiction means that the continuity of f at a point in E_1 implies the continuity of A on E_1.

For a fixed $x \in E_1$, if $f(tx)$ is continuous in t, then it follows from the above consideration that $A(tx)$ is continuous in t, hence A is linear. $\square$

The Hyers–Ulam stability result of Theorem 2.3 remains valid if E_1 is an abelian semigroup (ref. [105]).

The following corollary has been proved in the proof of Theorem 2.3.

Corollary 2.4. *Under the hypotheses of Theorem 2.3, if f is continuous at a single point of E_1, then A is continuous everywhere in E_1.*

As we see, we can explicitly construct the unique additive function satisfying (2.4) by means of the method expressed in (2.3). D. H. Hyers was the first person to suggest this method known as a *direct method* because it allows us to construct the additive function A satisfying (2.4) directly from the given function f in Theorem 2.3. It is the most powerful tool to study the stability of several functional equations and will be frequently used to construct certain function which is a solution of a given functional equation.

2.3   Hyers–Ulam–Rassias Stability

After Hyers gave an affirmative answer to Ulam's question, a large number of papers
have been published in connection with various generalizations of Ulam's problem
and Hyers's theorem.

There is no reason for the *Cauchy difference* $f(x + y) - f(x) - f(y)$ to be
bounded as in the expression of (2.2). Toward this point, Th. M. Rassias tried to
weaken the condition for the Cauchy difference and succeeded in proving what is
now known to be the *Hyers–Ulam–Rassias stability* for the additive Cauchy equation (see [159, 298, 301, 306]). This terminology is justified because the theorem of
Th. M. Rassias (Theorem 2.5 below) has strongly influenced mathematicians studying stability problems of functional equations. In fact, Th. M. Rassias [285] proved
the following:

Theorem 2.5 (Rassias). *Let E_1 and E_2 be Banach spaces, and let $f : E_1 \to E_2$
be a function satisfying the functional inequality*

$$\| f(x + y) - f(x) - f(y) \| \le \theta \big(\|x\|^p + \|y\|^p \big) \tag{2.5}$$

*for some $\theta > 0$, $p \in [0, 1)$, and for all $x, y \in E_1$. Then there exists a unique
additive function $A : E_1 \to E_2$ such that*

$$\| f(x) - A(x) \| \le \frac{2\theta}{2 - 2^p} \|x\|^p \tag{2.6}$$

*for any $x \in E_1$. Moreover, if $f(tx)$ is continuous in t for each fixed $x \in E_1$, then A
is linear.*

Proof. By induction on n, we prove that

$$\| 2^{-n} f(2^n x) - f(x) \| \le \theta \|x\|^p \sum_{m=0}^{n-1} 2^{m(p-1)} \tag{a}$$

for any $n \in \mathbb{N}$. Putting $y = x$ in (2.5) and dividing by 2 yield the validity of (a)
for $n = 1$. Assume now that (a) is true and we want to prove it for the case $n + 1$.
However, this is true because by (a) we obtain

$$\| 2^{-n} f(2^n 2x) - f(2x) \| \le \theta \|2x\|^p \sum_{m=0}^{n-1} 2^{m(p-1)},$$

therefore

$$\left\| 2^{-n-1} f\big(2^{n+1}x\big) - (1/2) f(2x) \right\| \le \theta \|x\|^p \sum_{m=1}^{n} 2^{m(p-1)}.$$

By the triangle inequality, we get

$$\|2^{-n-1} f\left(2^{n+1}x\right) - f(x)\|$$
$$\leq \|2^{-n-1} f\left(2^{n+1}x\right) - (1/2) f(2x)\| + \|(1/2) f(2x) - f(x)\|$$
$$\leq \theta \|x\|^p \sum_{m=0}^{n} 2^{m(p-1)},$$

which completes the proof of (a).

It then follows that

$$\|2^{-n} f(2^n x) - f(x)\| \leq \frac{2\theta}{2 - 2^p} \|x\|^p, \tag{b}$$

since $\sum_{m=0}^{\infty} 2^{m(p-1)}$ converges to $2/(2 - 2^p)$, as $0 \leq p < 1$. However, for $m > n > 0$, we have

$$\|2^{-m} f(2^m x) - 2^{-n} f(2^n x)\| = 2^{-n} \|2^{-(m-n)} f(2^{m-n} 2^n x) - f(2^n x)\|$$
$$\leq 2^{n(p-1)} \frac{2\theta}{2 - 2^p} \|x\|^p.$$

Therefore, the Rassias sequence $\{2^{-n} f(2^n x)\}$ is a Cauchy sequence for each $x \in E_1$. As E_2 is complete, we can define a function A by (2.3). It follows that

$$\|f\left(2^n (x + y)\right) - f(2^n x) - f(2^n y)\| \leq 2^{np} \theta \left(\|x\|^p + \|y\|^p\right).$$

Dividing by 2^n the last expression and letting $n \to \infty$, together with (2.3), yield that A is an additive function.

The inequality (2.6) immediately follows from (b) and (2.3).

We now want to prove that A is such a unique additive function. Assume that there exists another one, denoted by $A' : E_1 \to E_2$. Then there exists a constant $\varepsilon_1 \geq 0$ and q $(0 \leq q < 1)$ with

$$\|A'(x) - f(x)\| \leq \varepsilon_1 \|x\|^q. \tag{c}$$

By the triangle inequality, (2.6), and (c) we obtain

$$\|A(x) - A'(x)\| \leq \|A(x) - f(x)\| + \|f(x) - A'(x)\|$$
$$\leq \frac{2\theta}{2 - 2^p} \|x\|^p + \varepsilon_1 \|x\|^q.$$

Therefore,

$$\|A(x) - A'(x)\| = (1/n)\|A(nx) - A'(nx)\|$$

$$\leq \frac{1}{n}\left(\frac{2\theta}{2 - 2^p}\|nx\|^p + \varepsilon_1\|nx\|^q\right)$$

$$= n^{p-1}\frac{2\theta}{2 - 2^p}\|x\|^p + n^{q-1}\varepsilon_1\|x\|^q$$

for all $n \in \mathbb{N}$. By letting $n \to \infty$ we get $A(x) = A'(x)$ for any $x \in E_1$.

Assume that $f(tx)$ is continuous in t for any fixed $x \in E_1$. Since $A(x + y) = A(x) + A(y)$ for each $x, y \in E_1$, $A(qx) = qA(x)$ holds true for any rational number q. Fix x_0 in E_1 and ρ in E_2^* (the dual space of E_2). Define a function $\phi : \mathbb{R} \to \mathbb{R}$ by

$$\phi(t) = \rho(A(tx_0))$$

for all $t \in \mathbb{R}$. Then ϕ is additive. Moreover, ϕ is a Borel function because of the following reasoning: Let $\phi(t) = \lim_{n\to\infty} 2^{-n}\rho(f(2^n t x_0))$ and set $\phi_n(t) = 2^{-n}\rho(f(2^n t x_0))$. Then $\phi_n(t)$ are continuous functions. $\phi(t)$ is the pointwise limit of continuous functions, thus $\phi(t)$ is a Borel function. According to Theorem 2.1, ϕ is linear and hence it is continuous. Let $a \in \mathbb{R}$. Then $a = \lim_{n\to\infty} q_n$, where $\{q_n\}$ is a sequence of rational numbers. Hence,

$$\phi(at) = \phi\left(t \lim_{n\to\infty} q_n\right) = \lim_{n\to\infty} \phi(tq_n) = \lim_{n\to\infty} q_n\phi(t) = a\phi(t).$$

Therefore, $\phi(at) = a\phi(t)$ for any $a \in \mathbb{R}$. Thus, $A(ax) = aA(x)$ for any $a \in \mathbb{R}$. Hence, A is a linear function.   □

This theorem is a remarkable generalization of Theorem 2.3 and stimulated the concern of mathematicians toward the study of the stability problems of functional equations. T. Aoki [7] has provided a proof of a special case of Th. M. Rassias's theorem just for the stability of the additive function using the direct method. Aoki did not prove the last assertion of Rassias's Theorem 2.5 which provides the stability of the linear function.

Th. M. Rassias [289] noticed that the proof of this theorem also works for $p < 0$ and asked whether such a theorem can also be proved for $p \geq 1$. Z. Gajda [112] answered the *question of Rassias* for the case of $p > 1$ by a slight modification of the expression in (2.3). His idea to prove the theorem for this case is to replace n by $-n$ in the formula (2.3).

It turns out that 1 is the only critical value of p to which Theorem 2.5 cannot be extended. Z. Gajda [112] showed that this theorem is false for $p = 1$ by constructing a counterexample:

For a fixed $\theta > 0$ and $\mu = (1/6)\theta$ define a function $f : \mathbb{R} \to \mathbb{R}$ by

$$f(x) = \sum_{n=0}^{\infty} 2^{-n}\phi(2^n x),$$

where the function $\phi : \mathbb{R} \to \mathbb{R}$ is given by

$$\phi(x) = \begin{cases} \mu & \text{(for } x \in [1, \infty)), \\ \mu x & \text{(for } x \in (-1, 1)), \\ -\mu & \text{(for } x \in (-\infty, -1]). \end{cases}$$

Then the function f serves as a counterexample for $p = 1$ as presented in the following theorem.

Theorem 2.6 (Gajda). *The function f defined above satisfies*

$$|f(x + y) - f(x) - f(y)| \le \theta(|x| + |y|) \tag{2.7}$$

for all $x, y \in \mathbb{R}$, while there is no constant $\delta \ge 0$ and no additive function $A : \mathbb{R} \to \mathbb{R}$ satisfying the condition

$$|f(x) - A(x)| \le \delta|x| \tag{2.8}$$

for all $x, y \in \mathbb{R}$.

Proof. If $x = y = 0$, then (2.7) is trivially satisfied.

Now, we assume that $0 < |x| + |y| < 1$. Then there exists an $N \in \mathbb{N}$ such that

$$2^{-N} \le |x| + |y| < 2^{-(N-1)}.$$

Hence, $|2^{N-1}x| < 1$, $|2^{N-1}y| < 1$, and $|2^{N-1}(x + y)| < 1$, which implies that for each $n \in \{0, 1, \dots, N - 1\}$ the numbers $2^n x$, $2^n y$, and $2^n(x + y)$ belong to the interval $(-1, 1)$. Since ϕ is linear on this interval, we infer that

$$\phi(2^n(x + y)) - \phi(2^n x) - \phi(2^n y) = 0$$

for $n \in \{0, 1, \dots, N - 1\}$. As a result, we get

$$\frac{|f(x + y) - f(x) - f(y)|}{|x| + |y|} \le \sum_{n=N}^{\infty} \frac{|\phi(2^n(x + y)) - \phi(2^n x) - \phi(2^n y)|}{2^n(|x| + |y|)}$$

$$\le \sum_{k=0}^{\infty} \frac{3\mu}{2^k 2^N(|x| + |y|)}$$

$$\le \sum_{k=0}^{\infty} \frac{3\mu}{2^k}$$

$$= \theta.$$

Finally, assume $|x| + |y| \geq 1$. Then merely by means of the boundedness of f we have

$$\frac{|f(x + y) - f(x) - f(y)|}{|x| + |y|} \leq 6\mu = \theta,$$

since

$$|f(x)| \leq \sum_{n=0}^{\infty} 2^{-n}\mu = 2\mu.$$

Now, contrary to what we claim, suppose that there exist a constant $\delta \geq 0$ and an additive function $A : \mathbb{R} \to \mathbb{R}$ such that (2.8) holds true. Since f is defined by means of a uniformly convergent series of continuous functions, f itself is continuous. Hence, A is bounded on some neighborhood of zero. Then, by Theorem 2.1, there exists a real constant c such that $A(x) = cx$ for all $x \in \mathbb{R}$. Hence, it follows from (2.8) that

$$|f(x) - cx| \leq \delta|x|,$$

for any $x \in \mathbb{R}$, which implies

$$|f(x)|/|x| \leq \delta + |c|$$

for all $x \in \mathbb{R}$. On the other hand, we can choose an $N \in \mathbb{N}$ so large that $N\mu > \delta + |c|$. If we choose an $x \in \left(0, 2^{-(N-1)}\right)$, then we have $2^n x \in (0, 1)$ for each $n \in \{0, 1, \ldots, N - 1\}$. Consequently, for such an x we get

$$\frac{f(x)}{x} \geq \sum_{n=0}^{N-1} \frac{\phi(2^n x)}{2^n x} = \sum_{n=0}^{N-1} \frac{\mu 2^n x}{2^n x} = N\mu > \delta + |c|,$$

which leads to a contradiction.   □

Similarly, Th. M. Rassias and P. Šemrl [310] introduced a simple counterexample to Theorem 2.5 for $p = 1$ as follows:

The continuous real-valued function defined by

$$f(x) = \begin{cases} x \log_2(x + 1) & \text{(for } x \geq 0\text{)}, \\ x \log_2 |x - 1| & \text{(for } x < 0\text{)} \end{cases}$$

satisfies the inequality (2.7) with $\theta = 1$ and $|f(x) - cx|/|x| \to \infty$, as $x \to \infty$, for any real number c.

Furthermore, they also investigated the behaviors of functions which satisfy the inequality (2.7).

Theorem 2.7. *Let $f : \mathbb{R}^k \to \mathbb{R}^\ell$ be a function such that $f(tx)$ is continuous in t for each fixed x, where k and ℓ are given positive integers. Assume that f satisfies* *the inequality (2.7) for any $x, y \in \mathbb{R}^k$. Then for any $\varepsilon > 0$, there exists a real number M_ε such that*

$$|f(x)| \leq \begin{cases} M_\varepsilon |x|^{1+\varepsilon} & \text{(for } |x| \geq 1), \\ M_\varepsilon |x|^{1-\varepsilon} & \text{(for } |x| \leq 1). \end{cases} \tag{2.9}$$

Proof. Applying (2.7) and induction on n, we can prove

$$\begin{aligned} &\left| f(x_1 + \cdots + x_n) - f(x_1) - \cdots - f(x_n) \right| \\ &\leq \ \theta(n-1)\left(|x_1| + \cdots + |x_n| \right). \end{aligned} \tag{a}$$

Let $\{e_1, \ldots, e_k\}$ be the standard basis in $\mathbb{R}^k$. An arbitrary vector $x \in \mathbb{R}^k$ with $|x| \leq 1$ can be expressed in the form

$$x = \lambda_1 e_1 + \cdots + \lambda_k e_k,$$

where $|\lambda_i| \leq 1$ for $i \in \{1, 2, \ldots, k\}$. It follows from (a) that

$$\begin{aligned} &\left| f(\lambda_1 e_1 + \cdots + \lambda_k e_k) - f(\lambda_1 e_1) - \cdots - f(\lambda_k e_k) \right| \\ &\leq \ \theta(k-1)\left(|\lambda_1 e_1| + \cdots + |\lambda_k e_k| \right) \\ &\leq \ \theta(k-1)k. \end{aligned}$$

Then

$$\begin{aligned} \left| f(\lambda_1 e_1 + \cdots + \lambda_k e_k) \right| &\leq \theta(k-1)k + |f(\lambda_1 e_1)| + \cdots + |f(\lambda_k e_k)| \\ &\leq \theta(k-1)k + M_1 + \cdots + M_k, \end{aligned}$$

where

$$M_i = \max_{|\lambda| \leq 1} |f(\lambda e_i)|.$$

Hence, f is bounded on the unit ball in $\mathbb{R}^k$. Thus, there exists a real number c such that

$$|f(x)| \leq c|x| \tag{b}$$

for all x satisfying $1/2 \leq |x| \leq 1$.

  Claim that

$$\left| 2^{-n} f(2^n x) - f(x) \right| \leq n\theta|x| \tag{c}$$

for all $n \in \mathbb{N}$. By (2.7), (c) is true for $n = 1$. Assume now that (c) is true for some $n > 0$. Using the triangle inequality and (c), we get

$$\begin{aligned} &\left| 2^{-n-1} f(2^{n+1} x) - f(x) \right| \\ &\leq \left| 2^{-n-1} f(2^n 2x) - (1/2) f(2x) \right| + |(1/2) f(2x) - f(x)| \\ &\leq (n+1)\theta|x|, \end{aligned}$$

which ends the proof of (c).

For any x with $|x| > 1$, we can find an integer n such that the vector $y = 2^{-n}x$ satisfies $1/2 \leq |y| \leq 1$. Moreover, we have $n \leq \log_2 |x| + 1$. It follows from (c) that

$$|2^{-n} f(x) - f(y)| \leq n\theta|y|.$$

Therefore, by (b), we obtain

$$|f(x)| \leq 2^n\big(|f(y)| + n\theta|y|\big) \leq 2^n|y|(c + n\theta) \leq |x|(c + \theta(\log_2 |x| + 1)),$$

which proves the first part of (2.9).

A similar argument as in the proof of (c) yields

$$|2^n f(2^{-n}x) - f(x)| \leq n\theta|x|$$

for any $n \in \mathbb{N}$. For any x with $|x| \leq 1$, there exists an integer n such that the vector $y = 2^n x$ satisfies $1/2 \leq |y| \leq 1$. It follows that $n \leq -\log_2 |x|$. As in the previous case, we obtain

$$|2^n f(x) - f(y)| \leq n\theta|y|.$$

Thus,

$$|f(x)| \leq 2^{-n}\big(|f(y)| + n\theta|y|\big) \leq |x|(c - \theta \log_2 |x|).$$

Hence, the second part of (2.9) also holds true.   □

Th. M. Rassias [290] asked whether (2.6) gives the best possible estimate of the difference $\| f(x) - A(x) \|$ for $p \neq 1$. Th. M. Rassias and J. Tabor [313] answered the question for $p = 1/2$, and J. Brzdęk has partially answered the question for the case of $p > 0$ $(p \neq 1)$. As it is an interesting subject, we introduce the result of Brzdęk [30]:

Let $\theta > 0$ and $p > 0$ $(p \neq 1)$ be given, and let $A, f : \mathbb{R} \to \mathbb{R}$ be defined by

$$A(x) = 0$$

for all $x \in \mathbb{R}$, and

$$f(x) = \text{sign}(x) \frac{\theta}{|2^{p-1} - 1|}|x|^p$$

for all $x \in \mathbb{R}$, where sign : $\mathbb{R} \to \{-1, 0, 1\}$ is the *sign function*. Then A is an additive function and

$$|f(x) - A(x)| = \frac{\theta}{|2^{p-1} - 1|}|x|^p$$

for all $x \in \mathbb{R}$. We now aim to show that f also satisfies the inequality (2.5) for all $x, y \in \mathbb{R}$.

Let us start with the following lemma:

Lemma 2.8. *If f_1, $f_2 : [0, 1] \to \mathbb{R}$ are functions defined by*

$$f_1(x) = |(x + 1)^p - x^p - 1|/(x^p + 1),$$
$$f_2(x) = |(1 - x)^p + x^p - 1|/(x^p + 1)$$

for some $p > 0$ ($p \neq 1$) and for any $x \in [0, 1]$, then $f_1(x) \geq f_2(x)$ for all $x \in [0, 1]$.

Proof. Put $g_i(x) = f_i(x)(x^p + 1)$ for $x \in [0, 1]$ and $i \in \{1, 2\}$. Note that, in the case $p > 1$,

$$(1 - x)^p + x^p \leq 1 \quad \text{and} \quad (x + 1)^p - x^p \geq 1 \quad \text{for } x \in [0, 1]$$

and, in the case $0 < p < 1$,

$$(1 - x)^p + x^p \geq 1 \quad \text{and} \quad (x + 1)^p - x^p \leq 1 \quad \text{for } x \in [0, 1].$$

Hence, for every $x \in [0, 1]$,

$$g_1(x) = \begin{cases} (x + 1)^p - x^p - 1 & (\text{for } p > 1), \\ 1 + x^p - (x + 1)^p & (\text{for } 0 < p < 1) \end{cases} \qquad (a)$$

and

$$g_2(x) = \begin{cases} 1 - x^p - (1 - x)^p & (\text{for } p > 1), \\ (1 - x)^p + x^p - 1 & (\text{for } 0 < p < 1). \end{cases}$$

Consequently, for $x \in [0, 1]$,

$$g_1'(x) = \begin{cases} p(x + 1)^{p-1} - px^{p-1} & (\text{for } p > 1), \\ px^{p-1} - p(x + 1)^{p-1} & (\text{for } 0 < p < 1) \end{cases}$$

and

$$g_2'(x) = \begin{cases} p(1 - x)^{p-1} - px^{p-1} & (\text{for } p > 1), \\ px^{p-1} - p(1 - x)^{p-1} & (\text{for } 0 < p < 1). \end{cases}$$

Further, if $p > 1$, then

$$(x + 1)^{p-1} - x^{p-1} \geq (1 - x)^{p-1} - x^{p-1} \quad \text{for } x \in [0, 1],$$

and if $0 < p < 1$,

$$x^{p-1} - (x + 1)^{p-1} > x^{p-1} - (1 - x)^{p-1} \quad \text{for } x \in (0, 1).$$

Thus, $g_1'(x) \geq g_2'(x)$ for any $x \in (0, 1)$. Hence, $g_1(x) \geq g_2(x)$ for all $x \in [0, 1]$, since $g_1(0) = g_2(0) = 0$. That is, $f_1(x) \geq f_2(x)$ for each $x \in [0, 1]$.   □

Lemma 2.9. *Let* f_1 *be the same as in Lemma 2.8. Then*

$$\sup \{ f_1(x) \mid x \in [0, 1] \} = \left| 2^{p-1} - 1 \right|$$

for $p > 0$ *and* $p \neq 1$.

Proof. Suppose $f_1'(x) = 0$. Considering (a) in the proof of Lemma 2.8, it follows from the hypothesis that

$$\left(p(x + 1)^{p-1} - px^{p-1} \right)\left(x^p + 1 \right) - px^{p-1}\left((x + 1)^p - x^p - 1 \right) = 0.$$

The solution of this equation is $x = 1$ only.
    In this way, we have shown that

$$\sup \{ f_1(x) \mid x \in [0, 1] \} = \max \{ f_1(0),\ f_1(1) \} = \left| 2^{p-1} - 1 \right|,$$

which ends the proof.   □

    J. Brzdęk [30] proved that f satisfies the inequality (2.5):

Theorem 2.10 (Brzdęk). *The function* f *satisfies the functional inequality* (2.5) *for all* $x, y \in \mathbb{R}$.

Proof. First, let $x = y = 0$. Since $f(0) = 0$, it is clear that f satisfies (2.5) for this case.
    Now assume $(x, y) \neq (0, 0)$, and let

$$g(x, y) = \frac{\left| 2^{p-1} - 1 \right|}{\theta \left(|x|^p + |y|^p \right)} \left| f(x + y) - f(x) - f(y) \right|$$

for all $x, y \in \mathbb{R}$ with $x^2 + y^2 > 0$. Let us define

$$s = \sup \{ g(x, y) \mid x, y \in \mathbb{R},\ x^2 + y^2 > 0 \}.$$

Since $g(x, y) = g(y, x)$ for $x, y \in \mathbb{R}$ with $x^2 + y^2 > 0$, it is easily seen that

$$s = \sup \{ g(x, y) \mid x, y \in \mathbb{R},\ |x| \leq |y|,\ x^2 + y^2 > 0 \}.$$

Moreover, if $xy \geq 0$, $x^2 + y^2 > 0$ and $|x| \leq |y|$, then

$$g(x, y) = \left| |x + y|^p - |x|^p - |y|^p \right| / \left(|x|^p + |y|^p \right),$$

and if $xy < 0$ and $|x| \le |y|$, then

$$g(x, y) = \left| |x + y|^p + |x|^p - |y|^p \right| / \left(|x|^p + |y|^p \right).$$

Define

$$s_1 = \sup \left\{ g(x, y) \mid xy \ge 0, \ x^2 + y^2 > 0, \ |x| \le |y| \right\}$$

and

$$s_2 = \sup \left\{ g(x, y) \mid xy < 0, \ |x| \le |y| \right\}.$$

Then

$$s_1 = \sup \left\{ f_1(x) \mid x \in [0, 1] \right\}$$

and

$$s_2 = \sup \left\{ f_2(x) \mid x \in [0, 1] \right\}.$$

Therefore, by Lemmas 2.8 and 2.9, we get

$$s = \max \left\{ s_1, s_2 \right\} = \left| 2^{p-1} - 1 \right|$$

which implies that f satisfies (2.5) for all $x, y \in \mathbb{R}$ with $x^2 + y^2 > 0$. $\qquad\square$

Until now, we have proved that $\theta \|x\|^p / |1 - 2^{p-1}|$ gives the best possible upper bound for the norm of the difference $f(x) - A(x)$ in the case of $p > 0$ ($p \ne 1$).

Now, we return to the subject concerning the generalization of the bound condition for the norm of the Cauchy difference in (2.5).

A function $H : [0, \infty)^2 \to [0, \infty)$ is called *homogeneous* of degree p if it satisfies $H(tu, tv) = t^p H(u, v)$ for all $t, u, v \in [0, \infty)$. We can replace $\theta \left(\|x\|^p + \|y\|^p \right)$ with $H \left(\|x\|, \|y\| \right)$, where $H : [0, \infty)^2 \to [0, \infty)$ is a monotonically increasing symmetric homogeneous function of degree $p \ge 0$, $p \ne 1$, and still obtain a stability result. More precisely, Th. M. Rassias and P. Šemrl [311] generalized the result of Theorem 2.5 as follows:

Theorem 2.11 (Rassias and Šemrl). *Let E_1 and E_2 be a normed space and a real Banach space, respectively. Assume that $H : [0, \infty)^2 \to [0, \infty)$ is a monotonically increasing symmetric homogeneous function of degree p, where $p \in [0, \infty) \setminus \{1\}$. Let a function $f : E_1 \to E_2$ satisfy the inequality*

$$\| f(x + y) - f(x) - f(y) \| \le H \left(\|x\|, \|y\| \right) \tag{2.10}$$

for any $x, y \in E_1$. Then there exists a unique additive function $A : E_1 \to E_2$ such that

$$\| f(x) - A(x) \| \le \frac{H(1, 1)}{|2^p - 2|} \|x\|^p \tag{2.11}$$

for any $x \in E_1$. Moreover, the function A is linear if for each fixed $x \in E_1$ there exists a real number $\delta_x > 0$ such that $f(tx)$ is bounded on $[0, \delta_x]$.

Proof. The proof of the first part of the theorem is similar to Theorem 2.3 or 2.5. Therefore, we prove only the linearity of A under the condition that $f(tx)$ is locally bounded for each fixed x. So, let us assume that for every fixed $x \in E_1$ there exists a positive number δ_x such that the function $\| f(tx) \|$ is bounded on $[0, \delta_x]$. Fix $z \in E_1$ and $\varphi \in E_2^*$. Here, E_2^* denotes the dual space of E_2. Let us define

$$M_z = \sup \{ \| f(tz) \| \mid t \in [0, \delta_z] \}.$$

Consider the function $\phi : \mathbb{R} \to \mathbb{R}$ defined by $\phi(t) = \varphi(A(tz))$. It is obvious that ϕ is additive. For any real number $t \in [0, \delta_z]$, we have

$$|\phi(t)| = \left| \varphi(A(tz)) \right| \leq \| \varphi \| \, \| A(tz) \| \leq \| \varphi \| (\| A(tz) - f(tz) \| + \| f(tz) \|).$$

Using (2.11) we obtain

$$|\phi(t)| \leq \| \varphi \| \left(\frac{H(1, 1)}{|2^p - 2|} \delta_z^p \|z\|^p + M_z \right).$$

Since the additive function ϕ is bounded on an interval of positive length, in view of Theorem 2.1, it is of the form

$$\phi(t) = \phi(1) t$$

for all $t \in \mathbb{R}$. Therefore, $\varphi(A(tz)) = \varphi(tA(z))$ for any $t \in \mathbb{R}$, and consequently A is a linear function. $\square$

As expected, an analogue of Theorem 2.11 cannot be obtained in the case that H is a monotonically increasing symmetric homogeneous function of degree 1. It is not difficult to construct a counterexample from the following lemma presented in the paper [311].

Lemma 2.12. *Let $h : [0, \infty) \to [0, \infty)$ be a monotonically increasing function satisfying*

$$\lim_{t \to \infty} h(t) = \infty, \quad h(1) = 1, \quad and \quad h(t) = t h(1/t)$$

for all $t > 0$. Then there exists a continuous monotonically increasing function $g : [0, \infty) \to [0, \infty)$ such that

(i) $g(0) = 0$,
(ii) $g(t) \to \infty$ as $t \to \infty$,
(iii) $g(t + s) - g(t) \leq h(s/t)$ for all $s, t \in [0, \infty)$, $t \neq 0$.

Proof. Since $h(t) \to \infty$ as $t \to \infty$, we can find a monotonically increasing sequence $\{n_k\}$ of positive integers satisfying

$$n_1 = 1 \quad \text{and} \quad h(2^{n_k}) > k. \tag{a}$$

Let $\{a_k\}$ be a sequence given by

$$a_1 = 0, \quad a_2 = 2, \quad \text{and} \quad a_{k+1} = 2^{n_k} a_k \qquad (b)$$

for $k \in \{2, 3, \ldots\}$.
   We define

$$g(a_1) = 0 \quad \text{and} \quad g(a_k) = 1/3 + 1/4 + \cdots + 1/(k+1)$$

for $k \in \{2, 3, \ldots\}$. We extend g to $[0, \infty)$ such that g is piecewise linear. It is obvious that $g : [0, \infty) \to [0, \infty)$ is a continuous monotonically increasing function satisfying (i) and (ii).
   Let t be a real number with $0 < t \leq 1$. Since h is a monotonically increasing function satisfying $h(1) = 1$, we have $h(1/t) \geq 1$. It follows that $th(1/t) \geq t$. Using $h(t) = th(1/t)$ we finally get that

$$h(t) \geq t \qquad (c)$$

holds true for any $t \in (0, 1]$.
   In order to prove (iii) let us first assume that $t \geq s$ and choose $k \in \mathbb{N}$ such that $t \in [a_k, a_{k+1}]$. The definition of g implies

$$g(t + s) - g(t) \leq \frac{s}{(k+2)(a_{k+1} - a_k)}. \qquad (d)$$

For $k > 1$ we get using (b) that

$$g(t + s) - g(t) \leq \frac{s}{a_{k+1}} \frac{1}{(k+2)(1 - 2^{-n_k})} \leq \frac{s}{a_{k+1}} \leq \frac{s}{t}.$$

The relation (c) completes the proof in this case. The case, when $k = 1$, is an immediate consequence of (b), (c), and (d).
   Suppose now that $s > t$ and choose $k \in \mathbb{N}$ such that $s \in [a_k, a_{k+1}]$. Let us set $a_0 = 0$. We will first consider the case that $t \geq a_{k-1}$. Then we have

$$g(t + s) - g(t) \leq g(a_{k+2}) - g(a_{k-1}) \leq 1/3 + 1/4 + 1/5 < 1.$$

The desired relation (iii) follows now from $h(1) = 1$ and $s/t > 1$.
   It remains to consider the case when $t < a_{k-1}$. This implies $k \geq 3$. Therefore,

$$g(t + s) - g(t) \leq g(a_{k+2}) = 1/3 + 1/4 + \cdots + 1/(k+3) \leq k - 1.$$

Using (a) we have

$$g(t + s) - g(t) \leq h\big(a_k 2^{n_{k-1}} a_k^{-1}\big) \leq h\big(2^{n_{k-1}} s a_k^{-1}\big).$$

Applying $2^{n_k-1}t \leq 2^{n_k-1}a_{k-1} = a_k$ in the previous inequality, we obtain

$$g(t+s) - g(t) \leq h\big(2^{n_k-1}s2^{-n_k-1}t^{-1}\big) = h(s/t),$$

which ends the proof. $\Box$

Using the result of the last lemma, Th. M. Rassias and P. Šemrl [311] have succeeded in finding a counterexample to Theorem 2.11 for the case in which H is a monotonically increasing symmetric homogeneous function of degree 1.

Theorem 2.13. *Assume that* $H : [0, \infty)^2 \to [0, \infty)$ *is a symmetric monotonically increasing homogeneous function of degree* 1 *such that*

$$\lim_{s \to \infty} H(1, s) = \infty. \tag{2.12}$$

Then there exists a continuous function $f : \mathbb{R} \to \mathbb{R}$ *satisfying*

$$\big|f(t+s) - f(t) - f(s)\big| \leq H\big(|t|, |s|\big)$$

for all $t, s \in \mathbb{R}$ *and*

$$\sup_{t \neq 0} |f(t) - A(t)|/|t| = \infty \tag{2.13}$$

for any additive function $A : \mathbb{R} \to \mathbb{R}$.

Proof. For every real number $t > 1$ we have $H(1, t) \leq H(t, t) = tH(1, 1)$. This inequality yields $H(1, 1) \neq 0$ using (2.12). We can assume without loss of generality that $H(1, 1) = 1$. Let us define $h : [0, \infty) \to [0, \infty)$ by $h(t) = H(1, t)$. Obviously, h is a monotonically increasing function satisfying $h(t) \to \infty$, as $t \to \infty$, and $h(1) = 1$. Moreover, we have $h(t) = H(1, t) = tH(1/t, 1) = tH(1, 1/t) = th(1/t)$ for any $t > 0$. We choose a function g as in Lemma 2.12 and define $f(t) = (1/2)tg(t)$ for all $t \geq 0$. We extend f to $\mathbb{R}$ as an odd function.

We first prove the inequality

$$\big|f(t+s) - f(t) - f(s)\big| \leq H\big(|t|, |s|\big) \tag{a}$$

for all $s, t \in \mathbb{R}$. Clearly, (a) holds true in the case that $t = 0$ or $s = 0$. Next, we consider the case in which both numbers t and s are positive. Then we have

$$|f(t+s) - f(t) - f(s)| = (1/2)\big|t\big(g(t+s) - g(t)\big) + s\big(g(t+s) - g(s)\big)\big|.$$

As g is a monotonically increasing function, Lemma 2.12 (*iii*) yields

$$\begin{aligned}
|f(t+s) - f(t) - f(s)| &\leq (1/2)\big(th(s/t) + sh(t/s)\big) \\
&= (1/2)\big(H(t, s) + H(s, t)\big) \\
&= H\big(|t|, |s|\big).
\end{aligned}$$

Since f is an odd function, (a) holds true for $t, s < 0$ as well.

It remains to consider the case when $t > 0$ and $s < 0$. Let us first assume that $|t| > |s|$. The left side of (a) can be rewritten as $|f(t) - f(t + s) - f(-s)|$, but $t + s$ and $-s$ are positive real numbers. Thus,

$$|f(t + s) - f(t) - f(s)| \le H(t + s, -s) \le H(|t|, |s|).$$

The proof of (a) in the case when $|s| > |t|$ proceeds in a similar way.

Suppose now that there exists an additive function $A : \mathbb{R} \to \mathbb{R}$ such that

$$\sup_{t \ne 0} |f(t) - A(t)|/|t| < \infty. \tag{b}$$

As f is a continuous function, it is bounded on any finite interval. It follows from (b) that the additive function A is bounded on every finite interval $[a, b]$ of the real line with $0 \notin [a, b]$. This implies that A is of the form $A(t) = ct$ for some real number c. For $t \ge 0$ we have

$$|f(t) - A(t)|/|t| = |(1/2)g(t) - c|.$$

According to Lemma 2.12 (ii), this contradicts (b). $\qquad\square$

As seen in the above proof, the condition (2.12) is essential for the construction of a function f satisfying (2.13). The above theorem is a generalization of the result of Theorem 2.6 stating that the answer to Ulam's problem is negative in the special case that $H(|t|, |s|) = |t| + |s|$.

In the following theorem, we will introduce the behavior of functions satisfying the inequality (2.10), where H is a homogeneous function of degree 1 (ref. [311]).

Theorem 2.14. *Let E_1 and E_2 be a real normed space with $\dim E_1 > 1$ and a real Banach space, respectively. Suppose a function $f : E_1 \to E_2$ satisfies the inequality (2.10), where $H : [0, \infty)^2 \to [0, \infty)$ is a symmetric monotonically increasing homogeneous function of degree 1. Then the following conditions are equivalent:*

(i) $\displaystyle\sup_{\|x\| \le 1} \|f(x)\| < \infty$,

(ii) $\displaystyle\sup_{\|x\| = 1} \|f(x)\| < \infty$,

(iii) f is continuous at 0,

(iv) $\displaystyle\lim_{\|x\| \to \infty} \|f(x)\|/\|x\|^{1+\varepsilon} = 0$ *for any* $\varepsilon > 0$,

(v) $\displaystyle\lim_{\|x\| \to 0} \|f(x)\|/\|x\|^{1-\varepsilon} = 0$ *for any* $\varepsilon > 0$.

Proof. Let us set $H(1, 1) = \theta$. Claim that

$$\|2^{-n} f(2^n x) - f(x)\| \le (1/2)n\theta\|x\| \tag{a}$$

and

$$\|2^n f(2^{-n} x) - f(x)\| \le (1/2)n\theta\|x\| \tag{b}$$

for any $n \in \mathbb{N}$. The proof of (a) follows by induction on n. The case $n = 1$ is clear. Assume now that (a) is true for some $n > 0$ and we want to prove it for $n + 1$. Using the triangle inequality and (a), we get

$$
\begin{aligned}
\left\| 2^{-n-1} f\left(2^{n+1}x\right) - f(x) \right\| &\leq \left\| 2^{-n-1} f\left(2^{n+1}x\right) - (1/2)f(2x) \right\| \\
&\quad + \left\| (1/2)f(2x) - f(x) \right\| \\
&\leq (1/2)(n+1)\theta \|x\|.
\end{aligned}
$$

Replacing x with $2^{-n}x$ in (a) and multiplying the resulting inequality by 2^n, we obtain (b).

The implications $(i) \Rightarrow (ii)$ and $(v) \Rightarrow (iii)$ are easily seen. In order to prove $(ii) \Rightarrow (i)$, we choose a vector z such that $\|z\| \leq 1$. Since $\dim E_1 > 1$, the intersection of the unit spheres $S(0,1) = \{x \in E_1 \mid \|x\| = 1\}$ and $S(z,1) = \{x \in E_1 \mid \|x - z\| = 1\}$ is nonempty. Choose $w \in S(0,1) \cap S(z,1)$. Clearly, $z = w + (z - w)$. Since f satisfies (2.10), we have

$$
\begin{aligned}
\|f(z)\| &\leq H\left(\|w\|, \|z - w\|\right) + \|f(w)\| + \|f(z - w)\| \\
&\leq \theta + 2 \sup_{\|x\|=1} \|f(x)\|.
\end{aligned}
$$

Claim that $(iii) \Rightarrow (i)$. It is easy to see $f(0) = 0$. From (iii) it follows that there exist positive real numbers δ and M such that $\|x\| \leq \delta$ yields $\|f(x)\| \leq M$. We fix a positive integer n_0 satisfying $2^{-n_0} \leq \delta$. For every vector $z \in E_1$, $\|z\| \leq 1$, we get using (b) that

$$
\|f(z)\| \leq (1/2)n_0\theta\|z\| + 2^{n_0}\|f(2^{-n_0}z)\| \leq (1/2)n_0\theta + 2^{n_0}M.
$$

Claim that $(i) \Rightarrow (iv)$. It follows from (i) that there exists a real number c such that

$$
\|f(x)\| \leq c\|x\| \tag{c}
$$

for all x with $1/2 \leq \|x\| \leq 1$. For any x with norm greater than 1 we can find a positive integer n such that the vector $y = 2^{-n}x$ satisfies $1/2 \leq \|y\| \leq 1$. Moreover, we have $n \leq \log_2 \|x\| + 1$. It follows from (a) that

$$
\|2^{-n} f(x) - f(y)\| \leq (1/2)n\theta\|y\|.
$$

Therefore,

$$
\begin{aligned}
\|f(x)\| &\leq 2^n\left(\|f(y)\| + (1/2)n\theta\|y\|\right) \\
&\leq 2^n\|y\|\left(c + (1/2)n\theta\right) \\
&\leq \|x\|\left(c + (1/2)\theta(\log_2 \|x\| + 1)\right),
\end{aligned}
$$

which completes the proof of this implication.

Claim that $(i) \Rightarrow (v)$. For any x in the unit ball, $\|x\| \leq 1$, there exists an integer n such that the vector $y = 2^n x$ satisfies $1/2 \leq \|y\| \leq 1$. It follows that $n \leq -\log_2 \|x\|$. As before we get

$$\|2^n f(x) - f(y)\| \leq (1/2)n\theta\|y\|.$$

Thus,

$$\|f(x)\| \leq 2^{-n}\big(\|f(y)\| + (1/2)n\theta\|y\|\big) \leq \|x\|\big(c - (1/2)\theta \log_2 \|x\|\big).$$

Hence, (v) holds true.

Claim that $(iv) \Rightarrow (ii)$. It follows from (iv) with $\varepsilon = 1$ that there exist positive real numbers η and M such that $\|x\| \geq M$ implies $\|f(x)\| \leq \eta\|x\|^2$. Let us fix a positive integer n_0 satisfying $2^{n_0} \geq M$. Then for every $z \in E_1$ with $\|z\| = 1$, the inequality $\|f(2^{n_0}z)\| \leq 4^{n_0}\eta$ is true. A simple use of (a) completes the proof.

Applying a similar approach we can prove the implication $(v) \Rightarrow (ii)$. □

The assumption that $\dim E_1 > 1$ is indispensable in the above theorem. Every function $f : \mathbb{R} \to E_2$ satisfying (2.10), where H is a homogeneous function of degree 1, is bounded on the unit sphere $\{-1, 1\}$. However, such functions need not be bounded on the unit ball. The proof of the equivalence of the conditions (i), (iii), (iv), and (v) works also in the case that $\dim E_1 = 1$. A function $f : \mathbb{R}^2 \to \mathbb{R}$ defined by

$$f(t, s) = \begin{cases} t + s & (\text{for } t, s \in \mathbb{Q}), \\ 0 & (\text{for } t \in \mathbb{R}\backslash\mathbb{Q} \text{ or } s \in \mathbb{R}\backslash\mathbb{Q}) \end{cases}$$

is an example of a function satisfying (2.10), where H is a monotonically increasing symmetric homogeneous function of degree 1, of which point of continuity is only 0 (see [311]).

Under the additional assumption that $\sup_{s\geq 0} H(1, s) < \infty$, Th. M. Rassias and P. Šemrl [311] improved the previous result as follows:

Theorem 2.15. *Let E_1 and E_2 be those in Theorem 2.14. Suppose a function $f : E_1 \to E_2$ satisfies the inequality (2.10), where $H : [0, \infty)^2 \to [0, \infty)$ is a symmetric monotonically increasing homogeneous function of degree 1. Furthermore, assume that $\sup_{s\geq 0} H(1, s) < \infty$. Then the conditions (i), (ii), and (iii) given in Theorem 2.14 are equivalent to the following condition:*

(vi) there exists a real number M such that $\|f(x)\| \leq M\|x\|$ for all $x \in E_1$.

Proof. All we have to do is to prove that (i) implies (vi). Let us denote

$$\sup_{s\geq 0} H(1, s) = M_1.$$

Claim that

$$\|(1/k)f(kx) - f(x)\| \leq M_1 \|x\| \qquad (a)$$

for any integer $k > 1$. The verification of (a) follows by induction on k. In the case $k = 2$, we have

$$\|(1/2)f(2x) - f(x)\| \leq (1/2)H\big(\|x\|, \|x\|\big) \leq (1/2)M_1\|x\|.$$

Assume now that (a) is true for some $k > 1$ and we want to prove it for $k + 1$. Using the triangle inequality and (a), we get

$$
\begin{aligned}
\big\|(k &+ 1)^{-1} f\big((k+1)x\big) - f(x)\big\| \\
&\leq \big\|(k+1)^{-1} f\big((k+1)x\big) - (k+1)^{-1}f(kx) - (k+1)^{-1}f(x)\big\| \\
&\quad + \big\|(k+1)^{-1}f(kx) - \big(k/(k+1)\big)f(x)\big\| \\
&\leq (k+1)^{-1}H\big(\|x\|, k\|x\|\big) + \big(k/(k+1)\big)\|(1/k)f(kx) - f(x)\| \\
&\leq (k+1)^{-1}\big(M_1\|x\| + kM_1\|x\|\big) \\
&= M_1\|x\|.
\end{aligned}
$$

It follows from (i) that there exists a real number c such that (c) in the proof of Theorem 2.14 holds true for all x satisfying $1/2 \leq \|x\| \leq 1$. For any x with norm greater than 1 we can find an integer k (≥ 2) such that the vector $y = (1/k)x$ satisfies $1/2 \leq \|y\| \leq 1$. From (c) in the proof of Theorem 2.14 and (a) it follows

$$\|f(x)\| = \|f(ky)\| \leq k\|y\|(c + M_1) = \|x\|(c + M_1). \qquad (b)$$

A similar argument yields $\|f(x)\| \leq \|x\|(c + M_1)$ for any x having norm smaller than $1/2$. The relations (c) in the proof of Theorem 2.14 and (b) demonstrate that the assertion of the theorem holds true with $M = c + M_1$. □

In the following corollary, Rassias and Šemrl [311] generalized the result of Theorem 2.7, i.e., it was proved that if a function $f : E_1 \to E_2$ satisfies the inequality (2.10), where H is the same as in Theorem 2.14, and some suitable condition, then it behaves like that of Theorem 2.7.

Corollary 2.16. *Let E_1 and E_2 be a finite-dimensional real normed space with $\dim E_1 > 1$ and a real Banach space, respectively. Suppose a function $f : E_1 \to E_2$ satisfies the inequality (2.10), where $H : [0, \infty)^2 \to [0, \infty)$ is a symmetric monotonically increasing homogeneous function of degree 1. Moreover, assume that for every $x \in E_1$ there exists a positive real number δ_x such that the function $\| f(tx)\|$ is bounded on $[0, \delta_x]$. Then for every positive real number ε there exists a real number M_ε such that*

$$\|f(x)\| \leq \begin{cases} M_\varepsilon \|x\|^{1+\varepsilon} & (\text{for } \|x\| \geq 1), \\ M_\varepsilon \|x\|^{1-\varepsilon} & (\text{for } \|x\| \leq 1). \end{cases}$$

Further, assume that $\sup\limits_{s \geq 0} H(1, s) < \infty$. *Then there exists a real number* M *such that*

$$\|f(x)\| \leq M \|x\|$$

for all $x \in E_1$.

Proof. All norms on a finite-dimensional vector space are equivalent. Thus, without loss of generality, we can assume that E_1 is a Euclidean space $\mathbb{R}^k$. Let us set $\theta = H(1, 1)$. For an arbitrary pair $x, y \in E_1$ we have

$$\|f(x + y) - f(x) - f(y)\| \leq H\big(\|x\|, \|y\|\big)$$
$$\leq H\big(\|x\| + \|y\|, \|x\| + \|y\|\big)$$
$$= \theta\big(\|x\| + \|y\|\big).$$

Applying induction on n we can easily prove

$$\|f(x_1 + \cdots + x_n) - f(x_1) - \cdots - f(x_n)\|$$
$$\leq \theta(n - 1)\big(\|x_1\| + \cdots + \|x_n\|\big). \tag{a}$$

Let $\{e_1, \ldots, e_k\}$ be the standard basis in $\mathbb{R}^k$. According to the hypothesis, there exist positive real numbers $M_1, \delta_1, \ldots, \delta_k$ such that $|t| \leq \delta_i$ implies $\|f(te_i)\| \leq M_1$ for $i \in \{1, \ldots, k\}$. Choose a positive number K. Using (a) in the proof of Theorem 2.14 we can find a real number M_2 such that $|t| \leq K$ implies $\|f(te_i)\| \leq M_2$ for $i \in \{1, \ldots, k\}$.

An arbitrary vector $x \in \mathbb{R}^k$ with $\|x\| \leq K$ can be expressed in the form

$$x = t_1 e_1 + \cdots + t_k e_k,$$

where $|t_i| \leq K$ for $i \in \{1, \ldots, k\}$. It follows from (a) that

$$\|f(t_1 e_1 + \cdots + t_k e_k) - f(t_1 e_1) - \cdots - f(t_k e_k)\|$$
$$\leq \theta(k - 1)\big(\|t_1 e_1\| + \cdots + \|t_k e_k\|\big) \leq \theta(k - 1)kK.$$

Then

$$\|f(x)\| = \|f(t_1 e_1 + \cdots + t_k e_k)\|$$
$$\leq \theta(k - 1)kK + \|f(t_1 e_1)\| + \cdots + \|f(t_k e_k)\|$$
$$\leq \theta(k - 1)kK + kM_2.$$

Hence, we have proved that f is bounded on every bounded set in E_1. The assertion of our corollary is now a simple consequence of Theorems 2.14 and 2.15. $\qquad\square$

G. Isac and Th. M. Rassias [142] established a different generalization of Theorem 2.5 as follows:

Theorem 2.17 (Isac and Rassias). *Let E_1 and E_2 be a real normed space and a real Banach space, respectively. Let $\psi : [0, \infty) \to [0, \infty)$ be a function satisfying the following conditions:*

(i) $\lim\limits_{t \to \infty} \psi(t)/t = 0$,
(ii) $\psi(ts) \le \psi(t)\psi(s)$ *for all* $t, s \in [0, \infty)$,
(iii) $\psi(t) < t$ *for all* $t > 1$.

If a function $f : E_1 \to E_2$ satisfies the inequality

$$\| f(x + y) - f(x) - f(y) \| \le \theta \big(\psi(\|x\|) + \psi(\|y\|) \big) \tag{2.14}$$

for some $\theta \ge 0$ and for all $x, y \in E_1$, then there exists a unique additive function $A : E_1 \to E_2$ such that

$$\| f(x) - A(x) \| \le \frac{2\theta}{2 - \psi(2)} \psi(\|x\|) \tag{2.15}$$

for all $x \in E_1$. Moreover, if $f(tx)$ is continuous in t for each fixed x, then the function A is linear.

Proof. We will first prove that

$$\| 2^{-n} f(2^n x) - f(x) \| \le \theta \psi(\|x\|) \sum_{m=0}^{n-1} \big(\psi(2)/2 \big)^m \tag{a}$$

for any $n \in \mathbb{N}$ and for all $x \in E_1$. The proof of (a) follows by induction on n. The assertion for $n = 1$ is clear by (2.14). Assume now that (a) is true for $n > 0$ and we want to prove it for the case $n + 1$. Replacing x with $2x$ in (a) we obtain

$$\| 2^{-n} f(2^n 2x) - f(2x) \| \le \theta \psi(2\|x\|) \sum_{m=0}^{n-1} \big(\psi(2)/2 \big)^m .$$

By *(ii)* we get

$$\| 2^{-n} f(2^{n+1} x) - f(2x) \| \le \theta \psi(2) \psi(\|x\|) \sum_{m=0}^{n-1} \big(\psi(2)/2 \big)^m . \tag{b}$$

Multiplying both sides of (b) by $1/2$ we have

$$\| 2^{-n-1} f(2^{n+1} x) - (1/2) f(2x) \| \le \theta \psi(\|x\|) \sum_{m=1}^{n} \big(\psi(2)/2 \big)^m .$$

Using the triangle inequality, we now deduce

$$\|2^{-n-1}f(2^{n+1}x) - f(x)\|$$
$$\leq \|2^{-n-1}f(2^{n+1}x) - (1/2)f(2x)\| + \|(1/2)f(2x) - f(x)\|$$
$$\leq \theta\psi(\|x\|)\sum_{m=1}^{n}(\psi(2)/2)^m + \theta\psi(\|x\|)$$
$$= \theta\psi(\|x\|)\sum_{m=0}^{n}(\psi(2)/2)^m,$$

which ends the proof of (a).

Thus, it follows from (a) that

$$\|2^{-n}f(2^n x) - f(x)\| \leq \frac{2\theta\psi(\|x\|)}{2 - \psi(2)} \qquad (c)$$

for any $n \in \mathbb{N}$.

For $m > n > 0$ we obtain

$$\|2^{-m}f(2^m x) - 2^{-n}f(2^n x)\| = 2^{-n}\|2^{-(m-n)}f(2^m x) - f(2^n x)\|$$
$$= 2^{-n}\|2^{-r}f(2^r y) - f(y)\|,$$

where $r = m - n$ and $y = 2^n x$. Hence,

$$\|2^{-m}f(2^m x) - 2^{-n}f(2^n x)\| \leq 2^{-n}\theta\frac{2\psi(\|y\|)}{2 - \psi(2)}$$
$$\leq 2^{-n}\theta\frac{2\psi(2^n)\psi(\|x\|)}{2 - \psi(2)}$$
$$\leq \left(\frac{\psi(2)}{2}\right)^n\theta\frac{2\psi(\|x\|)}{2 - \psi(2)}.$$

However, by (iii), the Rassias sequence $\{2^{-n}f(2^n x)\}$ is a Cauchy sequence. Let us define

$$A(x) = \lim_{n\to\infty} 2^{-n}f(2^n x)$$

for all $x \in E_1$.

We will prove that A is additive. Let $x, y \in E_1$ be given. It then follows from (2.14) and (ii) that

$$\|f(2^n(x + y)) - f(2^n x) - f(2^n y)\| \leq \theta(\psi(\|2^n x\|) + \psi(\|2^n y\|))$$
$$\leq \theta\psi(2^n)(\psi(\|x\|) + \psi(\|y\|)),$$

which implies that

$$2^{-n} \left\| f\left(2^n(x+y)\right) - f(2^n x) - f(2^n y) \right\| \le \left(\psi(2)/2\right)^n \theta\left(\psi(\|x\|) + \psi(\|y\|)\right).$$

Using (iii) and letting $n \to \infty$, we conclude that A is additive.

By letting $n \to \infty$ in (c), we obtain the inequality (2.15).

Claim that A is such a unique additive function. Assume that there exists another one, denoted by $A' : E_1 \to E_2$, satisfying

$$\|f(x) - A'(x)\| \le \frac{2\theta'}{2 - \psi'(2)} \psi'(\|x\|), \tag{d}$$

where θ' (≥ 0) is a constant and $\psi' : [0, \infty) \to [0, \infty)$ is a function satisfying (i), (ii), and (iii). By (2.15) and (d), we get

$$\|A(x) - A'(x)\| \le \|A(x) - f(x)\| + \|f(x) - A'(x)\|$$
$$\le \frac{2\theta}{2 - \psi(2)} \psi(\|x\|) + \frac{2\theta'}{2 - \psi'(2)} \psi'(\|x\|).$$

Then,

$$\|A(x) - A'(x)\| = (1/n)\|A(nx) - A'(nx)\|$$
$$\le \frac{\psi(n)}{n} \frac{2\theta \psi(\|x\|)}{2 - \psi(2)} + \frac{\psi'(n)}{n} \frac{2\theta' \psi'(\|x\|)}{2 - \psi'(2)},$$

for every integer $n > 1$. In view of (i) and the last inequality, we conclude that $A(x) = A'(x)$ for all $x \in E_1$.

Because of the additivity of A it follows that $A(qx) = qA(x)$ for any $q \in \mathbb{Q}$. Using the same argument as in Theorem 2.5, we obtain that $A(ax) = aA(x)$ for all real numbers a. Hence, A is a linear function.   □

G. Isac and Th. M. Rassias [142] remarked that if $\psi(t) = t^p$ with $0 \le p < 1$, then from the last theorem we get the result of Theorem 2.5. If $p < 0$ and

$$\psi(t) = \begin{cases} 0 & (\text{for } t = 0), \\ t^p & (\text{for } t > 0), \end{cases}$$

then from Theorem 2.17 we obtain a generalization of Theorem 2.5 for $p < 0$ (ref. [112]). If $f(S)$ is bounded, where $S = \{x \in E_1 \mid \|x\| = 1\}$, then the A given in Theorem 2.17 is continuous. Indeed, this is a consequence of the inequalities

$$\|A(x)\| \le \|f(x)\| + \|A(x) - f(x)\|$$
$$\le \|f(x)\| + \frac{2\theta}{2 - \psi(2)} \psi(\|x\|)$$
$$\le \|f(x)\| + \frac{2\theta}{2 - \psi(2)} \psi(1)$$

for all $x \in S$.

The control functions H and ψ appearing in Theorems 2.11 and 2.17 were remarkably generalized and the Hyers–Ulam–Rassias stability with the generalized control function was also proved by P. Găvruta. In the following theorem, we will introduce his result [113].

Theorem 2.18 (Găvruta). *Let G and E be an abelian group and a Banach space, respectively, and let $\varphi : G^2 \to [0, \infty)$ be a function satisfying*

$$\Phi(x, y) = \sum_{k=0}^{\infty} 2^{-k-1} \varphi(2^k x, 2^k y) < \infty \tag{2.16}$$

for all $x, y \in G$. If a function $f : G \to E$ satisfies the inequality

$$\|f(x + y) - f(x) - f(y)\| \leq \varphi(x, y) \tag{2.17}$$

for any $x, y \in G$, then there exists a unique additive function $A : G \to E$ with

$$\|f(x) - A(x)\| \leq \Phi(x, x) \tag{2.18}$$

for all $x \in G$. Moreover, if $f(tx)$ is continuous in t for each fixed $x \in G$, then A is a linear function.

Proof. Putting $y = x$ in the inequality (2.17) yields

$$\|(1/2)f(2x) - f(x)\| \leq (1/2)\varphi(x, x) \tag{a}$$

for all $x \in G$. Applying an induction argument to n, we will prove

$$\|2^{-n} f(2^n x) - f(x)\| \leq \sum_{k=0}^{n-1} 2^{-k-1} \varphi(2^k x, 2^k x) \tag{b}$$

for any $x \in G$. Indeed,

$$\|2^{-n-1} f(2^{n+1} x) - f(x)\| \leq \|2^{-n-1} f(2^{n+1} x) - (1/2)f(2x)\| + \|(1/2)f(2x) - f(x)\|,$$

and by (a) and (b), we obtain

$$\|2^{-n-1} f(2^{n+1} x) - f(x)\|$$
$$\leq (1/2) \sum_{k=0}^{n-1} 2^{-k-1} \varphi(2^{k+1} x, 2^{k+1} x) + (1/2)\varphi(x, x)$$
$$= \sum_{k=0}^{n} 2^{-k-1} \varphi(2^k x, 2^k x),$$

which ends the proof of (b).

We will present that the Rassias sequence $\{2^{-n} f(2^n x)\}$ is a Cauchy sequence. Indeed, for $n > m > 0$, we have

$$\|2^{-n} f(2^n x) - 2^{-m} f(2^m x)\|$$

$$= 2^{-m} \|2^{-(n-m)} f(2^{n-m} 2^m x) - f(2^m x)\|$$

$$\leq 2^{-m} \sum_{k=0}^{n-m-1} 2^{-k-1} \varphi(2^{k+m} x, 2^{k+m} x)$$

$$= \sum_{k=m}^{n-1} 2^{-k-1} \varphi(2^k x, 2^k x).$$

Taking the limit as $m \to \infty$ and considering (2.16), we obtain

$$\lim_{m \to \infty} \|2^{-n} f(2^n x) - 2^{-m} f(2^m x)\| = 0.$$

Since E is a Banach space, it follows that the sequence $\{2^{-n} f(2^n x)\}$ converges. Let us denote

$$A(x) = \lim_{n \to \infty} 2^{-n} f(2^n x).$$

It follows from (2.17) that

$$\|f(2^n (x + y)) - f(2^n x) - f(2^n y)\| \leq \varphi(2^n x, 2^n y)$$

for all $x, y \in G$. Therefore,

$$\|2^{-n} f(2^n (x + y)) - 2^{-n} f(2^n x) - 2^{-n} f(2^n y)\| \leq 2^{-n} \varphi(2^n x, 2^n y). \qquad (c)$$

It follows from (2.16) that

$$\lim_{n \to \infty} 2^{-n} \varphi(2^n x, 2^n y) = 0.$$

Thus, (c) implies that $A : G \to E$ is an additive function.

Taking the limit in (b) as $n \to \infty$, we obtain the inequality (2.18).

It remains to show that A is uniquely defined. Let $A' : G \to E$ be another additive function satisfying (2.18). Then, we get

$$\|A(x) - A'(x)\| = \|2^{-n} A(2^n x) - 2^{-n} A'(2^n x)\|$$

$$\leq \|2^{-n} A(2^n x) - 2^{-n} f(2^n x)\|$$

$$+ \|2^{-n} f(2^n x) - 2^{-n} A'(2^n x)\|$$

$$\leq 2^{-n} \Phi(2^n x, 2^n x) + 2^{-n} \Phi(2^n x, 2^n x)$$

$$= 2^{-n} \sum_{k=0}^{\infty} 2^{-k} \varphi\left(2^{k+n}x, 2^{k+n}x\right)$$

$$= \sum_{k=n}^{\infty} 2^{-k} \varphi\left(2^{k}x, 2^{k}x\right)$$

for all $n \in \mathbb{N}$. Taking the limit in the above inequality as $n \to \infty$, we get

$$A(x) = A'(x)$$

for all $x \in G$.

From the additivity of A it follows that $A(qx) = qA(x)$ for any $q \in \mathbb{Q}$. Using the same argument as in Theorem 2.5, we obtain that $A(ax) = aA(x)$ for all real numbers a. Hence, A is a linear function. $\square$

Later, S.-M. Jung complemented Theorem 2.18 by proving a theorem which includes the following corollary as a special case (see [173, Theorem 4]).

Corollary 2.19. *Let E_1 and E_2 be a real normed space and a Banach space, respectively. Assume that a function $\varphi : E_1^2 \to [0, \infty)$ satisfies*

$$\Phi(x, y) = \sum_{k=1}^{\infty} 2^{k} \varphi\left(2^{-k}x, 2^{-k}y\right) < \infty$$

for all $x, y \in E_1$. If a function $f : E_1 \to E_2$ satisfies the inequality (2.17) for any $x, y \in E_1$, then there exists a unique additive function $A : E_1 \to E_2$ such that

$$\|f(x) - A(x)\| \leq (1/2)\Phi(x, x)$$

for all $x \in E_1$.

S.-M. Jung [156] has further generalized the result of Theorem 2.18 by making use of an idea from the previous theorem. In the following theorem, let G be an abelian group and E be a Banach space. Consider a function $\varphi : G^2 \to [0, \infty)$ satisfying $\varphi(x, y) = \varphi(y, x)$ for all $x, y \in G$. For all $n \in \mathbb{N}$ and all $x, y \in G$ define $\varphi_n^1(x, y) = \varphi(nx, y)$ and $\varphi_n^2(x, y) = \varphi(x, ny)$. By $a = (a_1, a_2, \ldots)$ we denote a sequence with $a_n \in \{1, 2\}$ for all $n \in \mathbb{N}$, and we define $\psi_k^a(x, y) = \varphi_1^{a_1}(x, y) + \cdots + \varphi_k^{a_k}(x, y)$ for a fixed integer $k > 1$. Suppose that there exists a sequence $a = (a_1, a_2, \ldots)$ with $a_n \in \{1, 2\}$ for all $n \in \mathbb{N}$ such that

$$\Psi_k(x, y) = \sum_{n=1}^{\infty} k^{-n} \psi_{k-1}^a\left(k^{n-1}x, k^{n-1}y\right) < \infty \tag{2.19}$$

for all $x, y \in G$. With these notations, Jung [156] proved the following theorem.

Theorem 2.20. *Suppose* $f : G \to E$ *is a function satisfying*

$$\|f(x + y) - f(x) - f(y)\| \leq \varphi(x, y) \tag{2.20}$$

for all $x, y \in G$. *Then there exists a unique additive function* $A : G \to E$ *such that*

$$\|f(x) - A(x)\| \leq \Psi_k(x, x) \tag{2.21}$$

for all $x \in G$. *Moreover, if* G *is a Banach space and* $f(tx)$ *is continuous in* t *for every fixed* $x \in G$, *then* A *is linear.*

Proof. We first claim

$$\|(1/n)f(nx) - f(x)\| \leq (1/n)\psi_{n-1}^a(x, x) \tag{a}$$

for each integer $n > 1$ and all $x \in G$. We verify it by induction on n. By putting $y = x$ in (2.20), we obtain

$$\|f(2x) - 2f(x)\| \leq \varphi(x, x) = \psi_1^a(x, x).$$

This implies the validity of (a) for the case $n = 2$. Assume now that the inequality (a) is valid for $n = m$ $(m \geq 2)$, i.e.,

$$\|f(mx) - mf(x)\| \leq \psi_{m-1}^a(x, x). \tag{b}$$

For the case $n = m + 1$, replacing y with mx in (2.20), we get

$$\|f(x + mx) - f(x) - f(mx)\| \leq \varphi(x, mx) = \varphi(mx, x). \tag{c}$$

It follows from (b) and (c) that

$$\begin{aligned}
\|f((m+1)x) &- (m+1)f(x)\| \\
&\leq \|f((m+1)x) - f(x) - f(mx)\| + \|f(mx) - mf(x)\| \\
&\leq \psi_m^a(x, x).
\end{aligned}$$

Accordingly, the assertion (a) is true for all integers $n > 1$ and all $x \in G$.

We claim

$$\|k^{-n}f(k^n x) - f(x)\| \leq \sum_{i=1}^{n} k^{-i}\psi_{k-1}^a(k^{i-1}x, k^{i-1}x) \tag{d}$$

for each $n \in \mathbb{N}$. We also prove it by induction on n. The validity of (d) for $n = 1$ follows from (a). By assuming the induction argument for $n = m$ and putting $n = 1$ and then substituting $k^m x$ for x in (d), we obtain

$$\left\| k^{-m-1} f\left(k^{m+1}x\right) - f(x) \right\|$$
$$\leq \left\| k^{-m-1} f\left(k^{m+1}x\right) - k^{-m} f(k^m x) \right\| + \left\| k^{-m} f(k^m x) - f(x) \right\|$$
$$\leq k^{-m} \left\| (1/k) f(k \cdot k^m x) - f(k^m x) \right\| + \sum_{i=1}^{m} k^{-i} \psi_{k-1}^a\left(k^{i-1}x, k^{i-1}x\right)$$
$$\leq k^{-m-1} \psi_{k-1}^a\left(k^m x, k^m x\right) + \sum_{i=1}^{m} k^{-i} \psi_{k-1}^a\left(k^{i-1}x, k^{i-1}x\right)$$
$$= \sum_{i=1}^{m+1} k^{-i} \psi_{k-1}^a\left(k^{i-1}x, k^{i-1}x\right).$$

Hence, the inequality (d) is true for all $n \in \mathbb{N}$.

We now claim that the Rassias sequence $\{k^{-n} f(k^n x)\}$ is a Cauchy sequence. Indeed, by (d), we have

$$\left\| k^{-n} f(k^n x) - k^{-m} f(k^m x) \right\|$$
$$= k^{-m} \left\| k^{-(n-m)} f(k^{n-m} \cdot k^m x) - f(k^m x) \right\|$$
$$\leq k^{-m} \sum_{i=1}^{n-m} k^{-i} \psi_{k-1}^a\left(k^{i-1} \cdot k^m x, \ k^{i-1} \cdot k^m x\right)$$
$$\leq \sum_{i=m+1}^{\infty} k^{-i} \psi_{k-1}^a\left(k^{i-1}x, k^{i-1}x\right)$$

for $n > m$. In view of (2.19), we can make the last term as small as possible by selecting sufficiently large m. Therefore, the given sequence is a Cauchy sequence. Since E is a Banach space, the sequence $\{k^{-n} f(k^n x)\}$ converges for every $x \in G$. Thus, we may define

$$A(x) = \lim_{n \to \infty} k^{-n} f(k^n x).$$

We claim that the function A is additive. By substituting $k^n x$ and $k^n y$ for x and y in (2.20), respectively, we get

$$\left\| f\left(k^n (x + y)\right) - f(k^n x) - f(k^n y) \right\| \leq \varphi(k^n x, k^n y). \qquad (e)$$

However, it follows from (2.19) that

$$\lim_{n \to \infty} k^{-n} \psi_{k-1}^a(k^n x, k^n y) = 0.$$

Therefore, dividing both sides of (e) by k^n and letting $n \to \infty$, we conclude that A is additive.

According to (d), the inequality (2.21) holds true for all $x \in G$.

We assert that A is uniquely determined. Let $A' : G \to E$ be another additive function with the property (2.21). Since A and A' are additive functions satisfying (2.21), we have

$$
\begin{aligned}
\|A(x) - A'(x)\| &= \|k^{-n} A(k^n x) - k^{-n} A'(k^n x)\| \\
&\leq k^{-n} \|A(k^n x) - f(k^n x)\| + k^{-n} \|f(k^n x) - A'(k^n x)\| \\
&\leq k^{-n} \Psi_k(k^n x, k^n x) + k^{-n} \Psi_k(k^n x, k^n x) \\
&= 2 \sum_{m=n+1}^{\infty} k^{-m} \psi_{k-1}^a\big(k^{m-1} x, k^{m-1} x\big).
\end{aligned}
$$

In view of (2.19), we can make the last term as small as possible by selecting sufficiently large n. Hence, it follows that $A = A'$.

Finally, it can also be proved that if G is a Banach space and $f(tx)$ is continuous in t for every fixed $x \in G$, then A is a linear function in the same way as in Theorem 2.5. $\qquad\qquad\qquad\qquad\qquad\qquad\qquad\qquad\qquad\qquad\qquad\qquad\qquad$ $\square$

It is worthwhile to note that P. Găvruta, M. Hossu, D. Popescu, and C. Căprău obtained the result of the previous theorem in the papers [118, 119] independently. Furthermore, Y.-H. Lee and K.-W. Jun generalized Theorem 2.20. Indeed, they replaced k with a rational number $a > 1$ in the condition (2.19) and proved Theorem 2.20 (see [233]).

There are still new valuable results for the Hyers–Ulam–Rassias stability of the additive Cauchy equation which were not cited above. Among them we have to state here an outstanding result of G. Isac and Th. M. Rassias [143] without proof:

Let E_1 and E_2 be a real normed space and a real Banach space, respectively. Assume that $f : E_1 \to E_2$ is a function such that $f(tx)$ is continuous in t for every fixed x in E_1. Assume that there exist $\theta \geq 0$ and $p_1, p_2 \in \mathbb{R}$ such that $p_2 \leq p_1 < 1$ or $1 < p_2 \leq p_1$. If f satisfies the inequality

$$
\|f(x + y) - f(x) - f(y)\| \leq \theta\big(\|x\|^{p_1} + \|y\|^{p_2}\big)
$$

for all $x, y \in E_1$, there exists a unique linear function $A : E_1 \to E_2$ such that

$$
\|f(x) - A(x)\| \leq
\begin{cases}
\theta(2 - 2^{p_1})^{-1}\big(\|x\|^{p_1} + \|x\|^{p_2}\big) & (\text{if } p_2 \leq p_1 < 1), \\
\theta(2^{p_2} - 2)^{-1}\big(\|x\|^{p_1} + \|x\|^{p_2}\big) & (\text{if } 1 < p_2 \leq p_1)
\end{cases}
$$

for any $x \in E_1$.

During the 31st International Symposium on Functional Equations, Th. M. Rassias [292] raised an open problem whether we can also expect a similar result for $p_2 < 1 < p_1$.

We here remark that J. M. Rassias [280] considered the case where the Cauchy difference $\|f(x + y) - f(x) - f(y)\|$ in (2.5) is bounded by $\theta \|x\|^p \|y\|^p$ ($\theta \geq 0$,

$0 \leq p < 1/2$) and obtained a similar result to that of Theorem 2.5 except the bound for the difference $\|f(x) - A(x)\|$ in (2.6) bounded by $\theta \|x\|^{2p}/(2 - 2^p)$ instead of $2\theta \|x\|^p/(2 - 2^p)$. Furthermore, he [281] also proved the Hyers–Ulam–Rassias stability of the additive Cauchy equation for the case where the norm of the difference in (2.5) is bounded by $\theta \|x\|^p \|y\|^q$ ($\theta \geq 0, 0 \leq p + q < 1$) with the modified approximation bound $\theta \|x\|^{p+q}/(2 - 2^{p+q})$ for $\|f(x) - A(x)\|$ in (2.6).

Several mathematicians have remarked interesting applications of the Hyers–Ulam–Rassias stability theory to various mathematical problems. We will now present some applications of this theory to the study of nonlinear analysis, especially in fixed point theory.

In nonlinear analysis it is well-known that finding the expression of the asymptotic derivative of a nonlinear operator can be a difficult problem. In this sense, we will explain how the Hyers–Ulam–Rassias stability theory can be used to evaluate the asymptotic derivative of some nonlinear operators.

The nonlinear problems considered in this book have been extensively studied by several mathematicians (cf. [6, 54, 226, 242, 364]). G. Isac and Th. M. Rassias were the first mathematicians to introduce the use of the Hyers–Ulam–Rassias stability theory for the study of these problems (see [144]).

Let E be a Banach space. A closed subset K of E is said to be a *cone* if it satisfies the following properties:

(C1) $K + K \subset K$;
(C2) $\lambda K \subset K$ for all $\lambda \geq 0$;
(C3) $K \cap (-K) = \{0\}$.

By K^* we denote the *dual* of K, i.e., $K^* = \{\phi \in E^* \mid \phi(x) \geq 0 \text{ for all } x \in K\}$. It is not difficult to see that each cone $K \subset E$ induces an *ordering* on E by the hypothesis that $x \leq y$ if and only if $y - x \in K$. If in E a cone is defined, then E is called an *ordered Banach space*. A cone $K \subset E$ is said to be *generating* if $E = K - K$ and it is said to be *normal* if there exists a $\delta \geq 1$ such that $\|x\| \leq \delta \|x + y\|$ for all $x, y \in K$. We say that a cone $K \subset E$ is *solid* if its topological interior is nonempty. We call a point $x_0 \in K$ a *quasi-interior point* if $\phi(x_0) > 0$ for any nonzero $\phi \in K^*$. If the cone $K \subset E$ is solid, then the quasi-interior points of K coincide precisely with its interior points.

We denote by $L(E, E)$ the space of linear bounded operators from E into E. It is well-known that for every $T \in L(E, E)$ the *spectral radius* $r(T)$ is well-defined, where $r(T) = \max\{|\lambda| \mid \lambda \in \sigma(T)\}$ and $\sigma(T)$ is the *spectrum* of T. We say that $T \in L(E, E)$ is *strictly monotone increasing* if for every pair $x, y \in E$ the relation $x < y$ (i.e., $x \leq y$ and $x \neq y$) implies $T(y) - T(x) \in K^\circ$, where K° denotes the interior of K.

Let $D \subset E$ be a bounded set. We define $\gamma(D)$, the *measure of noncompactness* of D, to be the minimum of all positive numbers δ such that D can be covered by finitely many sets of diameter less than δ. A function $f : E \to E$ is said to be a *k-set-contraction* if it is continuous and there exists a $k \geq 0$ such that $\gamma(f(D)) \leq k\gamma(D)$ for every bounded subset D of the domain of f. A function $f : E \to E$ is said to be a *strict-set-contraction* if it is a k-set-contraction for some

$k < 1$. A function $f : E \to E$ is called *compact* if it maps bounded subsets of the domain of f onto relatively compact subsets of E and f is said to be *completely continuous* if it is continuous and compact. Every completely continuous function is a strict-set-contraction.

Let K be a generating (or *total*, i.e., $E = \overline{K - K}$) cone in E. The function $f : K \to E$ is said to be *asymptotically differentiable along K* if there exists a $T \in L(E, E)$ such that

$$\lim_{\substack{x \in K \\ \|x\| \to \infty}} \|f(x) - T(x)\| / \|x\| = 0.$$

In this case, T is the unique function and we call it the *derivative at infinity along K of f*. We say that a function $f : E \to E$ is *asymptotically close to zero along K* if

$$\lim_{\substack{x \in K \\ \|x\| \to \infty}} \|f(x)\| / \|x\| = 0.$$

Let $\phi : [0, \infty) \to [0, \infty)$ be a function such that $\phi(t) > 0$ for all $t \geq \gamma$, where $\gamma \geq 0$. We say that $f : K \to E$ is *ϕ-asymptotically bounded along K* if there exist $b, c > 0$ such that for all $x \in K$, with $\|x\| \geq b$, we have $\|f(x)\| \leq c\phi(\|x\|)$. Every ϕ-asymptotically bounded function (along K) such that $\lim_{t \to \infty} \phi(t)/t = 0$ is asymptotically close to zero.

If K is a generating (or total) cone in E, then a function $f : K \to E$ is said to be *differentiable at $x_0 \in K$ along K* if there exists $f'(x_0) \in L(E, E)$ such that

$$\lim_{\substack{x \in K \\ x \to 0}} \|f(x_0 + x) - f(x_0) - f'(x_0)x\| / \|x\| = 0.$$

In this case, $f'(x_0)$ is the *derivative at x_0 along K of f* and it is uniquely determined.

To enlarge the class of the functions ψ such that the condition of Theorem 2.17 remains valid, we consider the following: Let F_ψ be the set of all functions $\psi : [0, \infty) \to [0, \infty)$ satisfying the conditions (*i*), (*ii*), and (*iii*) in Theorem 2.17. Let $P(\Psi)$ be the convex cone generated by the set F_ψ. We remark that a function $\psi \in P(\Psi)$ satisfies the condition (*i*) but generally does not satisfy the conditions (*ii*) and (*iii*) in Theorem 2.17. However, G. Isac and Th. M. Rassias [144] presented that Theorem 2.17 remains valid for each $\psi \in P(\Psi)$.

Theorem 2.21. *Let E_1 be a real normed space, E_2 a real Banach space, and $f : E_1 \to E_2$ a continuous function. Let $\psi \in P(\Psi)$ be given. If f satisfies the inequality (2.14) for some $\theta \geq 0$ and for all $x, y \in E_1$, then there exists a unique linear function $T : E_1 \to E_2$ satisfying the inequality (2.15) for all $x \in E_1$.*

Proof. We apply Theorem 2.18 for the function

$$\varphi(x, y) = \theta\big(\psi(\|x\|) + \psi(\|y\|)\big).$$

In this case, using the properties of the functions $\psi \in P(\Psi)$, we can show that $\Phi(x, y) < \infty$ for all $x, y \in E_1$, and the conclusion of our theorem follows.   □

The last theorem is significant because the class of functions satisfying (2.14) with $\psi \in P(\Psi)$ is strictly larger than the class of functions defined in Theorem 2.17. G. Isac and Th. M. Rassias [144] also proved the following theorem:

Theorem 2.22. *Let E be a Banach space ordered by a generating cone K and let $f : E \to E$, $g : K \to K$ be functions such that:*

(f1) *f is completely continuous, positive, and satisfies the inequality (2.14) for some $\psi \in P(\Psi)$ and $\theta > 0$ (i.e., $f(K) \subset K$);*

(f2) *there exist a quasi-interior point $x_0 \in K$ and $0 < \lambda_0 < 1$ such that $\lim\limits_{n\to\infty} 2^{-n} f(2^n x_0) \le \lambda_0 x_0$;*

(g1) *g is asymptotically close to zero along K;*

(h1) *$h = f + g$ is a strict-set-contraction from K to K.*

Then $h = f + g$ has a fixed point in K.

Proof. Let $S = \{x \in E \mid \|x\| = 1\}$. By assumption $(f1)$ and Theorem 2.21, we have that $T(x) = \lim\limits_{n\to\infty} 2^{-n} f(2^n x)$ is well defined for every $x \in E$ and T is the unique linear operator satisfying the inequality

$$\|f(x) - T(x)\| \le \frac{2\theta}{2 - \psi(2)} \psi(\|x\|) \tag{a}$$

for all $x \in E$. Since f is compact, we have that $f(S)$ is bounded, which implies that T is continuous. Indeed, the continuity of T is a consequence of the following inequalities:

$$\|T(x)\| \le \|f(x)\| + \|T(x) - f(x)\|$$
$$\le \|f(x)\| + \frac{2\theta}{2 - \psi(2)} \psi(\|x\|)$$
$$\le \|f(x)\| + \frac{2\theta}{2 - \psi(2)} \psi(1)$$

for all $x \in S$.

From the definition of T and the fact that $f(K) \subset K$, we deduce that T is positive (i.e., $T(K) \subset K$). From (a) and the properties of ψ, it follows that

$$\lim_{\substack{x \in K \\ \|x\| \to \infty}} \|f(x) - T(x)\| / \|x\| = 0,$$

i.e., T is the asymptotic derivative of f along K. In addition, from the principal theorem of [141] or [6], T is completely continuous (and it is also a strict-set-contraction). From $(g1)$ we have

$$\lim_{\substack{x \in K \\ \|x\| \to \infty}} \|h(x) - T(x)\|/\|x\|$$

$$\leq \lim_{\substack{x \in K \\ \|x\| \to \infty}} \|g(x)\|/\|x\| + \lim_{\substack{x \in K \\ \|x\| \to \infty}} \|f(x) - T(x)\|/\|x\|$$

$$= 0,$$

i.e., T is also the asymptotic derivative of h along K.

Since T is completely continuous, its spectrum consists of eigenvalues and zero. Suppose that $r(T) > 0$. From $(f2)$ we have that $T(x_0) \leq \lambda_0 x_0$ and using the Krein–Rutman theorem ([364, Proposition 7.26]) we have that there exists $\phi_0 \in K^* \setminus \{0\}$ such that $T^*(\phi_0) = r(T)\phi_0$ and $\phi_0(x_0) > 0$ (since x_0 is a quasi-interior point of K), where we denote by T^* the adjoint of T. Hence, we deduce

$$r(T) = \frac{\left(T^*(\phi_0)\right)(x_0)}{\phi_0(x_0)} = \frac{\phi_0\left(T(x_0)\right)}{\phi_0(x_0)} \leq \frac{\phi_0(\lambda_0 x_0)}{\phi_0(x_0)} = \lambda_0,$$

i.e., $r(T) < 1$.

Now, all the assumptions of [6, Theorem 1] are satisfied and hence $h = f + g$ has a fixed point in K. $\square$

Corollary 2.23. *Let E be a Banach space ordered by a generating cone K and let $f : E \to E$ be a function satisfying the conditions $(f1)$ and $(f2)$. Then f has a fixed point in K.*

In Theorem 2.22 and Corollary 2.23 we can replace $(f2)$ with the following condition:

$(f3)$ $\|f(x) - \lambda x\| > 2\theta\left(2 - \psi(2)\right)^{-1} \psi\left(\|x\|\right)$ *for all* $\lambda \geq 1$ *and* $x \in K \setminus \{0\}$.

G. Isac and Th. M. Rassias [144] investigated the existence of nonzero positive fixed points. We will express the result of G. Isac and Th. M. Rassias in the following theorem.

Theorem 2.24. *Let E be a Banach space ordered by a generating cone K and let $f : E \to E$, $g : K \to K$ be functions satisfying $(f1)$, $(g1)$, $(h1)$, and:*

$(f4)$ *there exist $\lambda_0 > 1$ and $x_0 \notin -K$ such that* $\lim_{n \to \infty} 2^{-n} f(2^n x_0) \geq \lambda_0 x_0$;

$(f5)$ $\|f(x) - x\| > 2\theta\left(2 - \psi(2)\right)^{-1} \psi\left(\|x\|\right)$ *for all* $x \in K \setminus \{0\}$;

$(h2)$ *h is differentiable at 0 along K and $h(0) = 0$;*

$(h3)$ *$h'(0)$ does not have a positive eigenvector belonging to an eigenvalue $\lambda \geq 1$.*

Then $h = f + g$ has a fixed point in $K \setminus \{0\}$.

Proof. As in the proof of Theorem 2.22, h is asymptotically differentiable along K and its derivative at infinity along K is $T(x) = \lim\limits_{n\to\infty} 2^{-n} f(2^n x)$ for each $x \in E$. Moreover, T is completely continuous and the inequality (a) in the proof of Theorem 2.22 is also satisfied. It follows from $(f5)$ that 1 is not an eigenvalue with corresponding positive eigenvector of T. From $(f4)$ we obtain $r(T) \geq \lambda_0$. Indeed, if $r(T) < \lambda_0$ and since $r(T) = \lim\limits_{n\to\infty} \|T^n\|^{1/n}$ (Gelfand's formula), $\lambda_0^{-n}\|T^n\| \leq k^n$ for sufficiently large n and some $k < 1$. From the fact that $T(x_0) \geq \lambda_0 x_0$ we deduce $\lambda_0^{-n} T^n(x_0) \geq x_0$ (since T is positive) and if we pass to the limit in the last relation, we obtain $x_0 \leq 0$, i.e., $x_0 \in -K$, which is a contradiction. Using the Krein–Rutman theorem again, we have that $r(T)$ is an eigenvalue of T with an eigenvector in K. Thus, all the assumptions of [54, Theorem 1] are satisfied and we conclude that h has a fixed point in $K \setminus \{0\}$. □

Let E be a Banach space ordered by a cone K, and let $L, A : E \to E$ be functions. We say that $\lambda > 0$ is an *asymptotic characteristic value* of (L, A) if L and λA are *asymptotically equivalent* (with respect to K), i.e.,

$$\lim_{\substack{x \in K \\ \|x\| \to 0}} \|L(x) - \lambda A(x)\|/\|x\| = 0.$$

An asymptotic characteristic value of (L, A) is a characteristic value of (L, A) in the sense of the definition provided by M. Mininni in [242], i.e., λ is a *characteristic value* of (L, A) *in Mininni's sense*, if there exists a sequence $\{x_n\}$ of elements of K such that

$$\lim_{n\to\infty} \|x_n\| = \infty \quad \text{and} \quad \lim_{n\to\infty} \|L(x_n) - \lambda A(x_n)\|/\|x_n\| = 0.$$

The following result is a consequence of Theorem 2.24.

Corollary 2.25. *Let E be a Banach space ordered by a generating cone K ($\subset E$), let $f : E \to E$ be a function such that $f(K) \subset K$, and let $L, A : E \to E$ be functions. Assume that the following conditions are satisfied:*

(i) f satisfies the conditions (f1), (f4), and (f5);
(ii) λ is an asymptotic characteristic value of (L, A);
(iii) $h = f + L - \lambda A$ satisfies the conditions (h1), (h2), and (h3).

Then the nonlinear eigenvalue problem, $L(x_*) + f(x_*) = x_* + \lambda A(x_*)$ with unknowns $x_* \in E \setminus \{0\}$ and $\lambda > 0$, has a solution.

It is well-known that the study of the nonlinear integral equation

$$x(u) = \int_\Omega G(u, v) f(v, x(v)) dv, \qquad (\alpha)$$

known as the *Hammerstein equation*, is of central importance in the study of several boundary-value problems (cf. [226, 364]).

In addition, some special interest is focused on the eigenvalue problem

$$x(u) = \lambda \int_{\Omega} G(u, v) f(v, x(v)) dv. \qquad (\beta)$$

If we denote by G the linear integral operator defined by the kernel $G(u, v)$ and by f the Nemyckii's nonlinear operator defined by $f(v, x(v))$, i.e., $f(u)(v) = f(v, x(v))$, then the equation (β) takes the abstract form

$$x = \lambda G f(x). \qquad (\gamma)$$

For the equation (γ) we consider the following hypotheses:

(H1) (E, K) and (F, P) are real ordered Banach spaces. The cone K is normal with nonempty interior.
(H2) The function $f : E \to F$ is continuous and the operator $G : F \to E$ is linear, compact, and positive.
(H3) G is strongly positive, i.e., $x < y$ implies $G(y) - G(x) \in K^{\circ}$.

We recall that $(\lambda_*, +\infty)$ is a bifurcation from infinity of equation (α) if $\lambda_* > 0$ and there is a sequence of solutions (λ_n, x_n) of (γ) such that $\lambda_n \to \lambda_*$ and $\|x_n\| \to \infty$ as $n \to \infty$.

Isac and Rassias [144] also contributed to the following theorem.

Theorem 2.26. *Consider equation (γ) and suppose that (H1), (H2), and (H3) are satisfied. In addition, assume that the function $f : E \to F$ satisfies the following conditions:*

(i) $f(K) \subset K$ and $f(S)$ is bounded, where $S = \{x \in E \mid \|x\| = 1\}$;
(ii) f satisfies the inequality (2.14) for some $\theta \geq 0$, $\psi \in P(\Psi)$, and for all $x, y \in E$;
(iii) $\lim_{n \to \infty} 2^{-n} f(2^n x) > 0$ for all $x \in K \setminus \{0\}$.

If we set $T(x) = \lim_{n \to \infty} 2^{-n} f(2^n x)$, for all $x \in E$, and $\lambda_* = r(GT)^{-1}$, then $(\lambda_*, +\infty)$ is the only bifurcation from infinity of equation (γ).

Proof. First, we note that T is the asymptotic derivative of f along K. Since by assumption *(iii)* we see that T is strictly positive on K, we remark that $r(GT) > 0$. We set

$$g(x) = \begin{cases} \|x^2\| f(x / \|x\|) & (\text{for } x \neq 0), \\ 0 & (\text{for } x = 0). \end{cases}$$

We know that $g'(0) = T$ and $(\lambda, +\infty)$ is a bifurcation point of $x = \lambda G f(x)$ if and only if $(\lambda, 0)$ is a bifurcation point of $x = \lambda G g(x)$. Our theorem now follows from a theorem of [364]. $\qquad \square$

Concerning Theorem 2.26, the two-sided estimates for the spectral radius obtained by V. Y. Stetsenko [337] are very essential. Stetsenko presented that if

$A : E \rightarrow E$ is a completely continuous operator and E is a Banach space ordered by a generating closed cone K with quasi-interior points and if some special assumptions are satisfied, then we can define the numbers λ_0, ρ, the vectors u_0, v_0, and a functional ϕ_0 such that

$$\lambda_0 - \rho \frac{\phi_0(v_0)}{\phi_0(u_0)} \leq r(A) \leq \lambda_0 - \frac{1}{\rho} \frac{\phi_0(v_0)}{\phi_0(u_0)}. \tag{δ}$$

G. Isac and Th. M. Rassias [144] raised an interesting problem to find new and more efficient two-sided estimates for the spectral radius of the operator GT with

$$T(x) = \lim_{n \to \infty} 2^{-n} f(2^n x)$$

when f satisfies the inequality (2.14) for some $\theta \geq 0$, $\psi \in P(\Psi)$, and for all $x, y \in E$.

Such an estimate of $r(GT)$, similar to the estimate (δ), is important for the approximation of the bifurcation point $(\lambda_*, +\infty)$ of the equation (γ).

2.4 Stability on a Restricted Domain

In previous sections, we have seen that the condition that a function f satisfies one of the inequalities (2.2), (2.5), (2.10), (2.14), (2.17), and (2.20) on the whole space, assures us of the existence of a unique additive function approximating f within a given distance.

It will also be interesting to study the stability problems of the additive Cauchy equation on a restricted domain. More precisely, we will study whether there exists a true additive function near a function satisfying one of those inequalities mentioned above only in a restricted domain.

F. Skof [330] and Z. Kominek [224] studied this question in the case of functions defined on certain subsets of $\mathbb{R}^N$ with values in a Banach space. First, we will introduce a lemma by Skof [330] which is necessary to prove the following propositions.

Lemma 2.27. *Let E be a Banach space. If a function $f : [0, \infty) \rightarrow E$ satisfies*

$$\| f(x + y) - f(x) - f(y) \| \leq \delta$$

for some $\delta \geq 0$ and for all $x, y \geq 0$, then there exists a unique additive function $A : \mathbb{R} \rightarrow E$ such that

$$\| f(x) - A(x) \| \leq \delta$$

for any $x \geq 0$.

Proof. Define a function $g : \mathbb{R} \to E$ by

$$g(x) = \begin{cases} f(x) & \text{(for } x \geq 0), \\ -f(-x) & \text{(for } x < 0). \end{cases}$$

It is not difficult to see that

$$\|g(x + y) - g(x) - g(y)\| \leq \delta$$

for all $x, y \in \mathbb{R}$. Therefore, by Theorem 2.3, there exists a unique additive function $A : \mathbb{R} \to E$ such that

$$\|g(x) - A(x)\| \leq \delta \qquad\qquad (a)$$

for any $x \in \mathbb{R}$. This function A is the unique additive function which satisfies (a) for any $x \geq 0$.   □

F. Skof [330] proved the following lemma for $N = 1$, and Z. Kominek [224] extended it for any positive integer N.

Lemma 2.28. *Let E be a Banach space and let N be a given positive integer. Given $c > 0$, let a function $f : [0, c)^N \to E$ satisfy the inequality*

$$\|f(x + y) - f(x) - f(y)\| \leq \delta \qquad\qquad (2.22)$$

for some $\delta \geq 0$ and for all $x, y \in [0, c)^N$ with $x + y \in [0, c)^N$. Then there exists an additive function $A : \mathbb{R}^N \to E$ such that

$$\|f(x) - A(x)\| \leq (4N - 1)\delta$$

for any $x \in [0, c)^N$.

Proof. First, we consider the case $N = 1$. We extend the function f to $[0, \infty)$ and then make use of Lemma 2.27 above. Let us represent an arbitrary $x \geq 0$ by $x = (1/2)nc + \mu$, where n is a nonnegative integer and $0 \leq \mu < (1/2)c$. Define a function $g : [0, \infty) \to E$ by

$$g(x) = f(\mu) + nf\big((1/2)c\big).$$

On the interval $[0, c)$ we have

$$\|g(x) - f(x)\| \leq \delta. \qquad\qquad (a)$$

In fact, when $0 \leq x < (1/2)c$, we have $g(x) = f(x)$. When $(1/2)c \leq x < c$, we get

$$\|g(x) - f(x)\| = \big\|f(\mu) + f\big((1/2)c\big) - f\big((1/2)c + \mu\big)\big\| \leq \delta,$$

since f satisfies (2.22).

We will now show that g satisfies

$$\|g(x + y) - g(x) - g(y)\| \leq 2\delta \tag{b}$$

for all $x, y \geq 0$. For given $x, y \geq 0$, let $x = (1/2)nc + \mu$ and $y = (1/2)mc + \omega$, where n and m are nonnegative integers and μ and ω belong to the interval $[0, (1/2)c)$. Assume $\mu + \omega \in [0, (1/2)c)$. Then

$$
\begin{aligned}
&\|g(x + y) - g(x) - g(y)\| \\
&= \big\| f(\mu + \omega) + (m + n)f\big((1/2)c\big) - f(\mu) \\
&\quad - nf\big((1/2)c\big) - f(\omega) - mf\big((1/2)c\big) \big\| \\
&= \| f(\mu + \omega) - f(\mu) - f(\omega) \| \\
&\leq \delta.
\end{aligned}
$$

Assume now that $\mu + \omega \in [(1/2)c, c)$. Put $\mu + \omega = (1/2)c + t$. Then, by (a), we have

$$
\begin{aligned}
&\|g(x + y) - g(x) - g(y)\| \\
&= \big\| f\big((1/2)c\big) + f(t) - f(\mu) - f(\omega) \big\| \\
&= \|g(\mu + \omega) - f(\mu + \omega) + f(\mu + \omega) - f(\mu) - f(\omega)\| \\
&\leq 2\delta.
\end{aligned}
$$

According to Lemma 2.27, (b) implies that there exists a unique additive function $A : \mathbb{R} \to E$ such that

$$\|g(x) - A(x)\| \leq 2\delta \tag{c}$$

for any $x \geq 0$. Therefore, by (a) and (c), we have

$$\|f(x) - A(x)\| \leq \|f(x) - g(x)\| + \|g(x) - A(x)\| \leq 3\delta$$

for all $x \in [0, c)$, which completes the proof for $N = 1$.

Assume now that $N > 1$. If we define the functions $f_i : [0, c) \to E$ for $i \in \{1, \ldots, N\}$ by

$$f_i(x_i) = f(0, \ldots, 0, x_i, 0, \ldots, 0),$$

then the functions f_i satisfy

$$\|f_i(x_i + y_i) - f_i(x_i) - f_i(y_i)\| \leq \delta$$

for all $x_i, y_i \in [0, c)$ with $x_i + y_i \in [0, c)$. Thus, the first part of this proof guarantees the existence of additive functions $A_i : \mathbb{R} \to E$ such that

$$\|f_i(x_i) - A_i(x_i)\| \leq 3\delta \tag{d}$$

for all $x_i \in [0, c)$.

Representing $x \in \mathbb{R}^N$ in the form $(x_1, \ldots, x_N)$ we see that the function $A : \mathbb{R}^N \to E$ given by

$$A(x) = \sum_{i=1}^{N} A_i(x_i)$$

is an additive function. Moreover, for each $x \in [0, c)^N$, by (d) and (2.22), we get

$$\| f(x) - A(x) \|$$
$$\leq \left\| f(x) - \sum_{i=1}^{N} f_i(x_i) \right\| + \sum_{i=1}^{N} \| f_i(x_i) - A_i(x_i) \|$$
$$\leq \left\| f(x_1, \ldots, x_{N-1}, 0) - \sum_{i=1}^{N-1} f_i(x_i) \right\|$$
$$\quad + \| -f(0, \ldots, 0, x_N) - f(x_1, \ldots, x_{N-1}, 0) + f(x) \| + 3N\delta$$
$$\leq \left\| f(x_1, \ldots, x_{N-1}, 0) - \sum_{i=1}^{N-1} f_i(x_i) \right\| + \delta + 3N\delta$$
$$\leq \cdots$$
$$\leq (N-1)\delta + 3N\delta.$$

This ends the proof. $\square$

Lemma 2.29. *Let E be a Banach space and let N be a positive integer. Given $c > 0$, let a function $f : (-c, c)^N \to E$ satisfy the inequality (2.22) for some $\delta \geq 0$ and for all $x, y \in (-c, c)^N$ with $x + y \in (-c, c)^N$. Then there exists an additive function $A : \mathbb{R}^N \to E$ such that*

$$\| f(x) - A(x) \| \leq (5N - 1)\delta$$

for any $x \in (-c, c)^N$.

Proof. First, we prove the assertion for $N = 1$. Put

$$g(x) = (1/2)\big(f(x) - f(-x)\big) \quad \text{and} \quad h(x) = (1/2)\big(f(x) + f(-x)\big)$$

for all $x \in (-c, c)$. We note that

$$\| h(x) \| \leq \delta \qquad\qquad\qquad (a)$$

for $x \in (-c, c)$ and

$$\| g(x + y) - g(x) - g(y) \| \leq \delta$$

for all $x, y \in (-c, c)$ with $x + y \in (-c, c)$. According to Lemma 2.28, there exists an additive function $A : \mathbb{R} \to E$ such that $\| g(x) - A(x) \| \leq 3\delta$ for each $x \in [0, c)$.

Since g and A are odd, $\|g(x) - A(x)\| \leq 3\delta$ holds true for any $x \in (-c, c)$. Hence, by (a), we get

$$\|f(x) - A(x)\| \leq \|h(x)\| + \|g(x) - A(x)\| \leq 4\delta$$

for all $x \in (-c, c)$. This completes the proof for the case $N = 1$.

Assume now that $N > 1$. The proof runs in the same way as in the proof of Lemma 2.28. We define functions $f_i : (-c, c) \to E$ by

$$f_i(x_i) = f(0, \ldots, 0, x_i, 0, \ldots, 0)$$

for $i \in \{1, \ldots, N\}$. Then, the functions f_i satisfy

$$\|f_i(x_i + y_i) - f_i(x_i) - f_i(y_i)\| \leq \delta$$

for all $x_i, y_i \in (-c, c)$ with $x_i + y_i \in (-c, c)$. Thus, the first part of our proof implies that there exist additive functions $A_i : \mathbb{R} \to E$ such that

$$\|f_i(x_i) - A_i(x_i)\| \leq 4\delta \qquad (b)$$

for $x_i \in (-c, c)$.

Expressing $x \in \mathbb{R}^N$ in the form $(x_1, \ldots, x_N)$ we see that the function $A : \mathbb{R}^N \to E$ defined by

$$A(x) = \sum_{i=1}^{N} A_i(x_i)$$

is additive. Let $x \in (-c, c)^N$ be given. Then, by (b) and (2.22), we have

$$\|f(x) - A(x)\| \leq \left\| f(x) - \sum_{i=1}^{N} f_i(x_i) \right\| + \sum_{i=1}^{N} \|f_i(x_i) - A_i(x_i)\|$$

$$\leq \|f(x) - f(0, \ldots, 0, x_N) - f(x_1, \ldots, x_{N-1}, 0)\|$$

$$+ \left\| f(x_1, \ldots, x_{N-1}, 0) - \sum_{i=1}^{N-1} f_i(x_i) \right\| + 4N\delta$$

$$\leq \left\| f(x_1, \ldots, x_{N-1}, 0) - \sum_{i=1}^{N-1} f_i(x_i) \right\| + \delta + 4N\delta$$

$$\leq \cdots$$

$$\leq (N-1)\delta + 4N\delta,$$

which ends the proof.   □

Using Corollary 2.23, Z. Kominek [224] proved a more generalized theorem concerning the stability of the additive Cauchy equation on a restricted domain.

Theorem 2.30 (Kominek). *Let E be a Banach space and let N be a positive integer. Suppose D is a bounded subset of $\mathbb{R}^N$ containing zero in its interior. Assume, moreover, that there exist a nonnegative integer n and a positive number $c > 0$ such that*

(i) $(1/2)D \subset D$,
(ii) $(-c, c)^N \subset D$,
(iii) $D \subset (-2^n c, 2^n c)^N$.

If a function $f : D \to E$ satisfies the functional inequality

$$\|f(x + y) - f(x) - f(y)\| \leq \delta \tag{2.23}$$

for some $\delta \geq 0$ and for all $x, y \in D$ with $x + y \in D$, then there exists an additive function $A : \mathbb{R}^N \to E$ such that

$$\|f(x) - A(x)\| \leq (2^n \cdot 5N - 1)\delta$$

for any $x \in D$.

Proof. On account of Lemma 2.29, there exists an additive function $A : \mathbb{R}^N \to E$ such that

$$\|f(x) - A(x)\| \leq (5N - 1)\delta \tag{a}$$

for any $x \in (-c, c)^N$. Taking an arbitrary $x \in D$ we observe, by (i), that $2^{-k}x \in D$ for $k \in \{1, \ldots, n\}$, and condition (iii) implies that $2^{-n}x \in (-c, c)^N$. It follows from (2.23) that for every $x \in D$ and each $k \in \{1, \ldots, n\}$

$$\left\| 2^{k-1} f\left(2^{-k+1} x\right) - 2^k f\left(2^{-k} x\right) \right\| \leq 2^{k-1}\delta,$$

therefore,

$$\|f(x) - 2^n f(2^{-n}x)\| \leq (2^n - 1)\delta. \tag{b}$$

Now, by (a), (b), and the additivity of A, we get

$$\begin{aligned} \|f(x) - A(x)\| &\leq \|f(x) - 2^n f(2^{-n}x)\| \\ &\quad + \|2^n f(2^{-n}x) - 2^n A(2^{-n}x)\| \\ &\leq (2^n - 1)\delta + 2^n(5N - 1)\delta \\ &= (2^n \cdot 5N - 1)\delta \end{aligned}$$

for any $x \in D$, which ends the proof. $\qquad\square$

L. Losonczi [238] proved the following theorem and applied it to the study of the Hyers–Ulam stability for Hosszú's equation (see Theorem 4.5).

Theorem 2.31. *Let E_1 and E_2 be a normed space and a Banach space, respectively. Suppose a function $f : E_1 \to E_2$ satisfies the inequality*

$$\| f(x + y) - f(x) - f(y) \| \leq \delta \qquad (2.24)$$

for some $\delta \geq 0$ and for all $(x, y) \in E_1^2 \setminus B$, where B is a subset of E_1^2 such that the first (or second) coordinates of the points of B form a bounded set. Then there exists a unique additive function $A : E_1 \to E_2$ such that

$$\| f(x) - A(x) \| \leq 5\delta \qquad (2.25)$$

for any $x \in E_1$.

Proof. Since the left-hand side of (2.24) is symmetric in x and y, we may assume that (2.24) holds true for all $(x, y) \in C = (E_1^2 \setminus B) \cup (E_1^2 \setminus B_s)$, where $B_s = \{(x, y) \in E_1^2 \mid (y, x) \in B\}$. Since

$$C = \left(E_1^2 \setminus B \right) \cup \left(E_1^2 \setminus B_s \right) = E_1^2 \setminus (B \cap B_s)$$

and both coordinates of the points from $B \cap B_s$ form a bounded set, we can find a number $a > 0$ such that $B \cap B_s \subset Q$, where

$$Q = \{(x, y) \in E_1^2 \mid \|x\| < a, \ \|y\| < a\}.$$

Choose a $t \in E_1$ such that $\|t\| = 2a$ and take any point $(u, v) \in B \cap B_s$. We then know that all the points

$$\left(u + v, 2t - (u + v) \right), \quad (u, t - u), \quad (t - v, v), \quad (t - u, t - v), \quad (t, t) \qquad (a)$$

are in $E_1^2 \setminus Q$ and hence also in $E_1^2 \setminus (B \cap B_s)$, since the inequalities

$$\|2t - (u + v)\| \geq \|2t\| - \|u + v\| \geq 4a - 2a > a,$$
$$\|t - u\| \geq \|t\| - \|u\| \geq a,$$
$$\|t - v\| \geq \|t\| - \|v\| \geq a,$$
$$\|t\| > a$$

are true. Thus, if $(u, v) \in B \cap B_s$, we get

$$
\begin{aligned}
\| f(u + v) &- f(u) - f(v) \| \\
&\leq \| f(u + v) + f(2t - (u + v)) - f(2t) \| \\
&\quad + \| - f(2t - (u + v)) + f(t - u) + f(t - v) \| \\
&\quad + \| - f(t - u) - f(u) + f(t) \| \\
&\quad + \| - f(v) - f(t - v) + f(t) \| \\
&\quad + \| - f(t) - f(t) + f(2t) \| \\
&\leq 5\delta,
\end{aligned}
$$

where we applied (2.24) for each of the points listed in (a).

If $(u, v) \in E_1^2 \setminus (B \cap B_s)$, then we obtain

$$\| f(u + v) - f(u) - f(v) \| \leq \delta$$

as we have already seen. Hence, the inequality

$$\| f(x + y) - f(x) - f(y) \| \leq 5\delta$$

holds true for all $x, y \in E_1$. According to Theorem 2.3, there exists a unique additive function $A : E_1 \to E_2$ such that the inequality (2.25) is true for all $x \in E_1$ because the assertion of Theorem 2.3 also holds true for the case when the domain E_1 of the function f is extended to an amenable semigroup and because each normed space is an amenable semigroup (see Corollary 2.42 below).   □

F. Skof [331] proved the following theorem and applied the result to the study of an asymptotic behavior of additive functions.

Theorem 2.32 (Skof). *Let E be a Banach space, and let $a > 0$ be a given constant. Suppose a function $f : \mathbb{R} \to E$ satisfies the inequality*

$$\| f(x + y) - f(x) - f(y) \| \leq \delta \qquad (2.26)$$

for some $\delta \geq 0$ and for all $x, y \in \mathbb{R}$ with $|x| + |y| > a$. Then there exists a unique additive function $A : \mathbb{R} \to E$ such that

$$\| f(x) - A(x) \| \leq 9\delta \qquad (2.27)$$

for all $x \in \mathbb{R}$.

Proof. Put

$$\varphi(x, y) = f(x + y) - f(x) - f(y)$$

for $x, y \in \mathbb{R}$. Suppose real numbers x and $y \neq 0$ are given. Let m and n be integers greater than 1. We then have

$$f(nx + my) = f(nx) + f(my) + \varphi(nx, my)$$

and

$$\begin{aligned}
f(nx + my) &= f\big((n - 1)x + (x + my)\big) \\
&= f\big((n - 1)x\big) + f(x) + f(my) \\
&\quad + \varphi(x, my) + \varphi\big((n - 1)x, x + my\big),
\end{aligned}$$

from which it follows that

$$f(nx) - f((n-1)x) - f(x)$$
$$= \varphi(x, my) + \varphi((n-1)x, x + my) - \varphi(nx, my).$$

If m is so large that $m > (a + |x|)/|y|$, the last equality implies

$$\| f(nx) - f((n-1)x) - f(x) \| \le 3\delta \qquad (a)$$

for any $x \in \mathbb{R}$ and any integer $n > 1$. The relation

$$f(nx) - nf(x) = \sum_{k=2}^{n} \big(f(kx) - f((k-1)x) - f(x) \big),$$

together with (a), yields

$$\| f(nx) - nf(x) \| \le 3(n-1)\delta \qquad (b)$$

for any real number x and $n > 1$. Obviously, it follows from (b) that

$$\| \varphi(nx, ny) - n\varphi(x, y) \| \le \| f(nx + ny) - nf(x + y) \|$$
$$+ \| f(nx) - nf(x) \| + \| f(ny) - nf(y) \|$$
$$\le 9(n-1)\delta$$

for all real numbers x and y with $(x, y) \ne (0, 0)$. Dividing both sides of the last inequality by n, then letting $n \to \infty$ and considering the fact that $(1/n)\varphi(nx, ny) \to 0$ as $n \to \infty$, we get

$$\| \varphi(x, y) \| \le 9\delta \qquad (c)$$

for all x, y with $(x, y) \ne (0, 0)$. Let x be a real number with $|x| > a$. Then, the inequality (2.26) with such an x and $y = 0$ yields $\| f(0) \| \le \delta$. Hence, we obtain $\| \varphi(0, 0) \| = \| f(0) \| \le \delta$. Therefore, (c) holds true for all $x, y \in \mathbb{R}$.

According to Theorem 2.3, there is a unique additive function $A : \mathbb{R} \to E$ such that the inequality (2.27) holds true for any x in $\mathbb{R}$. □

Analogously, we can easily generalize the last result. More precisely, we can extend the domain of the function f in the last theorem to an arbitrary normed space. Here, we remark that the domain E_1 of the function f in Theorem 2.3 can be extended to a normed space without reduction of the validity of the theorem. The proof of the following theorem given by F. Skof [331] is the same as that of the last theorem.

Theorem 2.33. *Let E_1 and E_2 be a normed space and a Banach space, respectively. Given $a > 0$, suppose a function $f : E_1 \to E_2$ satisfies*

$$\| f(x + y) - f(x) - f(y) \| \le \delta$$

for some $\delta \geq 0$ and for all $x, y \in E_1$ with $\|x\| + \|y\| > a$. Then there exists a unique additive function $A : E_1 \to E_2$ such that

$$\|f(x) - A(x)\| \leq 9\delta$$

for all $x \in E_1$.

Using this theorem, F. Skof [331] has studied an interesting asymptotic behavior of additive functions as we see in the following theorem.

Theorem 2.34 (Skof). *Let E_1 and E_2 be a normed space and a Banach space, respectively. Suppose z is a fixed point of E_1. For a function $f : E_1 \to E_2$ the following two conditions are equivalent:*

(i) $f(x + y) - f(x) - f(y) \to z$ as $\|x\| + \|y\| \to \infty$;
(ii) $f(x + y) - f(x) - f(y) = z$ for all $x, y \in E_1$.

Proof. It suffices to prove the implication $(i) \Rightarrow (ii)$ only because the reverse implication is a trivial case. Define $g(x) = f(x) + z$ for all $x \in E_1$. Then the condition (i) implies that

$$g(x + y) - g(x) - g(y) \to 0 \quad \text{as } \|x\| + \|y\| \to \infty. \tag{a}$$

Due to (a) there is a sequence $\{\delta_n\}$ monotonically decreasing to zero such that

$$\|g(x + y) - g(x) - g(y)\| \leq \delta_n$$

for all $x, y \in E_1$ with $\|x\| + \|y\| > n$. According to Theorem 2.33, there exists a unique additive function $A_n : E_1 \to E_2$ satisfying

$$\|g(x) - A_n(x)\| \leq 9\delta_n \tag{b}$$

for all $x \in E_1$. Let m and n be integers with $n > m > 0$. Then the additive function $A_n : E_1 \to E_2$ satisfies $\|g(x) - A_n(x)\| \leq 9\delta_m$ for all $x \in E_1$. The uniqueness argument implies $A_n = A_m$ for all integers n greater than $m > 0$. Therefore, $A_1 = A_2 = \ldots = A_n = \ldots$. If we define a function $A : E_1 \to E_2$ by $A(x) = A_n(x)$ for all $x \in E_1$ and for some $n > 1$, then A is an additive function. Letting $n \to \infty$ in (b), we conclude that g itself is an additive function. Thus,

$$0 = g(x + y) - g(x) - g(y) = f(x + y) - f(x) - f(y) - z$$

for all $x, y \in E_1$.   □

The following corollary is an immediate consequence of the last theorem (see [331]).

Corollary 2.35. *The function* $f : \mathbb{R} \to \mathbb{R}$ *is an additive function if and only if* $f(x+y) - f(x) - f(y) \to 0$ *as* $|x| + |y| \to \infty$.

Skof [334] dealt with functionals on a real Banach space and proved a theorem concerning the stability of the alternative equation $|f(x+y)| = |f(x) + f(y)|$:

Let E be a real Banach space, and let M be a given positive number. If a functional $f : E \to \mathbb{R}$ *satisfies the inequality*

$$\Big| |f(x+y)| - |f(x) + f(y)| \Big| \leq \delta$$

for some $\delta \geq 0$ and for all $x, y \in E$ with $\|x\| + \|y\| > M$, then there exists a unique additive functional $A : E \to \mathbb{R}$ such that

$$|f(x) - A(x)| \leq 40\delta + 11|f(0)| + 11 \max\{\delta, |f(0)|\}$$

for all $x \in E$.

On the basis of this result, Skof could obtain the following characterization of additive functionals (see [334]):

Let E be a real Banach space. The functional $f : E \to \mathbb{R}$ *is additive if and only if $f(0) = 0$ and $|f(x+y)| - |f(x) + f(y)| \to 0$ as $\|x\| + \|y\| \to \infty$.*

D. H. Hyers, G. Isac, and Th. M. Rassias [136] have proved a Hyers–Ulam–Rassias stability result of the additive Cauchy equation on a restricted domain and applied it to the study of the asymptotic derivability which is very important in nonlinear analysis.

Theorem 2.36 (Hyers, Isac, and Rassias). *Given a real normed space E_1 and a real Banach space E_2, let numbers $M > 0$, $\theta > 0$, and p with $0 < p < 1$ be chosen. Let a function* $f : E_1 \to E_2$ *satisfy the inequality*

$$\|f(x+y) - f(x) - f(y)\| \leq \theta(\|x\|^p + \|y\|^p) \tag{2.28}$$

for all $x, y \in E_1$ satisfying $\|x\|^p + \|y\|^p > M^p$. Then there exists an additive function $A : E_1 \to E_2$ such that

$$\|f(x) - A(x)\| \leq 2\theta(2 - 2^p)^{-1}\|x\|^p \tag{2.29}$$

for all $x \in E_1$ with $\|x\| > 2^{-1/p}M$.

Proof. When $\|x\| > 2^{-1/p}M$ or $2\|x\|^p > M^p$, we may put $y = x$ in (2.28) to obtain

$$\|(1/2)f(2x) - f(x)\| \leq \theta\|x\|^p. \tag{a}$$

Since $\|2x\|$ is also greater than $2^{-1/p}M$, we can replace x with $2x$ in (a). Thus, we can use the induction argument given in Theorem 2.5 to obtain the inequality

$$\|2^{-n}f(2^n x) - f(x)\| \leq 2\theta(2 - 2^p)^{-1}\|x\|^p \tag{b}$$

for all $x \in E_1$ with $\|x\| > 2^{-1/p}M$ and for any $n \in \mathbb{N}$. Hence, the limit

$$g(x) = \lim_{n \to \infty} 2^{-n} f(2^n x) \tag{c}$$

exists when $\|x\| > 2^{-1/p}M$. Therefore,

$$\|g(x) - f(x)\| \leq 2\theta(2 - 2^p)^{-1}\|x\|^p. \tag{d}$$

Clearly, when $\|x\| > 2^{-1/p}M$, we have

$$g(2x) = \lim_{n \to \infty} 2^{-n} f(2^{n+1}x) = 2 \lim_{n \to \infty} 2^{-n-1} f(2^{n+1}x),$$

so that

$$g(2x) = 2g(x) \tag{e}$$

for $\|x\| > 2^{-1/p}M$.

Assume now that $\|x\|$, $\|y\|$, and $\|x + y\|$ are all greater than $2^{-1/p}M$. Then, by (2.28), we find that

$$\|2^{-n}f(2^n(x + y)) - 2^{-n}f(2^n x) - 2^{-n}f(2^n y)\| \leq \theta 2^{-n(1-p)}(\|x\|^p + \|y\|^p)$$

for all $n \in \mathbb{N}$. It then follows from (c) that

$$g(x + y) = g(x) + g(y)$$

for all $x, y \in E_1$ with $\|x\|, \|y\|, \|x + y\| > 2^{-1/p}M$. Using an extension theorem of F. Skof [334], we will define a function $A : E_1 \to E_2$ to be an extension of the function g to the whole space E_1. Given any $x \in E_1$ with $0 < \|x\| \leq 2^{-1/p}M$, let $k = k(x)$ denote the largest integer such that

$$2^{-1/p}M < 2^k\|x\| \leq M. \tag{f}$$

Now, define a function $A : E_1 \to E_2$ by

$$A(x) = \begin{cases} 0 & \text{(for } x = 0\text{)}, \\ 2^{-k}g(2^k x) & \text{(for } 0 < \|x\| \leq 2^{-1/p}M, \text{ where } k = k(x)\text{)}, \\ g(x) & \text{(for } \|x\| > 2^{-1/p}M\text{)}. \end{cases}$$

If we take $x \in E_1$ with $0 < \|x\| \le 2^{-1/p}M$, then $k - 1$ is the largest integer satisfying

$$2^{-1/p}M < \|2^{k-1}(2x)\| \le M,$$

and we have

$$A(2x) = 2^{-(k-1)}g(2^{k-1}2x) = 2^{-k} \cdot 2g(2^k x) = 2A(x)$$

for $0 < \|x\| \le 2^{-1/p}M$. From the definition of A and (e), it follows that $A(2x) = 2A(x)$ for all $x \in E_1$. Given $x \in E_1$ with $x \ne 0$, choose a positive integer m so large that $\|2^m x\| > 2^{-1/p}M$. By the definition of A, we have

$$A(x) = 2^{-m}A(2^m x) = 2^{-m}g(2^m x),$$

and by (c) this implies that

$$A(x) = \lim_{n \to \infty} 2^{-(m+n)}f(2^{m+n}x),$$

which demonstrates

$$A(x) = \lim_{s \to \infty} 2^{-s}f(2^s x) \tag{g}$$

for $x \ne 0$. Since $A(0) = 0$, (g) holds true for all $x \in E_1$.

We will now present that

$$A(-x) = -A(x) \tag{h}$$

for all $x \in E_1$. It is obvious for $x = 0$. Take any $x \in E_1 \backslash \{0\}$ and choose an integer n so large that $\|2^n x\| > 2^{-1/p}M$. If we replace x and y in (2.28) with $2^n x$ and $-2^n x$, respectively, and divide the resulting inequality by 2^n, then we obtain

$$\|2^{-n}f(2^n x) + 2^{-n}f(-2^n x)\| \le 2\theta 2^{-n(1-p)}\|x\|^p + 2^{-n}\|f(0)\|.$$

If we let $n \to \infty$ in the last inequality, (h) follows from (g).

We note that the equation

$$A(x + y) = A(x) + A(y) \tag{i}$$

holds true when either x or y is zero. Assume then that $x \ne 0$ and $y \ne 0$. If $x + y = 0$, then (h) implies the validity of (i). The only remaining case is when x, y, and $x + y$ are all different from zero. In this case we may choose an integer n such that $\|2^n x\|$, $\|2^n y\|$, and $\|2^n(x + y)\|$ are all greater than $2^{-1/p}M$. Then, by (2.28), we get

$$\|f(2^n(x + y)) - f(2^n x) - f(2^n y)\| \le \theta 2^{np}(\|x\|^p + \|y\|^p).$$

If we divide both sides of this inequality by 2^n and then let $n \to \infty$, we find by (g) that (i) is true, thus A is additive.

By definition we have $A(x) = g(x)$ when $\|x\| > 2^{-1/p}M$, thus (2.29) follows from (d).   □

For convenience in applications, we give the following modified version of Theorem 2.36 (ref. [136]).

Theorem 2.37. *Given a real normed space E_1 and a real Banach space E_2, let numbers $m > 0$, $\theta > 0$, and p with $0 \leq p < 1$ be chosen. Suppose a function $f : E_1 \to E_2$ satisfies the inequality*

$$\|f(x + y) - f(x) - f(y)\| \leq \theta(\|x\|^p + \|y\|^p)$$

for all $x, y \in E_1$ such that $\|x\| > m$, $\|y\| > m$, and $\|x + y\| > m$. Then there exists an additive function $A : E_1 \to E_2$ such that

$$\|f(x) - A(x)\| \leq 2\theta(2 - 2^p)^{-1}\|x\|^p$$

for all $x \in E_1$ with $\|x\| > m$.

Proof. Assume that $\|x\| > m$. Then as in the proof of Theorem 2.36 we obtain $(a) - (e)$ (in the proof of Theorem 2.36) inclusive, but now all these formulas are satisfied for $\|x\| > m$. In particular,

$$g(x) = \lim_{n \to \infty} 2^{-n} f(2^n x)$$

when $\|x\| > m$. Also, if $\|x\| > m$, $\|y\| > m$, and $\|x + y\| > m$, then by hypothesis we see that

$$\left\|2^{-n} f(2^n(x + y)) - 2^{-n} f(2^n x) - 2^{-n} f(2^n y)\right\| \leq \theta 2^{-n(1-p)}(\|x\|^p + \|y\|^p)$$

and

$$g(x + y) = g(x) + g(y)$$

also hold true.

To apply Skof's extension procedure in the present case, let x in E_1 be given with $0 < \|x\| \leq m$ and define $k = k(x)$ to be the unique positive integer such that

$$m < 2^k \|x\| \leq 2m. \tag{j}$$

Now define a function $A : E_1 \to E_2$ by

$$A(x) = \begin{cases} 0 & \text{(for } x = 0), \\ 2^{-k} g(2^k x) & \text{(for } 0 < \|x\| \leq m), \\ g(x) & \text{(for } \|x\| > m). \end{cases}$$

The proofs, of (g) and (h), in the proof of the last theorem follow as before with the obvious changes. Indeed, we start with $x \in E_1$ satisfying $0 < \|x\| \leq m$ and let $k = k(x)$ as defined by (j), etc. Thus, (g) and (h) mentioned above hold true under the conditions of this theorem. The proof of the additivity of A also follows as before. □

We apply the result of Theorem 2.37 to the study of p-asymptotical derivatives: Let E_1 and E_2 be Banach spaces. Suppose $A : E_1 \to E_2$ is a function satisfying eventually a special property, for example, additivity, linearity, etc.

Let $0 < p < 1$ be arbitrary. A function $f : E_1 \to E_2$ is called p-asymptotically close to A if and only if $\|f(x) - A(x)\|/\|x\|^p \to 0$ as $\|x\| \to \infty$. Moreover, if $A \in L(E_1, E_2)$, then we say that A is a p-asymptotical derivative of f and if such an A exists, then f is called p-asymptotically derivable.

A function $f : E_1 \to E_2$ is called p-asymptotically additive if and only if for every $\theta > 0$ there exists a $\delta > 0$ such that

$$\|f(x + y) - f(x) - f(y)\| \leq \theta \big(\|x\|^p + \|y\|^p\big)$$

for all $x, y \in E_1$ such that $\|x\|^p, \|y\|^p, \|x + y\|^p > \delta$.

A function $A : E_1 \to E_2$ is called *additive outside a ball* if there exists an $r > 0$ such that $A(x + y) = A(x) + A(y)$ for all $x, y \in E_1$ such that $\|x\| \geq r$, $\|y\| \geq r$, and $\|x + y\| \geq r$.

Hyers, Isac, and Rassias [136] contributed to the following Theorems 2.38, 2.39, and Corollary 2.40.

Theorem 2.38. *If a function $f : E_1 \to E_2$ is p-asymptotically close to a function $A : E_1 \to E_2$ which is additive outside a ball, then f is p-asymptotically additive.*

Theorem 2.39. *If a function $f : E_1 \to E_2$ is p-asymptotically close to a function $A : E_1 \to E_2$ which is additive outside a ball, then f is q-asymptotically close to an additive function, where $0 < p < q < 1$.*

Corollary 2.40. *If a function $f : E_1 \to E_2$ is p-asymptotically close to a function $A : E_1 \to E_2$ which is additive outside a ball, then f has an additive q-asymptotical derivative, where $0 < p < q < 1$.*

2.5 Method of Invariant Means

So far we have dealt with generalizations of Theorem 2.3 in connection with the bounds for the Cauchy difference. Now, we will briefly introduce another generalization of the theorem from the point of view of the domain space of the functions involved.

Let $(G, \cdot)$ be a semigroup or group and let $B(G)$ denote the space of all bounded complex-valued functions on G with the norm

$$\|f\| = \sup \{f(x) \mid x \in G\}.$$

A linear functional m on $B(G)$ is called a *right invariant mean* if the following conditions are satisfied:

(i) $m(\overline{f}) = \overline{m(f)}$ *for* $f \in B(G)$,

(ii) $\inf \{f(x) \mid x \in G\} \leq m(f) \leq \sup \{f(x) \mid x \in G\}$ *for all real-valued* $f \in B(G)$,

(iii) $m(f_x) = m(f)$ *for all* $x \in G$ *and* $f \in B(G)$, *where* $f_x(t) = f(tx)$.

If *(iii)* in the above definition is replaced with $m(_x f) = m(f)$, where $_x f(t) = f(xt)$, then m is called a *left invariant mean*.

When a right (left) invariant mean exists on $B(G)$, we call G *right* (*left*) *amenable*. It is known that if G is a semigroup with both right and left invariant means, then there exists a two-sided invariant mean on $B(G)$ and in this case G is called *amenable*. It is also known that if G is a group, then either right or left amenability of G implies that G is amenable (ref. [127]). We remark that the norm of the functional m is one.

G. L. Forti [105] proved the following theorem.

Theorem 2.41 (Forti). *Assume that* $(G, \cdot)$ *is a right (left) amenable semigroup. If a function* $f : G \to \mathbb{C}$ *satisfies*

$$|f(x \cdot y) - f(x) - f(y)| \leq \delta$$

for some $\delta \geq 0$ *and for all* $x, y \in G$, *then there exists a homomorphism* $H : G \to \mathbb{C}$ *such that*

$$|f(x) - H(x)| \leq \delta$$

for all $x \in G$.

Proof. Let $m : B(G) \to \mathbb{C}$ be a right invariant mean. We use the notation m_x to indicate that the mean is to be applied with respect to the variable x. Define the function $H : G \to \mathbb{C}$ by

$$H(y) = m_x\big(f(x \cdot y) - f(x)\big).$$

Using the right invariance and the linearity of the functional m, we have

$$
\begin{aligned}
H(y) + H(z) &= m_x\big(f(x \cdot y) - f(x)\big) + m_x\big(f(x \cdot z) - f(x)\big) \\
&= m_x\big(f(x \cdot y \cdot z) - f(x \cdot z) + f(x \cdot z) - f(x)\big) \\
&= m_x\big(f(x \cdot y \cdot z) - f(x)\big) \\
&= H(y \cdot z),
\end{aligned}
$$

so that H is a homomorphism. We now get

$$
\begin{aligned}
|f(y) - H(y)| &= \left| f(y) - m_x\big(f(x \cdot y) - f(x)\big) \right| \\
&= \left| m_x\big(f(x \cdot y) - f(x) - f(y)\big) \right| \\
&\leq \|m_x\| \, |f(x \cdot y) - f(x) - f(y)| \\
&\leq \delta
\end{aligned}
$$

for all $y \in G$. The proof for the case of a left invariant mean is similar.   □

It should be remarked that L. Székelyhidi introduced the invariant mean method in [339, 340]. Also, he proved this theorem for the case when $(G, \cdot)$ is a right (left) amenable group (see [342]).

The method of invariant means does not provide a proof of uniqueness of the homomorphism H. However, it should be remarked that J. Rätz [316] proved the uniqueness of the homomorphism H.

Let us now introduce some terminologies. A vector space E is called a *topological vector space* if the set E is a topological space and if the vector space operations (vector addition and scalar multiplication) are continuous in the topology of E. A *local base* of a topological vector space E is a collection $\mathcal{B}$ of neighborhoods of 0 in E such that every neighborhood of 0 contains a member of $\mathcal{B}$. A topological vector space is called *locally convex* if there exists a local base of which members are convex. A topological space E is a *Hausdorff space* if distinct points of E have disjoint neighborhoods. A topological vector space E is called *sequentially complete* if every Cauchy sequence in E converges to a point of E.

L. Székelyhidi [343] presented that if the equation of homomorphism is stable for functions from a semigroup G into $\mathbb{C}$, then stability also holds true for functions $f : G \to E$, where E is a semi-reflexive complex locally convex Hausdorff topological vector space. Z. Gajda [111] significantly generalized this result, i.e., the stability result of Székelyhidi also holds true for functions $f : G \to E$, where E is a complex locally convex Hausdorff topological vector space which is sequentially complete.

Applying the above result of Gajda to Theorem 2.41, we obtain the following corollary (ref. [111]).

Corollary 2.42. *Let $(G, \cdot)$ be a right (left) amenable semigroup, and let E be a complex topological vector space which is locally convex, Hausdorff, and sequentially complete. If a function $f : G \to E$ satisfies*

$$
|f(x \cdot y) - f(x) - f(y)| \leq \delta
$$

for some $\delta \geq 0$ and for all $x, y \in G$, then there exists a unique homomorphism $H : G \to E$ such that

$$
|f(x) - H(x)| \leq \delta
$$

for all $x \in G$.

2.6 Fixed Point Method

In Theorems 2.3. 2.5, and 2.18, the relevant additive functions A are explicitly constructed from the given function f by means of

$$A(x) = \lim_{n \to \infty} 2^{-n} f(2^n x) \quad \text{or} \quad A(x) = \lim_{n \to \infty} 2^n f(2^{-n} x).$$

This method is called a direct method presented by D. H. Hyers [135] for the first time. On the other hand, another approach for proving the stability was introduced in Section 2.5. This approach is called the method of invariant means. In this section, we will deal with a new method, namely, the fixed point method.

For a nonempty set X, we introduce the definition of the generalized metric on X. A function $d : X^2 \to [0, \infty]$ is called a *generalized metric* on X if and only if d satisfies

(M1) $d(x, y) = 0$ *if and only if* $x = y$;
(M2) $d(x, y) = d(y, x)$ *for all* $x, y \in X$;
(M3) $d(x, z) \leq d(x, y) + d(y, z)$ *for all* $x, y, z \in X$.

We remark that the only difference between the generalized metric and the usual metric is that the range of the former is permitted to include the infinity.

We now introduce one of the fundamental results of the fixed point theory. For the proof, we refer to [240].

Theorem 2.43. *Let* (X, d) *be a generalized complete metric space. Assume that* $\Lambda : X \to X$ *is a strictly contractive operator with the Lipschitz constant* $L < 1$. *If there exists a nonnegative integer* n_0 *such that* $d\left(\Lambda^{n_0+1} x, \Lambda^{n_0} x\right) < \infty$ *for some* $x \in X$, *then the following statements are true:*

(i) The sequence $\{\Lambda^n x\}$ *converges to a fixed point* x^* *of* Λ;
(ii) x^* is the unique fixed point of Λ *in* $X^* = \{y \in X \mid d(\Lambda^{n_0} x, y) < \infty\}$;
(iii) If $y \in X^*$, *then*

$$d(y, x^*) \leq \frac{1}{1 - L} d(\Lambda y, y).$$

In 2003, V. Radu proved the Hyers–Ulam–Rassias stability of the additive Cauchy equation (2.1) by using the fixed point method (see [279] or [57]).

Theorem 2.44 (Cădariu and Radu). *Let* E_1 *and* E_2 *be a real normed space and a real Banach space, respectively. Let* p *and* θ *be nonnegative constants with* $p \neq 1$. *If a function* $f : E_1 \to E_2$ *satisfies the inequality*

$$\|f(x + y) - f(x) - f(y)\| \leq \theta \left(\|x\|^p + \|y\|^p\right) \tag{2.30}$$

for all $x, y \in E_1$, *then there exists a unique additive function* $A : E_1 \to E_2$ *such that*

$$\|f(x) - A(x)\| \leq \frac{2\theta}{|2 - 2^p|} \|x\|^p \tag{2.31}$$

for any $x \in E_1$.

Proof. We define the set

$$X = \{g : E_1 \to E_2 \mid p \cdot g(0) = 0\}$$

and introduce a generalized metric $d_p : X^2 \to [0, \infty]$ by

$$d_p(g, h) = \sup_{x \neq 0} \|g(x) - h(x)\| / \|x\|^p.$$

We know that (X, d_p) is complete.

Now, we define an operator $\Lambda : X \to X$ by

$$(\Lambda g)(x) = (1/q)g(qx),$$

where $q = 2$ if $p < 1$, and $q = 1/2$ if $p > 1$. Then, we have

$$
\begin{aligned}
\|(\Lambda g)(x) - (\Lambda h)(x)\| / \|x\|^p &= (1/q)\|g(qx) - h(qx)\| / \|x\|^p \\
&= q^{p-1} \|g(qx) - h(qx)\| / \|qx\|^p \\
&\leq q^{p-1} d_p(g, h)
\end{aligned}
$$

for any $g, h \in X$. Thus, we conclude that

$$d_p(\Lambda g, \Lambda h) \leq q^{p-1} d_p(g, h)$$

for all $g, h \in X$, i.e., Λ is a strictly contractive operator on X with the Lipschitz constant $L = q^{p-1}$.

If we put $y = x$ in (2.30), then we get

$$\|2f(x) - f(2x)\| \leq 2\theta \|x\|^p$$

for each $x \in E_1$, which implies that

$$d_p(f, \Lambda f) \leq \begin{cases} \theta & \text{(for } p < 1), \\ 2^{1-p}\theta & \text{(for } p > 1). \end{cases} \tag{a}$$

According to Theorem 2.43 (i), there exists a function $A : E_1 \to E_2$ which is a fixed point of Λ, i.e.,

$$A(2x) = 2A(x)$$

for all $x \in E_1$. In addition, A is uniquely determined in the set

$$X^* = \{g \in X \mid d_p(f, g) < \infty\}.$$

Moreover, due to Theorem 2.43 (i), we see that $d_p(\Lambda^n f, A) \to 0$ as $n \to \infty$, i.e.,

$$A(x) = \lim_{n \to \infty} q^{-n} f(q^n x) \qquad\qquad (b)$$

for any $x \in E_1$. In view of Theorem 2.43 (iii) and (a), we have

$$d_p(f, A) \le \frac{1}{1 - q^{p-1}} d_p(f, \Lambda f) \le \frac{2\theta}{|2 - 2^p|},$$

which implies the validity of the inequality (2.31).

If we replace x and y in (2.30) with $q^n x$ and $q^n y$, respectively, then we obtain

$$\left\| q^{-n} f\left(q^n(x + y)\right) - q^{-n} f(q^n x) - q^{-n} f(q^n y) \right\| \le L^n \theta \left(\|x\|^p + \|y\|^p \right)$$

for all $x, y \in E_1$. If we let $n \to \infty$ in the preceding inequality and consider (b), then we have

$$A(x + y) = A(x) + A(y)$$

for any $x, y \in E_1$. $\qquad\qquad\qquad\qquad\qquad\qquad\qquad\qquad\qquad\qquad\quad \Box$

2.7 Composite Functional Congruences

It is also interesting to study the stability problem when the values of the Cauchy difference $f(x + y) - f(x) - f(y)$ are forced to lie near integers, i.e.,

$$f(x + y) - f(x) - f(y) \in \mathbb{Z} + (-\varepsilon, \varepsilon), \qquad\qquad (2.32)$$

where ε is a small positive number.

R. Ger and P. Šemrl [123] proved that if a function $f : G \to \mathbb{R}$, where G is a cancellative abelian semigroup, satisfies the condition in (2.32) with $0 < \varepsilon < 1/4$, then there exists a function $p : G \to \mathbb{R}$ such that $p(x + y) - p(x) - p(y) \in \mathbb{Z}$ and $|f(x) - p(x)| \le \varepsilon$.

Such a property of functions satisfying (2.32) is called the *composite functional congruence*. It is a generalization of the functional congruence which was first studied by van der Corput [86].

Before stating the results of Ger and Šemrl, we introduce a theorem of M. Hosszú [134]. We may omit the proof because it is beyond the scope of this book.

Theorem 2.45. *Let G_1 and G_2 be a cancellative abelian semigroup and a divisible abelian group, respectively, in which the equation $nx = y$ has a unique solution $x \in G_2$ for each fixed $y \in G_2$ and any $n \in \mathbb{N}$. The most general form of solutions $f : G_1^2 \to G_2$ of the functional equation*

$$f(x + y, z) + f(x, y) = f(x, y + z) + f(y, z)$$

is $f(x, y) = B(x, y) + g(x + y) - g(x) - g(y)$, where $B : G_1^2 \to G_2$ is an arbitrary skew-symmetric biadditive function and where $g : G_1 \to G_2$ is an arbitrary function.

Let $(G, +)$ be a group and let $U, V \subset G$. We can define the addition and the subtraction between sets by

$$U + V = \{x + y \mid x \in U, \ y \in V\}$$

and

$$U - V = \{x - y \mid x \in U, \ y \in V\}.$$

For convenience, let us define $U^+ = U + U$ and $U^- = U - U$. We remark that if $(G, +)$ is an abelian group, then $(U^+)^- = (U^-)^+$.

A set of *generators* of a group G is a subset S of G such that each element of G can be represented (using the group operations) in terms of members of S, where the repetitions of members of S are allowed.

If G is an abelian group with a finite set of generators, then G is a Cartesian product of infinite cyclic groups $F_1, F_2, \ldots, F_m$ and cyclic groups $H_1, H_2, \ldots, H_n$ of finite order. If $n = 0$, G is called *torsion-free*.

Ger and Šemrl [123] proved the following theorem.

Theorem 2.46. *Let $(G_1, +)$ and $(G_2, +)$ be a cancellative abelian semigroup and a torsion-free divisible abelian group, respectively. Assume that U and V are nonempty subsets of G_2 with $(U^+)^- \cap (V^+)^- = \{0\}$. If a function $f : G_1 \to G_2$ satisfies*

$$f(x + y) - f(x) - f(y) \in U + V$$

for all $x, y \in G_1$, it can be represented by

$$f = u + v,$$

where $u, v : G_1 \to G_2$ satisfy the relations

$$u(x + y) - u(x) - u(y) \in U$$

and

$$v(x + y) - v(x) - v(y) \in V$$

for all $x, y \in G_1$. The functions u and v are determined uniquely up to an additive function.

Proof. There are functions $\psi : G_1^2 \to U$ and $\varphi : G_1^2 \to V$ such that

$$d(x, y) = f(x + y) - f(x) - f(y) = \psi(x, y) + \varphi(x, y)$$

for all $x, y \in G_1$. The commutativity of $(G_1, +)$ implies that d is symmetric.

Claim that ψ is symmetric. We have

$$\psi(x, y) - \psi(y, x) = d(x, y) - \varphi(x, y) - d(y, x) + \varphi(y, x) \in V^-.$$

On the other hand,

$$\psi(x, y) - \psi(y, x) \in U^-,$$

and since $0 \in U^- \cap V^-$, we infer that

$$\begin{aligned}
\psi(x, y) - \psi(y, x) &\in U^- \cap V^- \\
&\subset (U^- + U^-) \cap (V^- + V^-) \\
&= (U^+)^- \cap (V^+)^- \\
&= \{0\}.
\end{aligned}$$

Claim that ψ satisfies

$$\psi(x, y + z) + \psi(y, z) = \psi(x + y, z) + \psi(x, y) \qquad (a)$$

for all $x, y \in G_1$. A straightforward computation yields that d satisfies the same functional equation. Consequently,

$$\begin{aligned}
&\psi(x, y + z) + \psi(y, z) - \psi(x + y, z) - \psi(x, y) \\
&= \varphi(x + y, z) + \varphi(x, y) - \varphi(x, y + z) - \varphi(y, z) \\
&\in (U^+)^- \cap (V^+)^- \\
&= \{0\},
\end{aligned}$$

which proves (a).

According to Theorem 2.45, there exists a function $u : G_1 \to G_2$ such that

$$\psi(x, y) = u(x + y) - u(x) - u(y) \in U$$

for any $x, y \in G_1$ (since ψ is symmetric, we take $B(x, y) \equiv 0$ in Theorem 2.45). Now, define $v(x) = f(x) - u(x)$ for all $x \in G_1$. Then, we have

$$\varphi(x, y) = d(x, y) - \psi(x, y) = v(x + y) - v(x) - v(y) \in V$$

for all $x, y \in G_1$.

In order to prove the uniqueness, we assume that there are two representations

$$f = u + v = u' + v'$$

with the properties described above. Putting

$$a = u - u' = v' - v,$$

we get

$$a(x + y) - a(x) - a(y)$$
$$= \big(u(x + y) - u(x) - u(y)\big) - \big(u'(x + y) - u'(x) - u'(y)\big)$$
$$\in U^-.$$

Similarly,

$$a(x + y) - a(x) - a(y) \in V^-.$$

These two relations, together with $0 \in U^- \cap V^-$, imply the additivity of a. Moreover, we have

$$u' = u - a \quad \text{and} \quad v' = v + a$$

which ends the proof. $\square$

Let E be a vector space and let $V \subset E$. The intersection of all convex sets in E containing V is called the *convex hull* of V and denoted by $\mathrm{Co}\,V$. Thus, $\mathrm{Co}\,V$ is the smallest convex set containing V.

Let U be a subset of a topological vector space and let $\overline{U}$ denote the closure of U. Ger and Šemrl [123] also presented the following theorem.

Theorem 2.47 (Ger and Šemrl). *Let $(G, +)$ be a cancellative abelian semigroup, and let E be a Banach space. Assume that nonempty subsets $U, V \subset E$ satisfy $(U^+)^- \cap (V^+)^- = \{0\}$, $0 \in V$, and V is bounded. If a function $f : G \to E$ satisfies*

$$f(x + y) - f(x) - f(y) \in U + V$$

for all $x, y \in G$, then there exists a function $p : G \to E$ such that

$$p(x + y) - p(x) - p(y) \in U$$

for any $x, y \in G$, and

$$p(x) - f(x) \in \overline{\mathrm{Co}\,V}$$

for any $x \in G$. Moreover, if $U^- \cap 3\big(\overline{\mathrm{Co}\,V}\big)^- = \{0\}$, then the function p is unique.

Proof. According to Theorem 2.46, there exist functions $u, v : G \to E$ such that $f = u + v$, $u(x + y) - u(x) - u(y) \in U$, and $v(x + y) - v(x) - v(y) \in V$ for $x, y \in G$. It follows from [105, Theorem 4] that there exists an additive function $a : G \to E$ such that

$$v(x) - a(x) \in \overline{\mathrm{Co}(-V)} = -\overline{\mathrm{Co}\,V}$$

for all $x \in G$. Putting $p = u + a$ and applying $f = u + v$, we get the desired relation

$$p(x) - f(x) \in \overline{CoV}$$

for any $x \in G$.

Assume that $p_1, p_2 : G \to E$ are two functions such that

$$p_i(x + y) - p_i(x) - p_i(y) \in U$$

for any $x, y \in G$, and

$$p_i(x) - f(x) \in \overline{CoV}$$

for all $x \in G$ and $i \in \{1, 2\}$. Then, we have

$$r(x) = p_1(x) - p_2(x) = (p_1(x) - f(x)) - (p_2(x) - f(x)) \in (\overline{CoV})^-$$

for all $x \in G$. Consequently, using the notation $V_0 = \overline{CoV}$, we obtain

$$\begin{aligned} r(x + y) - r(x) - r(y) &\in (U - U) \cap (V_0^- - V_0^- - V_0^-) \\ &= U^- \cap 3V_0^- \\ &= \{0\}, \end{aligned}$$

since V_0^- is convex and symmetric with respect to zero. Thus, r is additive and bounded. An extended version of Theorem 2.1 implies that $r(x) = p_1(x) - p_2(x) \equiv 0$. This ends the proof.   □

Ger and Šemrl also gave the following corollary in the paper [123].

Corollary 2.48. *Let $(G, +)$ be a cancellative abelian semigroup, and let $\varepsilon \in (0, 1/4)$. If a function $f : G \to \mathbb{R}$ satisfies the congruence*

$$f(x + y) - f(x) - f(y) \in \mathbb{Z} + (-\varepsilon, \varepsilon)$$

for all $x, y \in G$, then there exists a function $p : G \to \mathbb{R}$ such that

$$p(x + y) - p(x) - p(y) \in \mathbb{Z}$$

for any $x, y \in G$, and

$$|f(x) - p(x)| \leq \varepsilon$$

for all $x \in G$.

Proof. Let $U = \mathbb{Z}$ and $V = (-\varepsilon, \varepsilon)$. Then, we have $(U^+)^- = \mathbb{Z}$ and $(V^+)^- = (-1, 1)$ and hence $(U^+)^- \cap (V^+)^- = \{0\}$. Therefore, the assertion is an immediate consequence of Theorem 2.47.   □

2.8   Pexider Equation

In this section, we prove the Hyers–Ulam–Rassias stability of the Pexider equation,

$$f(x + y) = g(x) + h(y).$$

Let G_1 and G_2 be abelian groups. It is well-known that functions $f, g, h : G_1 \rightarrow G_2$ satisfy the Pexider equation if and only if there exist an additive function $A :$ $G_1 \rightarrow G_2$ and constants $a, b \in G_2$ such that

$$f(x) = A(x) + a + b, \quad g(x) = A(x) + a, \quad h(x) = A(x) + b$$

for all $x \in G_1$.

In 1993, J. Chmieliński and J. Tabor [67] investigated the stability of the Pexider equation (ref. [126]). This paper seems to be the first one concerning the stability problem of the Pexider equation. We will introduce a theorem presented by K.-W. Jun, D.-S. Shin, and B.-D. Kim [155].

Theorem 2.49 (Jun, Shin, and Kim). *Let G and E be an abelian group and a Banach space, respectively. Let $\varphi : G^2 \rightarrow [0, \infty)$ be a function satisfying*

$$\Phi(x) = \sum_{i=1}^{\infty} 2^{-i} \left(\varphi\left(2^{i-1}x, 0\right) + \varphi\left(0, 2^{i-1}x\right) + \varphi\left(2^{i-1}x, 2^{i-1}x\right) \right) < \infty$$

and

$$\lim_{n \to \infty} 2^{-n} \varphi(2^n x, 2^n y) = 0$$

for all $x, y \in G$. If functions $f, g, h : G \rightarrow E$ satisfy the inequality

$$\| f(x + y) - g(x) - h(y) \| \leq \varphi(x, y) \tag{2.33}$$

for all $x, y \in G$, then there exists a unique additive function $A : G \rightarrow E$ such that

$$\| f(x) - A(x) \| \leq \|g(0)\| + \|h(0)\| + \Phi(x),$$

$$\| g(x) - A(x) \| \leq \|g(0)\| + 2\|h(0)\| + \varphi(x, 0) + \Phi(x), \tag{2.34}$$

$$\| h(x) - A(x) \| \leq 2\|g(0)\| + \|h(0)\| + \varphi(0, x) + \Phi(x)$$

for all $x \in G$.

Proof. If we put $y = x$ in (2.33), then we have

$$\| f(2x) - g(x) - h(x) \| \leq \varphi(x, x) \tag{a}$$

for all $x \in G$. Putting $y = 0$ in (2.33) yields that

$$\|f(x) - g(x) - h(0)\| \leq \varphi(x, 0) \qquad (b)$$

for any $x \in G$. It follows from (b) that

$$\|g(x) - f(x)\| \leq \|h(0)\| + \varphi(x, 0) \qquad (c)$$

for each $x \in G$. If we put $x = 0$ in (2.33), then we get

$$\|f(y) - g(0) - h(y)\| \leq \varphi(0, y)$$

for $y \in G$. Thus, we obtain

$$\|h(x) - f(x)\| \leq \|g(0)\| + \varphi(0, x) \qquad (d)$$

for all $x \in G$.

Let us define

$$u(x) = \|g(0)\| + \|h(0)\| + \varphi(0, x) + \varphi(x, 0) + \varphi(x, x).$$

Using the inequalities (a), (c), and (d), we have

$$\begin{aligned}
\|f(2x) &- 2f(x)\| \\
&\leq \|f(2x) - g(x) - h(x)\| + \|g(x) - f(x)\| + \|h(x) - f(x)\| \\
&\leq \|g(0)\| + \|h(0)\| + \varphi(0, x) + \varphi(x, 0) + \varphi(x, x) \\
&= u(x) \qquad (e)
\end{aligned}$$

for all $x \in G$. Replacing x with $2x$ in (e), we get

$$\left\|f\left(2^2 x\right) - 2f(2x)\right\| \leq u(2x) \qquad (f)$$

for any $x \in G$. It then follows from (e) and (f) that

$$\begin{aligned}
\left\|f\left(2^2 x\right) - 2^2 f(x)\right\| &\leq \left\|f\left(2^2 x\right) - 2f(2x)\right\| + 2\|f(2x) - 2f(x)\| \\
&\leq u(2x) + 2u(x)
\end{aligned}$$

for all $x \in G$.

Applying an induction argument on n, we will prove that

$$\|f(2^n x) - 2^n f(x)\| \leq \sum_{i=1}^{n} 2^{i-1} u(2^{n-i} x) \qquad (g)$$

for all $x \in G$ and $n \in \mathbb{N}$. In view of (e), the inequality (g) is true for $n = 1$. Assume that (g) is true for some $n > 0$. Substituting $2x$ for x in (g), we obtain

$$\left\| f\left(2^{n+1}x\right) - 2^n f(2x) \right\| \le \sum_{i=1}^{n} 2^{i-1} u\left(2^{n+1-i}x\right)$$

for any $x \in G$. Hence, it follows from (e) that

$$\left\| f\left(2^{n+1}x\right) - 2^{n+1} f(x) \right\| \le \left\| f\left(2^{n+1}x\right) - 2^n f(2x) \right\| + 2^n \left\| f(2x) - 2f(x) \right\|$$

$$\le \sum_{i=1}^{n} 2^{i-1} u\left(2^{n+1-i}x\right) + 2^n u(x)$$

$$= \sum_{i=1}^{n+1} 2^{i-1} u\left(2^{n+1-i}x\right)$$

for all $x \in G$, which proves the inequality (g).

By (g), we have

$$\left\| 2^{-n} f(2^n x) - f(x) \right\| \le \sum_{i=1}^{n} 2^{i-1-n} u(2^{n-i}x) \tag{h}$$

for all $x \in G$ and $n \in \mathbb{N}$. Moreover, if $m, n \in \mathbb{N}$ with $m < n$, then it follows from (e) that

$$\left\| 2^{-n} f(2^n x) - 2^{-m} f(2^m x) \right\|$$

$$\le \sum_{i=m}^{n-1} \left\| 2^{-i} f(2^i x) - 2^{-(i+1)} f\left(2^{i+1}x\right) \right\|$$

$$\le \sum_{i=m}^{n-1} 2^{-(i+1)} u(2^i x)$$

$$= \sum_{i=m}^{n-1} 2^{-(i+1)} \left(\|g(0)\| + \|h(0)\| + \varphi(0, 2^i x) + \varphi(2^i x, 0) + \varphi(2^i x, 2^i x) \right)$$

$$\le 2^{-m} \left(\|g(0)\| + \|h(0)\| \right)$$

$$+ \sum_{i=m}^{\infty} 2^{-(i+1)} \left(\varphi(0, 2^i x) + \varphi(2^i x, 0) + \varphi(2^i x, 2^i x) \right)$$

$$\to 0 \quad \text{as } m \to \infty$$

for all $x \in G$. Hence, $\{2^{-n} f(2^n x)\}$ is a Cauchy sequence for every $x \in G$. Since E is a Banach space, we can define a function $A : G \to E$ by

$$A(x) = \lim_{n \to \infty} 2^{-n} f(2^n x).$$

In view of (2.33), we obtain

$$\|2^{-n} f(2^n x + 2^n y) - 2^{-n} g(2^n x) - 2^{-n} h(2^n y)\| \leq 2^{-n} \varphi(2^n x, 2^n y)$$

for all $x \in G$ and $n \in \mathbb{N}$. It follows from (c) that

$$\|2^{-n} g(2^n x) - 2^{-n} f(2^n x)\| \leq 2^{-n} (\|h(0)\| + \varphi(2^n x, 0)) \tag{i}$$

for any $x \in G$ and $n \in \mathbb{N}$. Since

$$2^{-n} \varphi(2^n x, 0) \leq 2 \sum_{i=n}^{\infty} 2^{-(i+1)} (\varphi(0, 2^i x) + \varphi(2^i x, 0) + \varphi(2^i x, 2^i x))$$
$$\to 0 \quad \text{as } n \to \infty,$$

it follows from (i) that

$$\lim_{n \to \infty} 2^{-n} g(2^n x) = \lim_{n \to \infty} 2^{-n} f(2^n x) = A(x) \tag{j}$$

for each $x \in G$. Also, by (d), we have

$$\|2^{-n} h(2^n x) - 2^{-n} f(2^n x)\| \leq 2^{-n} (\|g(0)\| + \varphi(0, 2^n x)) \tag{k}$$

for all $x \in G$ and $n \in \mathbb{N}$. Similarly, it follows from (k) that

$$\lim_{n \to \infty} 2^{-n} h(2^n x) = \lim_{n \to \infty} 2^{-n} f(2^n x) = A(x) \tag{l}$$

for each $x \in G$. Thus, by (2.33), (j), (l), and the commutativity of G, we get

$$0 = \left\| \lim_{n \to \infty} \left(2^{-n} f(2^n x + 2^n y) - 2^{-n} g(2^n x) - 2^{-n} h(2^n y) \right) \right\|$$
$$= \|A(x + y) - A(x) - A(y)\|$$

for all $x, y \in G$.

Taking the limit in (h) as $n \to \infty$ yields

$$\|A(x) - f(x)\|$$

$$\leq \lim_{n \to \infty} \sum_{i=1}^{n} 2^{i-1-n} u(2^{n-i} x)$$

$$= \lim_{n \to \infty} (1 - 2^{-n})(\|g(0)\| + \|h(0)\|)$$

$$+ \lim_{n \to \infty} \sum_{i=1}^{n} 2^{-i} (\varphi(0, 2^{i-1} x) + \varphi(2^{i-1} x, 0) + \varphi(2^{i-1} x, 2^{i-1} x))$$

$$= \|g(0)\| + \|h(0)\| + \Phi(x)$$

for each $x \in G$, which proves (2.34).

It remains to prove the uniqueness of A. Assume that $A' : G \to E$ is another additive function which satisfies the inequalities in (2.34). Then we have

$$\|A(x) - A'(x)\|$$

$$\leq 2^{-n} \|A(2^n x) - f(2^n x)\| + 2^{-n} \|f(2^n x) - A'(2^n x)\|$$

$$\leq 2^{-n+1} (\|g(0)\| + \|h(0)\| + \Phi(2^n x))$$

$$= 2^{-(n-1)} (\|g(0)\| + \|h(0)\|)$$

$$+ 2 \sum_{i=n+1}^{\infty} 2^{-i} (\varphi(0, 2^{i-1} x) + \varphi(2^{i-1} x, 0) + \varphi(2^{i-1} x, 2^{i-1} x))$$

$$\to 0 \quad \text{as } n \to \infty$$

for each $x \in G$, which implies that $A(x) = A'(x)$ for all $x \in G$. $\square$

Corollary 2.50. *Let E_1 and E_2 be Banach spaces and let $\theta \geq 0$ and $p \in [0, 1)$ be constants. If functions $f, g, h : E_1 \to E_2$ satisfy the inequality*

$$\|f(x + y) - g(x) - h(y)\| \leq \theta(\|x\|^p + \|y\|^p)$$

for all $x, y \in E_1$, then there exists a unique linear function $A : E_1 \to E_2$ such that

$$\|f(x) - A(x)\| \leq \|g(0)\| + \|h(0)\| + \frac{4\theta}{2 - 2^p} \|x\|^p,$$

$$\|g(x) - A(x)\| \leq \|g(0)\| + 2\|h(0)\| + \frac{6 - 2^p}{2 - 2^p} \theta \|x\|^p,$$

$$\|h(x) - A(x)\| \leq 2\|g(0)\| + \|h(0)\| + \frac{6 - 2^p}{2 - 2^p} \theta \|x\|^p$$

for any $x \in E_1$.

In 2000, Y.-H. Lee and K.-W. Jun [232] investigated the Hyers–Ulam–Rassias stability of the Pexider equation on the restricted domains (ref. [274]).

2.9 Remarks

T. Aoki [7] appears to be the first to extend the theorem of Hyers (Theorem 2.3) for additive functions. Indeed, Aoki provided a proof of the special case of the theorem of Th. M. Rassias when the given function is additive.

It was Th. M. Rassias [285] who was the first to prove the stability of the linear function in Banach spaces.

D. G. Bourgin [28] stated the following result without proof, which is similar to the theorem of Găvruta (Theorem 2.18).

Let E_1 and E_2 be Banach spaces and let $\varphi : [0, \infty)^2 \to [0, \infty)$ be a monotone nondecreasing function such that

$$\Phi(x) = \sum_{k=1}^{\infty} 2^{-k} \varphi(2^k \|x\|, 2^k \|x\|) < \infty$$

for all $x \in E_1$. If a function $f : E_1 \to E_2$ satisfies

$$\|f(x + y) - f(x) - f(y)\| \leq \varphi(\|x\|, \|y\|)$$

for all $x, y \in E_1$, then there exists a unique additive function $A : E_1 \to E_2$ such that

$$\|f(x) - A(x)\| \leq \Phi(x)$$

for any $x \in E_1$.

G. L. Forti [104] used a similar idea in proving his stability theorem for a class of functional equations of the form

$$f(F(x, y)) = H(f(x), f(y))$$

with f as the unknown function, which includes the additive Cauchy equation as a special case. In the special case of the additive Cauchy equation, it proves that the result of Bourgin with an abelian semigroup $(G, +)$ instead of the Banach space E_1 holds true under an additional condition such as

$$\lim_{k \to \infty} 2^{-k} \varphi(2^k \|x\|, 2^k \|y\|) = 0$$

for all $x, y \in G$.

Recently, S.-M. Jung and S. Min [193] have proved the Hyers–Ulam–Rassias stability of the functional equations of the type $f(x + y) = H(f(x), f(y))$ by using the fixed point method, where H is a bounded linear transformation.

Chapter 3
Generalized Additive Cauchy Equations

It is very natural for one to try to transform the additive Cauchy equation into other forms. Some typically *generalized additive Cauchy equations* will be introduced. The functional equation $f(ax + by) = af(x) + bf(y)$ appears in Section 3.1. The Hyers–Ulam stability problem is discussed in connection with a *question of Th. M. Rassias and J. Tabor*. In Section 3.2, the functional equation (3.3) is introduced, and the Hyers–Ulam–Rassias stability for this equation is also studied. The stability result for this equation will be used to answer the question of Rassias and Tabor cited above. The last section deals with the functional equation $f(x + y)^2 = (f(x) + f(y))^2$. The continuous solutions and the Hyers–Ulam stability for this functional equation will be investigated.

3.1 Functional Equation $f(ax + by) = af(x) + bf(y)$

It is a very natural thing to generalize the additive Cauchy equation (2.1) into the functional equation $f(ax + by) = af(x) + bf(y)$ and study the stability problems for this equation.

In the paper [312], Th. M. Rassias and J. Tabor asked whether the functional equation $f(ax + by + c) = Af(x) + Bf(y) + C$ with $abAB \neq 0$ is stable in the sense of Hyers, Ulam, and Rassias.

C. Badea [8] answered this question of Rassias and Tabor for the case when $c = C = 0$, $a = A$, and $b = B$.

Theorem 3.1 (Badea). *Let a and b be nonnegative real numbers with $\alpha = a + b > 0$. Let $H : [0, \infty)^2 \to [0, \infty)$ be a function for which there exists a positive number $k < \alpha$ such that $H(\alpha s, \alpha t) \leq kH(s, t)$ for all $s, t \in [0, \infty)$. Given a real normed space E_1 and a real Banach space E_2, assume that a function $f : E_1 \to E_2$ satisfies the inequality*

$$\| f(ax + by) - af(x) - bf(y) \| \leq H(\|x\|, \|y\|) \tag{3.1}$$

for all $x, y \in E_1$. Then there exists a unique function $A : E_1 \to E_2$ such that

$$A(ax + by) = aA(x) + bA(y)$$

for any x and y in E_1, and

$$\|f(x) - A(x)\| \le (\alpha - k)^{-1} H(\|x\|, \|x\|) \tag{3.2}$$

for all $x \in E_1$.

Proof. Let x be a fixed point of E_1. Putting $y = x$ in the inequality (3.1) yields

$$\|f(\alpha x) - \alpha f(x)\| \le H(\|x\|, \|x\|)$$

which implies

$$\|(1/\alpha) f(\alpha x) - f(x)\| \le (1/\alpha) H(\|x\|, \|x\|). \tag{a}$$

We now use an induction argument to prove

$$\|\alpha^{-n} f(\alpha^n x) - f(x)\| \le \frac{1 - (k/\alpha)^n}{\alpha - k} H(\|x\|, \|x\|) \tag{b}$$

for any $n \in \mathbb{N}$. Assume that the inequality (b) holds true for a fixed $n \ge 1$. If we substitute αx for x in (b) and divide the resulting inequality by α, then we obtain

$$\|\alpha^{-n-1} f(\alpha^{n+1} x) - (1/\alpha) f(\alpha x)\| \le \frac{1 - (k/\alpha)^n}{\alpha(\alpha - k)} H(\alpha\|x\|, \alpha\|x\|)$$

$$\le \frac{1 - (k/\alpha)^n}{\alpha(\alpha - k)} k H(\|x\|, \|x\|).$$

This, together with (a), leads to the inequality

$$\|\alpha^{-n-1} f(\alpha^{n+1} x) - f(x)\| \le \left(\frac{1 - (k/\alpha)^n}{\alpha(\alpha - k)} k + \frac{1}{\alpha}\right) H(\|x\|, \|x\|).$$

It can be easily seen that

$$\frac{1 - (k/\alpha)^{n+1}}{\alpha - k} = \frac{1 - (k/\alpha)^n}{\alpha(\alpha - k)} k + \frac{1}{\alpha}$$

which implies the validity of the inequality (b) for every $n \in \mathbb{N}$.

Let m and n be integers with $m > n > 0$. Then we have

$$\|\alpha^{-m} f(\alpha^m x) - \alpha^{-n} f(\alpha^n x)\| = \alpha^{-n} \|\alpha^{-(m-n)} f(\alpha^m x) - f(\alpha^n x)\|$$

$$= \alpha^{-n} \|\alpha^{-r} f(\alpha^r y) - f(y)\|,$$

where $r = m - n$ and $y = \alpha^n x$. Therefore, using this equality and the inequality (b), we get

$$\|\alpha^{-m} f(\alpha^m x) - \alpha^{-n} f(\alpha^n x)\| \leq \frac{1 - (k/\alpha)^r}{\alpha^n (\alpha - k)} H(\|\alpha^n x\|, \|\alpha^n x\|).$$

However,

$$
\begin{aligned}
H(\|\alpha^n x\|, \|\alpha^n x\|) &= H(\alpha \|\alpha^{n-1} x\|, \alpha \|\alpha^{n-1} x\|) \\
&\leq k H(\|\alpha^{n-1} x\|, \|\alpha^{n-1} x\|) \\
&\leq \cdots \\
&\leq k^n H(\|x\|, \|x\|).
\end{aligned}
$$

Hence,

$$\|\alpha^{-m} f(\alpha^m x) - \alpha^{-n} f(\alpha^n x)\| \leq (k/\alpha)^n \frac{1 - (k/\alpha)^r}{\alpha - k} H(\|x\|, \|x\|).$$

As $k < \alpha$, the right-hand side of the inequality tends to zero if n tends to infinity. Therefore, the sequence $\{\alpha^{-n} f(\alpha^n x)\}$ is a Cauchy sequence. We may use the direct method to define

$$A(x) = \lim_{n \to \infty} \alpha^{-n} f(\alpha^n x)$$

for all $x \in E_1$. The inequality (3.2) follows immediately from (b).

We will show that A satisfies the equality $A(ax + by) = a A(x) + b A(y)$ for all $x, y \in E_1$. Let x and y be points of E_1. Then (3.1) implies

$$
\begin{aligned}
\|f(\alpha^n (ax + by)) - af(\alpha^n x) - bf(\alpha^n y)\| &\leq H(\|\alpha^n x\|, \|\alpha^n y\|) \\
&\leq k^n H(\|x\|, \|y\|)
\end{aligned}
$$

and hence

$$\alpha^{-n} \|f(\alpha^n (ax + by)) - af(\alpha^n x) - bf(\alpha^n y)\| \leq (k/\alpha)^n H(\|x\|, \|y\|).$$

We get $A(ax + by) = a A(x) + b A(y)$, since $(k/\alpha)^n$ tends to zero as n tends to infinity.

Suppose there exists another function $A' : E_1 \to E_2$ such that

$$\|f(x) - A'(x)\| \leq (\alpha - k')^{-1} H'(\|x\|, \|x\|)$$

for a certain function H' with the corresponding number $k' < \alpha$ and such that

$$A'(ax + by) = a A'(x) + b A'(y)$$

for all $x, y \in E_1$. Since $A(\alpha^n x) = \alpha^n A(x)$ and $A'(\alpha^n x) = \alpha^n A'(x)$, we have

$$\|A(x) - A'(x)\| \leq \|A(x) - f(x)\| + \|f(x) - A'(x)\|$$
$$\leq (\alpha - k)^{-1} H(\|x\|, \|x\|) + (\alpha - k')^{-1} H'(\|x\|, \|x\|)$$

and thus

$$\|A(x) - A'(x)\| = \|\alpha^{-n} A(\alpha^n x) - \alpha^{-n} A'(\alpha^n x)\|$$
$$\leq (k/\alpha)^n (\alpha - k)^{-1} H(\|x\|, \|x\|)$$
$$+ (k'/\alpha)^n (\alpha - k')^{-1} H'(\|x\|, \|x\|).$$

Both terms of the right-hand side of the above formula tend to zero for n tending to infinity. Thus, A' coincides with A.   □

As Gajda [111] extended the result of Theorem 2.5, Badea [8] also modified the result of the previous theorem by using a similar method. Hence, we omit the proof.

Theorem 3.2. *Let a and b be nonnegative real numbers such that $\alpha = a + b > 0$. Let $H : [0, \infty)^2 \to [0, \infty)$ be a function for which there exists a positive number $k < 1/\alpha$ such that $H(s/\alpha, t/\alpha) \leq k H(s, t)$ for all $s, t \in [0, \infty)$. Given a real normed space E_1 and a real Banach space E_2, assume that a function $f : E_1 \to E_2$ satisfies the inequality*

$$\|f(ax + by) - af(x) - bf(y)\| \leq H(\|x\|, \|y\|)$$

for all $x, y \in E_1$. Then there exists a unique function $A : E_1 \to E_2$ such that

$$A(ax + by) = a A(x) + b A(y)$$

for any $x, y \in E_1$, and

$$\|f(x) - A(x)\| \leq (k/(1 - \alpha k)) H(\|x\|, \|x\|)$$

for all $x \in E_1$.

Theorems 3.1 and 3.2 generalize the results of Theorems 2.11 and 2.17, respectively. In the case when the function H is given by

$$H(s, t) = \theta(\psi(s) + \psi(t)),$$

Badea [8] proved the following corollary.

Corollary 3.3. *Let a and b be nonnegative real numbers such that $\alpha = a + b > 0$. Let $\psi : [0, \infty) \to [0, \infty)$ be a function for which there exists a positive number*

$k < \alpha$ such that $\psi(\alpha s) \leq k\psi(s)$ for all $s \geq 0$. Given a real normed space E_1 and a real Banach space E_2, assume that a function $f : E_1 \to E_2$ satisfies the inequality

$$\|f(ax + by) - af(x) - bf(y)\| \leq \theta\big(\psi(\|x\|) + \psi(\|y\|)\big)$$

for some $\theta \geq 0$ and for all $x, y \in E_1$. Then there exists a unique function $A : E_1 \to E_2$ such that

$$A(ax + by) = aA(x) + bA(y)$$

for any $x, y \in E_1$, and

$$\|f(x) - A(x)\| \leq \big(2\theta/(\alpha - k)\big)\psi(\|x\|)$$

for all $x \in E_1$.

3.2 Additive Cauchy Equations of General Form

Throughout this section, let E_1 and E_2 be a complex normed space and a complex Banach space, respectively, let $m \geq 2$ be a fixed integer, and let (a_{ij}) be a matrix in $\mathbb{C}^{m \times m}$ of which determinant, denoted by ω, is nonzero. By ω_{ij}, $i, j \in \{1, \ldots, m\}$, we denote the *cofactor* of the matrix (a_{ij}) corresponding to the entry a_{ij}. For any $i \in \{1, \ldots, m\}$ we denote by ω_i the determinant of the matrix that remains after all entries of the i-th column in (a_{ij}) are replaced with 1. Let a_i and b_i, $i \in \{1, \ldots, m\}$, be complex numbers for which there exist some $j, k \in \{1, \ldots, m\}$ with $a_j \neq 0$ and $b_k \neq 0$.

S.-M. Jung [162] investigated the stability problem for a generalized additive Cauchy functional equation

$$f\left(x_0 + \sum_{j=1}^{m} a_j x_j\right) = \sum_{i=1}^{m} b_i f\left(\sum_{j=1}^{m} a_{ij} x_j\right) \tag{3.3}$$

for all $x_1, \ldots, x_m \in E_1$, where x_0 is a fixed point of E_1. In the case of $x_0 = 0$ in (3.3), the Hyers–Ulam–Rassias stability problem is treated in Theorem 3.6. Further, this result will be applied to the study of a question on the stability for a special form of generalized additive Cauchy equation suggested by Rassias and Tabor [312].

Let us define

$$r = \sum_{i=1}^{m} a_i \omega_i \omega^{-1} \quad \text{and} \quad B = \sum_{i=1}^{m} b_i \tag{3.4}$$

and assume $|r|, |B| \notin \{0, 1\}$. For a fixed $x_0 \in E_1$ and an $x \in E_1$ we define

$$s_0(x) = x, \quad s_1(x) = rx + x_0, \quad \text{and} \quad s_{n+1}(x) = s_n(s_1(x))$$

for any $n \in \mathbb{N}$. It is not difficult to show

$$s_{m+n}(x) = s_m(s_n(x)) \tag{3.5}$$

for any $m, n \in \mathbb{N}$. Further, suppose that a function $\varphi : E_1^m \to [0, \infty)$ satisfies

$$\Phi(x_1, \ldots, x_m) = \sum_{i=0}^{\infty} |B|^{-i-1} \varphi\big(\omega_1 \omega^{-1} s_i(x_1), \ldots, \omega_m \omega^{-1} s_i(x_m)\big) < \infty \tag{3.6}$$

and

$$\Phi\big(s_n(x), \ldots, s_n(x)\big) = o\big(|B|^n\big) \quad \text{as} \quad n \to \infty \tag{3.7}$$

for all $x, x_1, \ldots, x_m \in E_1$.

Lemma 3.4. *It holds true that*

$$\omega = \sum_{j=1}^{m} a_{ij} \omega_j$$

for all $i \in \{1, \ldots, m\}$.

Proof. Consider the following system of inhomogeneous linear equations with m unknowns $z_1, z_2, \ldots, z_m$ in $\mathbb{C}$

$$\begin{cases} a_{11}z_1 + a_{12}z_2 + \cdots + a_{1m}z_m = 1, \\ a_{21}z_1 + a_{22}z_2 + \cdots + a_{2m}z_m = 1, \\ \cdots \quad\quad \cdots \quad\quad\quad \cdots \\ a_{m1}z_1 + a_{m2}z_2 + \cdots + a_{mm}z_m = 1. \end{cases}$$

Since $\omega \neq 0$, the solution of this system is uniquely determined by $z_j = \omega_j \omega^{-1}$, $j \in \{1, \ldots, m\}$. Hence, it holds true that

$$\sum_{j=1}^{m} a_{ij} \omega_j \omega^{-1} = 1,$$

for any $i \in \{1, \ldots, m\}$, which ends the proof.   □

We now investigate the stability problem for a generalized additive Cauchy equation, i.e., the stability problem for the functional inequality

$$\left\| f\left(x_0 + \sum_{j=1}^{m} a_j x_j\right) - \sum_{i=1}^{m} b_i f\left(\sum_{j=1}^{m} a_{ij} x_j\right) \right\| \leq \varphi(x_1, \ldots, x_m) \tag{3.8}$$

for all $x_1, \ldots, x_m \in E_1$.

S.-M. Jung [162] contributed to the following theorem:

Theorem 3.5. *If a function* $f : E_1 \rightarrow E_2$ *satisfies the inequality* (3.8) *for all* $x_1, \ldots, x_m \in E_1$, *then there exists a unique function* $A : E_1 \rightarrow E_2$ *which satisfies*

$$A(rx + x_0) = BA(x) \tag{3.9}$$

and

$$\|f(x) - A(x)\| \leq \Phi(x, \ldots, x) \tag{3.10}$$

for all $x \in E_1$.

Proof. Let x be an arbitrary point of E_1. Using Lemma 3.4 and putting $x_i = \omega_i \omega^{-1} x$, $i \in \{1, \ldots, m\}$, it follows from (3.8) that

$$\|f(s_1(x)) - Bf(x)\| \leq \varphi(\omega_1 \omega^{-1} x, \ldots, \omega_m \omega^{-1} x). \tag{a}$$

We will now prove

$$\|f(s_n(x)) - B^n f(x)\| \leq \sum_{i=0}^{n-1} |B|^{n-1-i} \varphi(\omega_1 \omega^{-1} s_i(x), \ldots, \omega_m \omega^{-1} s_i(x)) \tag{b}$$

for all $n \in \mathbb{N}$. In view of (a) it is easy to see the validity of (b) in the case of $n = 1$. Let us assume that (b) holds true for some integer $n > 0$. Replacing x in (b) with $s_1(x)$ and using (3.5) we obtain

$$\begin{aligned}
\|f(s_{n+1}(x)) &- B^n f(s_1(x))\| \\
&= \|f(s_{n+1}(x)) - B^n(f(s_1(x)) - Bf(x)) - B^{n+1} f(x)\| \\
&\leq \sum_{i=0}^{n-1} |B|^{n-1-i} \varphi(\omega_1 \omega^{-1} s_{i+1}(x), \ldots, \omega_m \omega^{-1} s_{i+1}(x)).
\end{aligned}$$

Hence, it follows from (a) that

$$\|f(s_{n+1}(x)) - B^{n+1} f(x)\| \leq \sum_{i=0}^{n} |B|^{n-i} \varphi(\omega_1 \omega^{-1} s_i(x), \ldots, \omega_m \omega^{-1} s_i(x))$$

which implies the validity of (b) for all $n \in \mathbb{N}$.

If we divide both sides in (b) by $|B|^n$ we get

$$\|B^{-n} f(s_n(x)) - f(x)\| \leq \sum_{i=0}^{n-1} |B|^{-i-1} \varphi(\omega_1 \omega^{-1} s_i(x), \ldots, \omega_m \omega^{-1} s_i(x)). \tag{c}$$

Next, we present that the sequence $\{B^{-n} f(s_n(x))\}$ is a Cauchy sequence. In view of (3.5), (c), and (3.6), for any positive integers p and q with $p < q$, the following estimation is possible:

$$\|B^{-q} f(s_q(x)) - B^{-p} f(s_p(x))\|$$
$$= |B|^{-p} \| B^{-(q-p)} f\left(s_{q-p}(s_p(x))\right) - f\left(s_p(x)\right) \|$$
$$\leq |B|^{-p} \sum_{i=0}^{q-p-1} |B|^{-i-1} \varphi\left(\omega_1 \omega^{-1} s_{p+i}(x), \ldots, \omega_m \omega^{-1} s_{p+i}(x)\right)$$
$$\leq \sum_{i=p}^{q-1} |B|^{-i-1} \varphi\left(\omega_1 \omega^{-1} s_i(x), \ldots, \omega_m \omega^{-1} s_i(x)\right)$$
$$\to 0 \quad \text{as} \quad p \to \infty.$$

Since E_2 is a Banach space, we can define

$$A(x) = \lim_{n \to \infty} B^{-n} f\left(s_n(x)\right) \qquad (d)$$

for all $x \in E_1$. On account of (3.6), (c), and (d), the validity of (3.10) is clear. Further, replacing x in (d) with $s_1(x)$ and in view of (3.5), the validity of (3.9) is also obvious.

It only remains to prove the uniqueness of A. Assume that $A' : E_1 \to E_2$ is another function that satisfies (3.9) and (3.10). From (3.9) and (3.5) it follows that

$$A(s_n(x)) = B^n A(x) \quad \text{and} \quad A'(s_n(x)) = B^n A'(x)$$

for all $n \in \mathbb{N}$. Hence, by (3.10) and (3.7), we get

$$\|A(x) - A'(x)\|$$
$$= |B|^{-n} \|A(s_n(x)) - A'(s_n(x))\|$$
$$\leq |B|^{-n} \left(\|A(s_n(x)) - f(s_n(x))\| + \|f(s_n(x)) - A'(s_n(x))\|\right)$$
$$\leq 2|B|^{-n} \Phi\left(s_n(x), \ldots, s_n(x)\right)$$
$$\to 0 \quad \text{as} \quad n \to \infty,$$

which implies the uniqueness of A.   □

On the other hand, if $x_0 = 0$ is assumed in the functional inequality (3.8), we can prove the Hyers–Ulam–Rassias stability of the equation (3.8) (ref. [162]).

Theorem 3.6. *Assume that $x_0 = 0$ in the functional inequality (3.8). If a function $f : E_1 \to E_2$ satisfies the inequality (3.8) for all $x_1, \ldots, x_m \in E_1$, then there exists a unique function $A : E_1 \to E_2$ which satisfies (3.10) and*

$$A\left(\sum_{j=1}^{m} a_j x_j\right) = \sum_{i=1}^{m} b_i A\left(\sum_{j=1}^{m} a_{ij} x_j\right) \tag{3.11}$$

for all $x_1, \ldots, x_m \in E_1$.

Proof. Let us define A as in (d) in the proof of the last theorem. On account of Theorem 3.5, it only remains to prove (3.11). Since $s_n(x) = r^n x$ $(x_0 = 0)$, by (3.8) with $x_0 = 0$ and by (3.6), we obtain

$$\left\| A\left(\sum_{j=1}^{m} a_j x_j\right) - \sum_{i=1}^{m} b_i A\left(\sum_{j=1}^{m} a_{ij} x_j\right) \right\|$$

$$= \lim_{n \to \infty} |B|^{-n} \left\| f\left(r^n \sum_{j=1}^{m} a_j x_j\right) - \sum_{i=1}^{m} b_i f\left(r^n \sum_{j=1}^{m} a_{ij} x_j\right) \right\|$$

$$\le \lim_{n \to \infty} |B|^{-n} \varphi(r^n x_1, \ldots, r^n x_m)$$

$$= |B| \lim_{n \to \infty} |B|^{-n-1} \varphi\left(\omega_1 \omega^{-1} s_n(\omega_1^{-1} \omega x_1), \ldots, \omega_m \omega^{-1} s_n(\omega_m^{-1} \omega x_m)\right)$$

$$= 0,$$

which completes our proof.   □

Since $\omega \neq 0$ is assumed, the set of vectors

$$\{(a_{11}, a_{12}, \ldots, a_{1m}), (a_{21}, a_{22}, \ldots, a_{2m}), \ldots, (a_{m1}, a_{m2}, \ldots, a_{mm})\}$$

establishes a basis for $\mathbb{C}^m$. Therefore, we can uniquely determine the complex numbers $d_1, d_2, \ldots, d_m$ such that

$$(a_1, a_2, \ldots, a_m) = \sum_{i=1}^{m} d_i (a_{i1}, a_{i2}, \ldots, a_{im}).$$

Jung [162] introduced some properties of the general solution of the generalized additive Cauchy equation (3.11).

Theorem 3.7. *Assume that $\omega_{kk} \neq 0$ for some $k \in \{1, \ldots, m\}$. If a function $f : E_1 \to E_2$ satisfies the generalized additive Cauchy equation (3.11) for all $x_1, \ldots, x_m \in E_1$ and, in addition, this satisfies $f(0) = 0$, then it holds true that $f(d_k x) = b_k f(x)$ for all $x \in E_1$.*

Proof. Putting $x_i = \omega_{ki} \omega_{kk}^{-1} x_k$, $i \in \{1, \ldots, m\}$, and using this well-known fact

$$\sum_{j=1}^{m} a_{ij} \omega_{kj} = \delta_{ik} \omega$$

yield

$$\sum_{j=1}^{m} a_j x_j = \sum_{j=1}^{m}\sum_{i=1}^{m} d_i a_{ij} x_j = \sum_{i=1}^{m} d_i \sum_{j=1}^{m} a_{ij} x_j = d_k \omega \omega_{kk}^{-1} x_k$$

and

$$\sum_{j=1}^{m} a_{ij} x_j = \sum_{j=1}^{m} a_{ij}\omega_{kj}\omega_{kk}^{-1} x_k = \delta_{ik}\omega\omega_{kk}^{-1} x_k.$$

Hence, the functional equation (3.11) can be transformed into

$$f\left(d_k \omega\omega_{kk}^{-1} x_k\right) = b_k f\left(\omega\omega_{kk}^{-1} x_k\right),$$

since $f(0) = 0$. Putting $x = \omega\omega_{kk}^{-1} x_k$ ends the proof.   □

As already introduced in Section 3.1, Rassias and Tabor [312] raised a question concerning the stability of a generalized additive Cauchy functional inequality of the form

$$\|f(a_1 x + a_2 y + v) - b_1 f(x) - b_2 f(y) - w\| \le \theta\left(\|x\|^p + \|y\|^q\right), \qquad (3.12)$$

where $a_1 a_2 b_1 b_2 \ne 0$, $\theta \ge 0$, $p, q \in \mathbb{R}$, v and w are fixed points of E_1 and E_2, respectively. Jung [162] partially answered this question as stated in the following theorem.

Theorem 3.8. *Assume that* $0 < |a_1 + a_2| < 1$, $0 < p, q \le 1$, *and* $|b_1 + b_2| > 1$. *If a function* $f : E_1 \to E_2$ *satisfies the inequality* (3.12) *for all* $x, y \in E_1$, *then there exists a unique function* $A : E_1 \to E_2$ *which satisfies*

$$A\left((a_1 + a_2)x + v\right) = (b_1 + b_2)A(x)$$

and

$$\|f(x) - A(x)\| \le M_1 + M_2\|x\|^p + M_3\|x\|^q$$

for all $x \in E_1$, *where* M_1, M_2, *and* M_3 *are appropriate constants.*

Proof. The functional inequality (3.12) can be transformed into

$$\|f(a_1 x + a_2 y + v) - b_1 f(x) - b_2 f(y)\| \le \varphi(x, y),$$

where $\varphi(x, y) = \|w\| + \theta\left(\|x\|^p + \|y\|^q\right)$ for all $x, y \in E_1$. Since $a_{11} = 1$, $a_{12} = 0$, $a_{21} = 0$, $a_{22} = 1$, and $\omega = \omega_1 = \omega_2 = 1$, we have $r = a_1 + a_2$ and $B = b_1 + b_2$. Therefore, using the triangle inequality for the norm and using the condition $0 < p, q \le 1$, we get

$$\|s_n(x)\|^t \leq |a_1 + a_2|^{tn} \|x\|^t + \sum_{j=0}^{n-1} |a_1 + a_2|^{tj} \|v\|^t \qquad (a)$$

for $t \in \{p, q\}$, since

$$s_n(x) = (a_1 + a_2)^n x + \sum_{j=0}^{n-1} (a_1 + a_2)^j v.$$

It follows from (a) and the given hypotheses that

$$\sum_{i=0}^{\infty} |b_1 + b_2|^{-i-1} \|s_i(x)\|^t$$

$$\leq \|x\|^t |b_1 + b_2|^{-1} \sum_{i=0}^{\infty} \left(|a_1 + a_2|^t |b_1 + b_2|^{-1} \right)^i$$

$$+ \|v\|^t |b_1 + b_2|^{-1} \sum_{i=0}^{\infty} |b_1 + b_2|^{-i} \sum_{j=0}^{i-1} |a_1 + a_2|^{tj}$$

$$\leq c_1 \|x\|^t + c_2 \|v\|^t$$

for $t \in \{p, q\}$, where c_1 and c_2 are given constants. Hence, we obtain

$$\Phi(x, y) = \sum_{i=0}^{\infty} |b_1 + b_2|^{-i-1} \varphi(s_i(x), s_i(y))$$

$$= \sum_{i=0}^{\infty} |b_1 + b_2|^{-i-1} \left(\|w\| + \theta \|s_i(x)\|^p + \theta \|s_i(y)\|^q \right) \qquad (b)$$

$$\leq M_1 + M_2 \|x\|^p + M_3 \|y\|^q,$$

where M_1, M_2, and M_3 are appropriate constants. Therefore,

$$\Phi(x, y) < \infty.$$

From (a) and (b), it follows that

$$\Phi(s_n(x), s_n(x)) \leq M_1 + M_2 \|s_n(x)\|^p + M_3 \|s_n(x)\|^q$$

$$\leq M_1 + M_2 \left(|a_1 + a_2|^{pn} \|x\|^p + \sum_{j=0}^{n-1} |a_1 + a_2|^{pj} \|v\|^p \right)$$

$$+ M_3 \left(|a_1 + a_2|^{qn} \|x\|^q + \sum_{j=0}^{n-1} |a_1 + a_2|^{qj} \|v\|^q \right).$$

Hence, we achieve

$$\Phi\big(s_n(x), s_n(x)\big) = o\big(|b_1 + b_2|^n\big) \quad \text{as } n \to \infty. \tag{c}$$

In view of (b), (c), and Theorem 3.5, the assertions are true.   □

3.3 Functional Equation $f(x+y)^2 = (f(x) + f(y))^2$

The functional equation $f(x+y)^2 = \big(f(x) + f(y)\big)^2$ may be regarded as a special form of the functional equation

$$f(x+y)^2 = af(x)f(y) + bf(x)^2 + cf(y)^2 \quad (a,b,c \in \mathbb{C}) \tag{3.13}$$

of which solutions and the Hyers–Ulam stability problem were proved by S.-M. Jung [157].

Throughout this section, let E be a real normed space. We first describe the solutions of the functional equation (3.13) briefly:

(i) The case of $b \neq 1$ or $c \neq 1$.
   The function $f : E \to \mathbb{C}$ is a solution of the functional equation (3.13) with $b \neq 1$ or $c \neq 1$ if and only if there exists a complex number α such that $f(x) = \alpha$ for all $x \in E$, where $\alpha = 0$ for the case of $a + b + c \neq 1$.
(ii) The case of $a = -1$ and $b = c = 1$.
   The function $f : E \to \mathbb{C}$ is a solution of the functional equation

$$f(x+y)^2 = -f(x)f(y) + f(x)^2 + f(y)^2$$

   if and only if there exists a complex number α such that $f(x) = \alpha$ for all $x \in E$.
(iii) The case of $a \notin \{-1,2\}$ and $b = c = 1$.
   The function $f(x) \equiv 0$ $(x \in E)$ is the unique solution of the functional equation

$$f(x+y)^2 = af(x)f(y) + f(x)^2 + f(y)^2,$$

   where $a \notin \{-1,2\}$.
(iv) The case of $a = 2$ and $b = c = 1$.
   In this case the functional equation (3.13) can be rewritten as

$$f(x+y)^2 = \big(f(x) + f(y)\big)^2. \tag{3.14}$$

In the following theorem, solutions of the functional equation (3.14) which are continuous at a point will be studied (see [157]).

Theorem 3.9. *Let $f : \mathbb{R} \to \mathbb{C}$ be a function which is continuous at 0. The function f is a solution of the functional equation (3.14) if and only if f is linear, i.e., there exists a complex number c such that $f(x) = cx$ for all $x \in \mathbb{R}$.*

Proof. If we set $x = y = 0$ in (3.14), then we get $f(0) = 0$. By putting $y = -x$ in (3.14) and using $f(0) = 0$ we can show the oddness of f. Now, we will prove that

$$f(nx) = nf(x) \qquad (a)$$

for all integers n and any $x \in \mathbb{R}$. By putting $y = x$ in (3.14) we get

$$f(2x) = 2f(x) \quad \text{or} \quad f(2x) = -2f(x). \qquad (b)$$

Replacing y in (3.14) with $2x$ yields

$$f(3x)^2 = \begin{cases} 9f(x)^2 & (\text{if } f(2x) = 2f(x)), \\ f(x)^2 & (\text{if } f(2x) = -2f(x)). \end{cases} \qquad (c)$$

In view of (3.14) and (b), it follows

$$f(4x)^2 = \big(f(2x) + f(2x)\big)^2 = 16f(x)^2. \qquad (d)$$

On the other hand, replacing y in (3.14) with $3x$ and using (c), we obtain

$$f(4x)^2 = \begin{cases} 16f(x)^2 & (\text{if } f(3x) = 3f(x)), \\ 4f(x)^2 & (\text{if } f(3x) = -3f(x) \text{ or } f(3x) = f(x)), \\ 0 & (\text{if } f(3x) = -f(x)). \end{cases} \qquad (e)$$

Comparing (d) with (e) and taking (c) into consideration, we get $f(3x) = 3f(x)$. Hence, it follows from (b) and (c) that

$$f(2x) = 2f(x).$$

Thus, (a) holds true for $n = 2$. Assume that (a) is true for all positive integers $\leq n$ $(n \geq 2)$. Then, putting $y = nx$ in (3.14) yields

$$f\big((n + 1)x\big) = (n + 1)f(x) \quad \text{or} \quad f\big((n + 1)x\big) = -(n + 1)f(x). \qquad (f)$$

Replacing y in (3.14) with $(n + 1)x$ yields

$$f\big((n + 2)x\big)^2 = \begin{cases} (n + 2)^2 f(x)^2 & (\text{if } f((n + 1)x) = (n + 1)f(x)), \\ n^2 f(x)^2 & (\text{if } f((n + 1)x) = -(n + 1)f(x)). \end{cases} \qquad (g)$$

On the other hand, by substituting $2x$ and nx for x and y in (3.14), respectively, and by using induction hypothesis, it holds true that

$$f\left((n+2)x\right)^2 = (n+2)^2 f(x)^2. \tag{h}$$

Comparing (g) with (h) and considering (f) yield

$$f\left((n+1)x\right) = (n+1)f(x),$$

which ends the proof of (a).

By substituting x/n $(n \neq 0)$ for x in (a), we achieve

$$f(x/n) = (1/n)f(x). \tag{i}$$

Hence, by (a) and (i), we have

$$f(qx) = qf(x)$$

for every rational number q. If we put $x = 1$ in the above equation, then we attain

$$f(q) = f(1)q. \tag{j}$$

The continuity of f at 0, together with (3.14), implies that $f(x)^2$ is continuous at each $x \in \mathbb{R}$. From this fact and (j), it follows that

$$f(r) = f(1)r \quad \text{or} \quad f(r) = -f(1)r \tag{k}$$

for all irrational numbers r. Assume that f satisfies $f(r) = -f(1)r$ for some irrational number r. Then, by (k), it holds true that

$$f(q+r) = f(1)(q+r) \quad \text{or} \quad f(q+r) = -f(1)(q+r) \tag{l}$$

for any rational number $q \neq 0$. On the other hand, by (3.14), (j), and the assumption, we reach

$$f(q+r)^2 = f(1)^2(q-r)^2.$$

By comparing this equation with (l), we conclude that $f(1) = 0$. Hence, if $f(r) = -f(1)r$ holds true for some irrational number r, then it follows that

$$f(x) = 0$$

for all $x \in \mathbb{R}$. Assume now that $f(r) = f(1)r$ for all irrational numbers r. Then this assumption, together with (j), yields

$$f(x) = f(1)x$$

for all $x \in \mathbb{R}$.

Conversely, every complex-valued function f defined on $\mathbb{R}$ of the form $f(x)=cx$ with a constant c satisfies the functional equation (3.14). □

Remark. If f is a real-valued function, the functional equation (3.14) is equivalent to

$$|f(x + y)| = |f(x) + f(y)| \quad \text{for all} \ x, y \in E.$$

Since $(\mathbb{R}, |\cdot|)$ is a strictly normed space, according to F. Skof [333], the function f is a solution of the functional equation (3.14) if and only if f is an additive function.

Using ideas from Theorems 2.3 and 2.5, Jung [157] proved the Hyers–Ulam stability of the functional equation (3.14).

Theorem 3.10. *Suppose* $f : E \to \mathbb{R}$ *is a function which satisfies*

$$\left| f(x + y)^2 - \big(f(x) + f(y)\big)^2 \right| \le \delta \tag{3.15}$$

for some $\delta \ge 0$ *and for any* $x, y \in E$. *Then there exists an additive function* $A :$ $E \to \mathbb{R}$ *which satisfies*

$$\left| f(x)^2 - A(x)^2 \right| \le (1/3)\delta \tag{3.16}$$

for all $x \in E$. *Moreover, if* $A' : E \to \mathbb{R}$ *is another additive function which satisfies* (3.16), *then*

$$A(x)^2 = A'(x)^2 \tag{3.17}$$

for any $x \in E$.

Proof. By using induction on n we first prove that

$$\left| f\big(2^n x\big)^2 - \big(2^n f(x)\big)^2 \right| \le \delta \sum_{i=0}^{n-1} 2^{2i} \tag{a}$$

for any $n \in \mathbb{N}$. For $n = 1$, it is trivial by (3.15). Assume that (a) holds true for some n. Then, by substituting $2^n x$ for x and y in (3.15) and by using (a), we show

$$\begin{aligned}
\left| f\big(2^{n+1}x\big)^2 - \big(2^{n+1} f(x)\big)^2 \right| &\le \left| f\big(2^{n+1}x\big)^2 - \big(2f(2^n x)\big)^2 \right| \\
&\quad + 2^2 \left| f(2^n x)^2 - \big(2^n f(x)\big)^2 \right| \\
&\le \delta + 2^2 \delta \sum_{i=0}^{n-1} 2^{2i} \\
&\le \delta \sum_{i=0}^{n} 2^{2i},
\end{aligned}$$

which ends the proof of (a).

Dividing both sides in (a) by 2^{2n} yields

$$\left|2^{-2n} f(2^n x)^2 - f(x)^2\right| \le (1/3)\delta \tag{b}$$

for all $x \in E$ and $n \in \mathbb{N}$. It follows from (b) that, for $n \ge m > 0$,

$$\left|\left(2^{-n} f(2^n x)\right)^2 - \left(2^{-m} f(2^m x)\right)^2\right|$$
$$= 2^{-2m}\left|\left(2^{-(n-m)} f(2^{n-m} 2^m x)\right)^2 - f(2^m x)^2\right|$$
$$\le 2^{-2m}(1/3)\delta.$$

Hence, it holds true that

$$\left|\left(2^{-n} f(2^n x)\right)^2 - \left(2^{-m} f(2^m x)\right)^2\right| \to 0 \quad \text{as } m \to \infty. \tag{c}$$

For each $x \in E$ we define

$$I_x^+ = \left\{n \in \mathbb{N} \mid f(2^n x) \ge 0\right\} \quad \text{and} \quad I_x^- = \left\{n \in \mathbb{N} \mid f(2^n x) < 0\right\}.$$

In view of (c), we know that if I_x^+ or I_x^- is an infinite set, then the sequence $\{2^{-n} f(2^n x)\}_{n \in I_x^+}$ or $\{2^{-n} f(2^n x)\}_{n \in I_x^-}$ is a Cauchy sequence, respectively. Now, let us define

$$A(x) = \begin{cases} \displaystyle\lim_{\substack{n \to \infty \\ n \in I_x^+}} 2^{-n} f(2^n x) & \text{(if } I_x^+ \text{ is infinite)}, \\[2ex] \displaystyle\lim_{\substack{n \to \infty \\ n \in I_x^-}} 2^{-n} f(2^n x) & \text{(otherwise)}. \end{cases}$$

It is clear that if both I_x^+ and I_x^- are infinite sets, then

$$A(x) = -\lim_{\substack{n \to \infty \\ n \in I_x^-}} 2^{-n} f(2^n x). \tag{d}$$

The definition of A and (b) imply the validity of (3.16).

Let $x, y \in E$ be given arbitrarily. It is not difficult to prove that there is at least one infinite set among the sets $I_x^+ \cap I_y^+ \cap I_{x+y}^+, I_x^+ \cap I_y^+ \cap I_{x+y}^-, \ldots, I_x^- \cap I_y^- \cap I_{x+y}^-$. We may choose such an infinite set and denote it by I. Let $n \in I$ be given. Replacing x and y in (3.15) with $2^n x$ and $2^n y$, respectively, and then dividing the resulting inequality by 2^{2n}, we obtain

$$\left|\left(2^{-n} f(2^n (x + y))\right)^2 - \left(2^{-n} f(2^n x) + 2^{-n} f(2^n y)\right)^2\right| \le 2^{-2n}\delta. \tag{e}$$

By letting $n \rightarrow \infty$ through I in (e) and taking (d) into consideration, we immediately achieve

$$A(x + y)^2 = \big(A(x) + A(y)\big)^2 \quad \text{or} \quad A(x + y)^2 = \big(A(x) - A(y)\big)^2 \qquad (f)$$

for all $x, y \in E$. The second equality in (f) can take place, for example, when both I_y^+ and I_y^- are infinite sets and $I = I_x^+ \cap I_y^- \cap I_{x+y}^+$ for some $x, y \in E$, because (e) and $I = I_x^+ \cap I_y^- \cap I_{x+y}^+$ lead to

$$\lim_{\substack{n \to \infty \\ n \in I}} 2^{-n} f\big(2^n (x + y)\big) = \lim_{\substack{n \to \infty \\ n \in I_{x+y}^+}} 2^{-n} f\big(2^n (x + y)\big) = A(x + y),$$

$$\lim_{\substack{n \to \infty \\ n \in I}} 2^{-n} f(2^n x) = \lim_{\substack{n \to \infty \\ n \in I_x^+}} 2^{-n} f(2^n x) = A(x),$$

and

$$\lim_{\substack{n \to \infty \\ n \in I}} 2^{-n} f(2^n y) = \lim_{\substack{n \to \infty \\ n \in I_y^-}} 2^{-n} f(2^n y) = -A(y).$$

$A(x) \equiv 0 \ (x \in E)$ is not only the unique solution of the second functional equation but also a solution of the first equation in (f). Hence, it holds true that

$$A(x + y)^2 = \big(A(x) + A(y)\big)^2 \quad \text{or} \quad |A(x + y)| = |A(x) + A(y)|$$

for all $x, y \in E$. According to the above Remark, A is an additive function.

Now, suppose $A' : E \rightarrow \mathbb{R}$ is another additive function which satisfies (3.16). Since A and A' are additive functions, it is clear that

$$A(nx) = n A(x) \quad \text{and} \quad A'(nx) = n A'(x)$$

for all $n \in \mathbb{N}$ and any $x \in E$. Hence, by (3.16), we get

$$\begin{aligned}
\big|A(x)^2 - A'(x)^2\big| &= n^{-2}\big|A(nx)^2 - A'(nx)^2\big| \\
&\leq n^{-2}\big(|A(nx)^2 - f(nx)^2| + |f(nx)^2 - A'(nx)^2|\big) \\
&\leq (2/3)n^{-2}\delta \\
&\to 0 \quad \text{as } n \to \infty,
\end{aligned}$$

which ends the proof of (3.17). □

Chapter 4
Hosszú's Functional Equation

In 1967, M. Hosszú introduced the functional equation $f(x + y - xy) = f(x) + f(y) - f(xy)$ in a presentation at a meeting on functional equations held in Zakopane, Poland. In honor of M. Hosszú, this equation is called *Hosszú's functional equation*. As one can easily see, Hosszú's functional equation is a kind of generalized form of the additive Cauchy functional equation. In Section 4.1, it will be proved that Hosszú's equation is stable in the sense of C. Borelli. We discuss the Hyers–Ulam stability problem of Hosszú's equation in Section 4.2. In Section 4.3, Hosszú's functional equation will be generalized, and the stability (in the sense of Borelli) of the generalized equation will be proved. It is surprising that Hosszú's functional equation is not stable on the unit interval. It will be discussed in Section 4.4. In the final section, we will survey the Hyers–Ulam stability of Hosszú's functional equation of Pexider type.

4.1 Stability in the Sense of Borelli

We have seen in Chapter 3 that the additive Cauchy functional equation can be generalized in various forms. *Hosszú's functional equation*

$$f(x + y - xy) = f(x) + f(y) - f(xy) \qquad (4.1)$$

is the most famous among generalized forms of the additive Cauchy equation. Hence, this equation will be surveyed separately from Chapter 3. Every solution of Hosszú's functional equation is said to be a *Hosszú function*. A function f is called *affine* if it can be represented by $f(x) = A(x) + c$, where A is an additive function and c is a constant. According to T. M. K. Davison [92], the function $f : K \to G$ is a Hosszú function if and only if it is affine, where K is a field with at least five elements and G is an abelian group.

Throughout this section, let $f : \mathbb{R} \to \mathbb{R}$ be a function, and let g and h denote the odd and the even part of a corresponding function f, respectively.

We now start with the proof of Borelli's theorem concerning the stability of Hosszú's functional equation (see [23]).

Lemma 4.1. *If the inequality*

$$\left| h(x + y - xy) - h(x) - h(y) + h(xy) \right| \le \varepsilon \tag{4.2}$$

holds true for some $\varepsilon \ge 0$ *and for all* $x, y \in \mathbb{R}$, *then* $|h(x)| \le 2\varepsilon + |h(1)|$ *for all* $x \in \mathbb{R}$.

Proof. If we replace x and y in (4.2) with $-x$ and $-y$, respectively, we get

$$\left| h(-x - y - xy) - h(x) - h(y) + h(xy) \right| \le \varepsilon,$$

since h is even. Thus,

$$
\begin{aligned}
& |h(x + y - xy) - h(-x - y - xy)| \\
& \quad \le |h(x + y - xy) - h(x) - h(y) + h(xy)| \\
& \quad\quad + |-h(-x - y - xy) + h(x) + h(y) - h(xy)| \\
& \quad \le 2\varepsilon
\end{aligned}
$$

and setting $y = -1$ yield $|h(2x - 1) - h(1)| \le 2\varepsilon$. $2x - 1$ spans the whole $\mathbb{R}$, which ends the proof. $\qquad\square$

C. Borelli [23] also provided the following lemmas.

Lemma 4.2. *Let a function* $f : \mathbb{R} \to \mathbb{R}$ *satisfy the inequality*

$$\left| f(x + y - xy) - f(x) - f(y) + f(xy) \right| \le \delta \tag{4.3}$$

for some $\delta \ge 0$ *and for all* $x, y \in \mathbb{R}$, *and let the even part h of f satisfy the inequality* (4.2) *for some* $\varepsilon \ge 0$ *and for all* $x, y \in \mathbb{R}$. *Define* $\mu = \delta + \varepsilon$ *and* $\gamma = \max\{2\mu, |g(2) - 2g(1)|\}$. *Then the odd part g of f can be represented by* $g(x) = p(x) + s(x)$ *for any real number x, where p and s are odd functions, p satisfies the equation $p(2x) = 2p(x)$, and $|s(x)| \le \gamma$ for all $x \in \mathbb{R}$.*

Proof. It is obvious that

$$\left| g(x + y - xy) - g(x) - g(y) + g(xy) \right| \le \mu$$

and

$$\left| g(x - y + xy) - g(x) - g(-y) + g(-xy) \right| \le \mu.$$

The oddness of g implies that

$$\left| g(x - y + xy) - g(x) + g(y) - g(xy) \right| \le \mu,$$

therefore

$$\begin{aligned}
\left| g(x + y - xy) + g(x - y + xy) - 2g(x) \right| \\
\leq \left| g(x + y - xy) - g(x) - g(y) + g(xy) \right| \\
+ \left| g(x - y + xy) - g(x) + g(y) - g(xy) \right| \\
\leq 2\mu.
\end{aligned}$$

Taking $y = x/(1 - x)$ for $x \neq 1$ yields

$$\left| g(2x) + g(0) - 2g(x) \right| = \left| g(2x) - 2g(x) \right| \leq 2\mu,$$

for $x \neq 1$, and so

$$\left| g(2x) - 2g(x) \right| \leq \gamma$$

for all $x \in \mathbb{R}$. Using induction on n we can verify that

$$\left| 2^{-n} g(2^n x) - g(x) \right| \leq \sum_{k=1}^{n} 2^{-k} \gamma \leq \gamma \qquad (a)$$

for all $x \in \mathbb{R}$ and $n \in \mathbb{N}$. For $m > n > 0$ we get

$$\begin{aligned}
\left| 2^{-m} g(2^m x) - 2^{-n} g(2^n x) \right| &= 2^{-n} \left| 2^{-(m-n)} g(2^{m-n} 2^n x) - g(2^n x) \right| \\
&\leq 2^{-n} \gamma \\
&\to 0 \quad \text{as } n \to \infty.
\end{aligned}$$

Hence, the sequence $\{2^{-n} g(2^n x)\}$ is a Cauchy sequence. Let us define

$$p(x) = \lim_{n \to \infty} 2^{-n} g(2^n x)$$

for all $x \in \mathbb{R}$. We can then conclude that

$$p(2x) = \lim_{n \to \infty} 2^{-n} g(2^n 2x) = 2 \lim_{n \to \infty} 2^{-n-1} g(2^{n+1} x) = 2p(x),$$

p is odd and by (a)

$$\left| g(x) - p(x) \right| = \left| s(x) \right| \leq \gamma$$

for all $x \in \mathbb{R}$. $\qquad \square$

Lemma 4.3. *Let a function* $f : \mathbb{R} \to \mathbb{R}$ *satisfy the inequality* (4.3) *for some* $\delta \geq 0$ *and for all* $x, y \in \mathbb{R}$. *If the even part* h *of* f *satisfies the inequality* (4.2) *for some* $\varepsilon \geq 0$ *and for all* $x, y \in \mathbb{R}$, *then the function* $p : \mathbb{R} \to \mathbb{R}$ *defined in Lemma 4.2 is additive.*

Proof. From the preceding lemma it follows that

$$
\begin{aligned}
&\left| p(x + y - xy) - p(x) - p(y) + p(xy) \right| \\
&\leq \left| g(x + y - xy) - g(x) - g(y) + g(xy) \right| \\
&\quad + \left| -s(x + y - xy) + s(x) + s(y) - s(xy) \right| \\
&\leq \mu + 4\gamma,
\end{aligned}
$$

for all $x, y \in \mathbb{R}$, and hence

$$
\left| p(-x - y - xy) - p(-x) - p(-y) + p(xy) \right| \leq \mu + 4\gamma.
$$

Since p is odd, from the last inequality we get

$$
\left| -p(x + y + xy) + p(x) + p(y) + p(xy) \right| \leq \mu + 4\gamma.
$$

These inequalities and the 2-homogeneity of p (see Lemma 4.2) imply

$$
\begin{aligned}
&\left| p(x + y + xy) - p(x + y - xy) - p(2xy) \right| \\
&= \left| p(x + y - xy) - p(x + y + xy) + 2p(xy) \right| \\
&\leq \left| p(x + y - xy) - p(x) - p(y) + p(xy) \right| \\
&\quad + \left| -p(x + y + xy) + p(x) + p(y) + p(xy) \right| \\
&\leq 2\mu + 8\gamma.
\end{aligned}
$$

In the last inequality we make the following change of variables:

$$
u = -x - y + xy \quad \text{and} \quad v = -2xy. \tag{a}
$$

Since for all $u \in \mathbb{R}$ and $v \geq 0$ there exists at least one pair (x, y) satisfying (a), we have

$$
|p(u + v) - p(u) - p(v)| \leq 2\mu + 8\gamma
$$

for all $u \in \mathbb{R}$ and $v \geq 0$. According to Lemma 2.27, there exists an additive function $A : \mathbb{R} \to \mathbb{R}$ such that

$$
|p(u) - A(u)| \leq 2\mu + 8\gamma
$$

for all $u \geq 0$. Indeed, we have $A = p$: Assume there exists $w \geq 0$ such that $A(w) \neq p(w)$. By the properties of A and p we have

$$
|p(2^n w) - A(2^n w)| = 2^n |p(w) - A(w)| \leq 2\mu + 8\gamma,
$$

for all $n \in \mathbb{N}$, which leads to a contradiction. Thus, p is additive on $[0, \infty)$. Since it is odd, p is additive on the whole $\mathbb{R}$. $\square$

We now introduce the main theorem of Borelli (see [23]).

Theorem 4.4 (Borelli). *Let a function* $f : \mathbb{R} \to \mathbb{R}$ *satisfy the inequality* (4.3) *for some* $\delta \geq 0$ *and for all* $x, y \in \mathbb{R}$. *There exists an additive function* $A : \mathbb{R} \to \mathbb{R}$ *such that the difference* $f - A$ *is bounded if and only if the even part* h *of* f *satisfies the inequality* (4.2) *for some* $\varepsilon \geq 0$ *and for all* $x, y \in \mathbb{R}$.

Proof. Assume that h satisfies (4.2). By the previous lemmas we have

$$f(x) = g(x) + h(x) = p(x) + s(x) + h(x),$$

where p is additive and $|s(x) + h(x)| \leq \gamma + 2\varepsilon + |h(1)|$, i.e., f differs from a solution of Hosszú's equation by a bounded function.

Conversely, assume $f(x) = A(x) + b(x)$, where A is additive and b is bounded. We then have

$$h(x) = (1/2)\big(f(x) + f(-x)\big) = (1/2)\big(b(x) + b(-x)\big),$$

so h satisfies (4.2). $\qquad\qquad\qquad\qquad\qquad\qquad\qquad\qquad\qquad\qquad$ □

Borelli raised a question as to whether one can also prove the Hyers–Ulam stability for Hosszú's functional equation. Section 4.2 is wholly devoted to the answer to this question of Borelli.

4.2 Hyers–Ulam Stability

L. Losonczi affirmatively answered the *question of Borelli*. More precisely, he proved in his paper [238] that Hosszú's functional equation (4.1) is stable in the sense of Hyers and Ulam.

Theorem 4.5 (Losonczi). *Let* E *be a Banach space and suppose that a function* $f : \mathbb{R} \to E$ *satisfies the functional inequality* (4.3) *for some* $\delta \geq 0$ *and for all* $x, y \in \mathbb{R}$. *Then there exist a unique additive function* $A : \mathbb{R} \to E$ *and a unique constant* $b \in E$ *such that*

$$\| f(x) - A(x) - b \| \leq 20\delta$$

for all $x \in \mathbb{R}$.

Proof. Let us define

$$(Hf)(x, y) = f(x + y - xy) - f(x) - f(y) + f(xy)$$

and

$$(Gf)(x, y) = f(xy) - f(x) - f(y)$$

for all $x, y \in \mathbb{R}$. Then we have

$$(Gf)(xy, z) + (Gf)(x, y) - (Gf)(x, yz) - (Gf)(y, z) = 0 \qquad (a)$$

for all $x, y \in \mathbb{R}$. Since

$$f(x + y - xy) = (Hf)(x, y) - (Gf)(x, y),$$

it follows from (4.3) and (a) that

$$\begin{aligned}
&\| f(xy + z - xyz) + f(x + y - xy) - f(x + yz - xyz) - f(y + z - yz) \| \\
&\quad = \| (Hf)(xy, z) + (Hf)(x, y) - (Hf)(x, yz) - (Hf)(y, z) \| \\
&\quad \leq 4\delta.
\end{aligned}$$

Substituting in the last expression

$$-3 = y + z - yz, \qquad (b)$$

$$u - 3 = xy + z - xyz, \qquad (c)$$

and

$$v - 3 = x + y - xy \qquad (d)$$

for $x, y, z \in \mathbb{R}$ yield

$$\| f(u - 3) + f(v - 3) - f(u + v - 3) - f(-3) \| \leq 4\delta \qquad (e)$$

for $(u, v) \in D$, where D is the set of all pairs $(u, v) \in \mathbb{R}^2$ for which the system (b)–(d) has at least one real solution triple x, y, z.

It is clear from (b) that $y \neq 1$. Thus, from (b) and (d) we get

$$x = 1 + \frac{v - 4}{1 - y}, \quad z = 1 - \frac{4}{1 - y}.$$

Substituting these into (c), we obtain after a simple calculation that

$$uy^2 - 2(u + 2v - 8)y + u = 0. \qquad (f)$$

The image of the set $\{(x, y, 1 - 4/(1 - y)) \mid x \in \mathbb{R}, \ y \neq 1\}$ under the transformations (b)–(d) is the set of all pairs (u, v) for which equation (f) has at least one real solution $y \neq 1$.

If $u = 0, v \in \mathbb{R}$, then (f) has at least one real solution $y \neq 1$, namely, $y = 0$ if $v \neq 4$ and $y = $ arbitrary real if $v = 4$.

If $u \neq 0$, then $y = 1$ is a solution of (f) exactly if $v = 4$. Thus, (f) has at least one real solution $y \neq 1$ if and only if $v \neq 4$ and the discriminant of (f) is nonnegative, i.e., if $4(u + 2v - 8)^2 - 4u^2 \geq 0$ or $(v - 4)(u + v - 4) \geq 0$. We can write these conditions in the form

$$\begin{cases} u + v \geq 4 & \text{(for } v > 4), \\ u + v \leq 4 & \text{(for } v < 4). \end{cases} \tag{g}$$

In summary, we obtained that

$$D = \{(u, v) \in \mathbb{R}^2 \mid v \neq 4 \text{ and } (g) \text{ holds true}\}.$$

From (e) with F defined by $F(u) = f(u - 3) - f(-3)$ $(u \in \mathbb{R})$, we get

$$\|F(u + v) - F(u) - F(v)\| \leq 4\delta \tag{h}$$

for all $(u, v) \in D$. Since the left-hand side of (h) is symmetric in u and v, (h) also holds true on

$$D^* = \{(u, v) \mid (u, v) \in D \text{ or } (v, u) \in D\}.$$

It is easy to check that $D^* = \mathbb{R}^2 \setminus \triangle$, where $\triangle$ is the triangle

$$\triangle = \{(u, v) \mid u, v \in (0, 4], \ u + v \in (4, 8]\}.$$

According to Theorem 2.31, there exists a unique additive function $A : \mathbb{R} \to E$ such that

$$\|F(u) - A(u)\| \leq 20\delta.$$

Using the definition of F we get from the last inequality with $x = u - 3$, $A(x - 3) = A(x) - A(3)$, and $b = A(3) + f(-3)$ that

$$\|f(x) - A(x) - b\| \leq 20\delta$$

for all $x \in \mathbb{R}$. □

Later, Theorem 4.5 was generalized by P. Găvruta [117] and further by P. Volkmann [357]:

Theorem 4.6. *Let E be a Banach space. Suppose a function $f : \mathbb{R} \to E$ satisfies the functional inequality (4.3) for some $\delta \geq 0$ and for all $x, y \in \mathbb{R}$. Then there exist a unique additive function $A : \mathbb{R} \to E$ and a unique constant $b \in E$ such that*

$$\|f(x) - A(x) - b\| \leq 4\delta$$

for all $x \in \mathbb{R}$.

4.3 Generalized Hosszú's Functional Equation

In Chapter 3, we have seen various generalizations of the additive Cauchy functional equation. The following functional equation

$$f(x + y + qxy) = f(x) + f(y) + qf(xy)$$

can be regarded as a generalized form of Hosszú's equation (4.1). A function $f : \mathbb{R} \to \mathbb{R}$ will be called a solution of the generalized q-Hosszú's equation if and only if it satisfies

$$(H_q f)(x, y) = f(x + y + qxy) - f(x) - f(y) - qf(xy) = 0$$

for all $x, y \in \mathbb{R}$.

According to a paper [213] of Pl. Kannappan and P. K. Sahoo, every solution $f : \mathbb{R} \to \mathbb{R}$ of the generalized q-Hosszú's equation is given by

$$\begin{cases} f(x) = A(x) & \text{(for } q \in \mathbb{Q} \setminus \{-1, 0\}), \\ f(x) = A(x) + c & \text{(for } q = -1), \end{cases}$$

where $A : \mathbb{R} \to \mathbb{R}$ is an additive function and c is a real constant.

In this section, the stability problem of the generalized Hosszú's equation will be discussed (in the sense of Borelli) for the case of $q \in \mathbb{Q} \setminus \{-1/2, 0, 1/2\}$. We denote by g and h the odd and even part of a corresponding function $f : \mathbb{R} \to \mathbb{R}$, respectively.

We will now start this section with the following three lemmas presented by S.-M. Jung and Y.-H. Kye [191].

Lemma 4.7. *If h satisfies the functional inequality $|(H_q h)(x, y)| \le \varepsilon$ for all $x, y \in \mathbb{R}$ and for some $\varepsilon > 0$, then*

$$|h(x)| \le 2\varepsilon + |h(1/q)|$$

for each $x \in \mathbb{R}$.

Proof. Since h is the even part of f, by the hypothesis we have

$$\begin{aligned} |h(x + y + qxy) - h(x) - h(y) - qh(xy)| &\le \varepsilon, \\ |h(-x - y + qxy) - h(x) - h(y) - qh(xy)| &\le \varepsilon \end{aligned} \tag{a}$$

for all $x, y \in \mathbb{R}$. It follows from (a) that

$$|h(x + y + qxy) - h(-x - y + qxy)| \le 2\varepsilon$$

for all $x, y \in \mathbb{R}$. Replacing x and y with $(1/2)(u - 1/q)$ and $1/q$, respectively, in the last inequality, we obtain

$$|h(u) - h(1/q)| \leq 2\varepsilon.$$

Since u can take every real number as its value, the assertion is true.   □

Lemma 4.8. *Let ε and δ be given positive numbers. Assume that a function $f : \mathbb{R} \to \mathbb{R}$ and its even part h satisfy the inequalities*

$$|(H_q f)(x, y)| \leq \delta \quad and \quad |(H_q h)(x, y)| \leq \varepsilon \tag{4.4}$$

for all $x, y \in \mathbb{R}$. Furthermore, suppose there exists a $\xi > 0$ such that the odd part g satisfies

$$\max \{(1/n)|g(n/q) - ng(1/q)| \mid n \in \mathbb{N}\} < \xi. \tag{4.5}$$

It then holds true that

$$g(x) = g_1(x) + g_2(x),$$

where g_1 and g_2 are odd functions such that $g_1(2qx) = 2qg_1(x)$ for each $x \in \mathbb{R}$, and g_2 is bounded.

The condition (4.5) is a kind of approximate homogeneity (additivity) of g which is much weaker than an ordinary approximate homogeneity.

Proof. It follows from $|(H_q f)(x, y)| = |(H_q g)(x, y) + (H_q h)(x, y)|$ that

$$|(H_q g)(x, y)| \leq \delta + \varepsilon \tag{a}$$

for all $x, y \in \mathbb{R}$. Let $\eta = \max\{\delta + \varepsilon, \xi\}$. We now assert that

$$|g(nx) - ng(x)| \leq |n|\eta \tag{b}$$

for any real x and every integer $n \neq 0$.

First, we assume that n is a natural number. The assertion (b) is true for $n = 1$. By (a), we get

$$|(H_q g)(x, -y)| = \big|g(x - y - qxy) - g(x) + g(y) + qg(xy)\big| \leq \eta. \tag{c}$$

According to (a) and (c), we obtain

$$\big|g(x + y + qxy) + g(x - y - qxy) - 2g(x)\big| \leq 2\eta. \tag{d}$$

Substituting $x/(1 + qx)$, $x \neq -1/q$, for y in (d) yields

$$|g(2x) - 2g(x)| \leq 2\eta,$$

since $g(0) = 0$. Hence, by (4.5), $|g(2x) - 2g(x)| \leq 2\eta$ for all $x \in \mathbb{R}$.

Assume that $|g(nx) - ng(x)| \leq n\eta$ for $n \in \{1, \ldots, m\}$ $(m \geq 2)$ as induction hypothesis. Then, replacing y in (d) with $mx/(1 + qx)$, $x \neq -1/q$, yields

$$\begin{aligned}
\big|g\big((m + 1)x\big) &- g\big((m - 1)x\big) - 2g(x)\big| \\
&= \big|g\big((m + 1)x\big) - (m + 1)g(x) - g\big((m - 1)x\big) + (m - 1)g(x)\big| \\
&\leq 2\eta,
\end{aligned}$$

and hence

$$\big|g\big((m + 1)x\big) - (m + 1)g(x)\big| \leq (m + 1)\eta.$$

Thus, it holds true that $|g(nx) - ng(x)| \leq n\eta$ for all $x \in \mathbb{R}$ and any $n \in \mathbb{N}$.

Now, it is obvious that the assertion (b) is also true for any negative integer n, since g is an odd function. Let m be any nonzero integer. It then follows from (b) that

$$|g(mx) - mg(x)| \leq |m|\eta \tag{e}$$

for all $x \in \mathbb{R}$. Dividing by $|n|$ both sides of (e), we have

$$\big|(1/n)g(mx) - (m/n)g(x)\big| \leq \big(|m|/|n|\big)\eta. \tag{f}$$

Replacing x with $(m/n)x$ and dividing by $|n|$ both sides of (b) yield

$$\big|(1/n)g(mx) - g\big((m/n)x\big)\big| \leq \eta \tag{g}$$

for all real x and nonzero integers m and n. By using (f) and (g), we conclude that for all $x \in \mathbb{R}$ and nonzero integers m and n

$$\big|g\big((m/n)x\big) - (m/n)g(x)\big| \leq \big(|m|/|n| + 1\big)\eta. \tag{h}$$

Now, for any real x, let

$$g_1(x) = \begin{cases} \lim\limits_{n \to \infty} (2q)^{-n} g\big((2q)^n x\big) & (\text{if } |q| > 1/2), \\ \lim\limits_{n \to \infty} (2q)^n g\big((2q)^{-n} x\big) & (\text{if } |q| < 1/2). \end{cases}$$

By using (h), we show that if $|q| > 1/2$, then $\{(2q)^{-n}g((2q)^n x)\}$ is a Cauchy sequence for each real x:

$$
\begin{aligned}
&\left|(2q)^{-n}g\big((2q)^n x\big) - (2q)^{-m}g\big((2q)^m x\big)\right| \\
&= |2q|^{-n}\left|g\big((2q)^{n-m}(2q)^m x\big) - (2q)^{n-m}g\big((2q)^m x\big)\right| \\
&\le |2q|^{-n}\big(|2q|^{n-m} + 1\big)\eta \\
&\to 0 \quad \text{as } m, n \to \infty.
\end{aligned}
$$

Analogously, the sequence $\{(2q)^n g((2q)^{-n} x)\}$ is also a Cauchy sequence, when $|q| < 1/2$. Thus, $g_1(x)$ exists for any $x \in \mathbb{R}$. In view of the definitions of g_1 and g_2 we conclude that $g_1(2qx) = 2qg_1(x)$ for each real x, and g_1, g_2 are odd functions, and g_2 is bounded, since it follows from (h) that

$$
|g_2(x)| = |g(x) - g_1(x)| \le \eta \tag{i}
$$

for all $x \in \mathbb{R}$. $\qquad\qquad\square$

Lemma 4.9. *Let ε and δ be given positive numbers. Suppose a function $f : \mathbb{R} \to \mathbb{R}$ and its even part h satisfy the inequalities in (4.4) for all $x, y \in \mathbb{R}$. Furthermore, assume that there exists a $\xi > 0$ satisfying the condition (4.5). Then, the function g_1 defined in Lemma 4.8 is additive on $\mathbb{R}$.*

Proof. Set $\eta = \max\{\delta + \varepsilon, \xi\}$. It follows from (a) and (i) in the proof of Lemma 4.8 that

$$
\begin{aligned}
|(H_q g_1)(x, y)| &= |(H_q g)(x, y) - (H_q g_2)(x, y)| \\
&\le |(H_q g)(x, y)| + |(H_q g_2)(x, y)| \\
&\le (|q| + 4)\eta \tag{a}
\end{aligned}
$$

for all $x, y \in \mathbb{R}$. Therefore, we also have

$$
|(H_q g_1)(-x, -y)| \le (|q| + 4)\eta. \tag{b}
$$

By using (a) and (b), we obtain

$$
\begin{aligned}
&\left|g_1(x + y + qxy) + g_1(-x - y + qxy) - 2qg_1(xy)\right| \\
&= \left|g_1(x + y + qxy) - g_1(x + y - qxy) - g_1(2qxy)\right| \\
&\le 2(|q| + 4)\eta \tag{c}
\end{aligned}
$$

for all $x, y \in \mathbb{R}$, since g_1 is odd and $g_1(2qx) = 2qg_1(x)$.

For given u and v, consider a system of the following equations,

$$
x + y - qxy = u \quad \text{and} \quad 2qxy = v. \tag{d}
$$

If we assume $y \neq 0$, it follows from (d) that

$$y = \frac{q(2u + v) \pm \sqrt{q^2(2u + v)^2 - 8qv}}{4q}.$$

Hence, there exists at least one pair (x, y) satisfying (d) either for all $u \in \mathbb{R}$ and all $v < 0$ (when $q > 0$) or for all $u \in \mathbb{R}$ and all $v > 0$ (when $q < 0$). Thus, by using (c) and (d), we have

$$\left| g_1(u + v) - g_1(u) - g_1(v) \right| \leq 2\left(|q| + 4\right)\eta \qquad\qquad (e)$$

either for all $u, v < 0$ (and for $q > 0$) or for all $u, v > 0$ (and for $q < 0$.)

For the case when $u, v < 0$, $q > 1/2$, and when n is an arbitrary natural number, we get

$$|2q|^n \left| g_1(u + v) - g_1(u) - g_1(v) \right| \leq 2\left(|q| + 4\right)\eta$$

by substituting $(2q)^n u$ and $(2q)^n v$ for u and v in (e), respectively, and using Lemma 4.8. Similarly, if $u, v < 0$, $0 < q < 1/2$, and if n is a natural number, we replace u and v in (e) with $(2q)^{-n} u$ and $(2q)^{-n} v$, respectively. Then

$$|2q|^{-n} \left| g_1(u + v) - g_1(u) - g_1(v) \right| \leq 2\left(|q| + 4\right)\eta,$$

since Lemma 4.8 also implies $g_1\left((2q)^{-1}x\right) = (2q)^{-1}g_1(x)$. These together imply that if $q > 0$, then

$$g_1(u + v) = g_1(u) + g_1(v)$$

for all $u, v < 0$.

When $q < 0$, we analogously obtain the same equation for all $u, v > 0$. However, the fact that g_1 is odd implies that g_1 is additive on the whole $\mathbb{R}$. $\qquad\qquad \square$

Now, we introduce the main theorem of this section proved by Jung and Kye [191]. This theorem says that the generalized Hosszú's equation is stable in the sense of Hyers, Ulam, and Borelli.

Theorem 4.10. *Suppose $q \notin \{-1/2, 0, 1/2\}$ is a fixed rational number. Let $\delta > 0$ be given and let $f : \mathbb{R} \to \mathbb{R}$ be a function satisfying the first inequality in (4.4) for all $x, y \in \mathbb{R}$. Then there exists an additive function $A : \mathbb{R} \to \mathbb{R}$ such that $f - A$ is bounded if and only if there exist positive constants ε and ξ such that the second inequality in (4.4) and the condition (4.5) hold true for all $x, y \in \mathbb{R}$.*

Proof. First, assume that $|(H_q f)(x, y)| \leq \delta$ and $|(H_q h)(x, y)| \leq \varepsilon$ for all $x, y \in \mathbb{R}$. Moreover, suppose there exists a $\xi > 0$ such that g satisfies the condition (4.5). From the above lemmas and (i) in the proof of Lemma 4.8, we obtain

$$f(x) = g(x) + h(x) = g_1(x) + g_2(x) + h(x),$$

where g_1 is additive and $|f(x) - g_1(x)| = |g_2(x) + h(x)| \leq \eta + 2\varepsilon + |h(1/q)|$ for all $x \in \mathbb{R}$.

Conversely, assume that $f(x) = a(x) + b(x)$, where a is additive and b is bounded. Then

$$h(x) = (1/2)\big(f(x) + f(-x)\big) = (1/2)\big(b(x) + b(-x)\big).$$

Hence, $|(H_q h)(x, y)|$ is bounded. Moreover, we have

$$g(x) = (1/2)\big(f(x) - f(-x)\big) = a(x) + (1/2)\big(b(x) - b(-x)\big)$$

and

$$
\begin{aligned}
(1/n)&\big|g(n/q) - ng(1/q)\big| \\
&= \big(1/(2n)\big)\big|b(n/q) - b(-n/q) - nb(1/q) + nb(-1/q)\big| \\
&\leq \big(1/(2n)\big)\big|b(n/q) - b(-n/q)\big| + (1/2)\big|b(1/q) - b(-1/q)\big|.
\end{aligned}
$$

Since b is assumed to be bounded, we can choose a $\xi > 0$ such that the condition (4.5) is true. $\quad\square$

It is not clear whether the generalized Hosszú's equation is also stable in the sense of Hyers, Ulam, and Borelli for the case when $|q| = 1/2$. Naturally, one can raise a question as to whether the generalized Hosszú's equation is stable in the sense of Hyers and Ulam.

4.4 Hosszú's Equation is not Stable on the Unit Interval

As we have seen in Theorem 2.30, the additive Cauchy equation is stable on various bounded intervals. However, Hosszú's functional equation is not stable on the unit interval, where we denote by the unit interval any interval with ends 0 and 1. J. Tabor was the first person to prove this surprising fact. We first introduce a lemma necessary to prove the main theorem of J. Tabor (ref. [348]).

Lemma 4.11. *Let $\varepsilon \geq 0$ be given. Suppose that a function $F : [0, \infty) \to \mathbb{R}$ satisfies the inequalities*

$$|F(x + y) - F(x)| \leq \varepsilon \tag{4.6}$$

for all $x, y \geq 0$ with $x \geq y$, and

$$|F(x)| \leq \varepsilon \tag{4.7}$$

for any $x \in [0, 6]$. Let us define a function $f : (0, 1) \to \mathbb{R}$ by

$$f(x) = F\big(g(x)\big),$$

where

$$g(x) = \begin{cases} -\ln x & \text{(for } x \in (0, 1/2]), \\ -\ln(1-x) & \text{(for } x \in [1/2, 1)). \end{cases}$$

Then, the inequality

$$\left| f(x + y - xy) - f(x) - f(y) + f(xy) \right| \le 4\varepsilon \tag{4.8}$$

holds true for every $x, y \in (0, 1)$.

Proof. Let $x, y \in (0, 1)$. Claim that

$$g(x + y - xy) = g(x) + g(y) \quad \text{for } x \ge y \ge 1/2. \tag{a}$$

If $1/2 \le u < 1$, then $0 < 1 - u \le 1/2$, and hence $(1 - x)(1 - y) < 1/2$. Since $1 - x - y + xy = (1 - x)(1 - y) < 1/2$, we have $x + y - xy > 1/2$. Thus, we have

$$g(x + y - xy) = -\ln\big((1 - x)(1 - y)\big) = g(x) + g(y)$$

for $x \ge y \ge 1/2$.

Claim that

$$|g(xy) - g(y)| \le 3 \quad \text{for } x \ge y \text{ and } x \ge 1/2. \tag{b}$$

Suppose that $y \ge 1/2$. If $xy \le 1/2$, then $1/2 \le y \le 1/\sqrt{2}$ and $1/4 \le xy \le 1/2$, thus

$$|g(xy) - g(y)| \le |g(xy)| + |g(y)| \le \ln 4 + \left| \ln\big(1 - 1/\sqrt{2}\big) \right| \le 3.$$

If $xy \ge 1/2$, then

$$|g(xy) - g(y)| = \left| \ln(1 - xy) - \ln(1 - y) \right| = \left| \ln\left(1 + y\frac{1 - x}{1 - y}\right) \right| \le 3.$$

Assume now that $y < 1/2$. Then,

$$|g(xy) - g(y)| = |\ln xy - \ln y| \le 3.$$

We claim that

$$|g(x + y - xy) - g(x)| \le 3 \quad \text{for } x \ge y \text{ and } y \le 1/2. \tag{c}$$

Let $x^* = 1 - x$. Then, we have $g(x) = g(x^*)$. Making use of the equality

$$g(x + y - xy) = g\big((x^* y^*)^*\big) = g(x^* y^*)$$

and interchanging the role of x and y, we obtain (c) from (b).

Let us show that

$$|F(x) - F(y)| \leq 2\varepsilon \qquad (d)$$

for any $x, y \geq 0$ with $|x - y| \leq 3$. The inequality (d) follows from (4.7) in the case when $x, y \in [0, 6]$. In the other case, we may assume that $x \geq y$ and $x > 6$. Then, $x - y \leq 3 < y$ and due to (4.6) we have

$$|F(x) - F(y)| = |F(y + (x - y)) - F(y)| \leq \varepsilon \leq 2\varepsilon.$$

We will prove that (4.8) holds true. Without loss of generality, we may assume that $x \geq y$. Suppose that $x \geq y \geq 1/2$. Then, by (a), (b), (d), and (4.6), we get

$$\begin{aligned}
&\left| f(x + y - xy) - f(x) - f(y) + f(xy) \right| \\
&= \left| F\big(g(x + y - xy)\big) - F\big(g(x)\big) - F\big(g(y)\big) + F\big(g(xy)\big) \right| \\
&\leq \left| F\big(g(x + y - xy)\big) - F\big(g(x) + g(y)\big) \right| \\
&\quad + \left| F\big(g(x) + g(y)\big) - F\big(g(x)\big) \right| + \left| F\big(g(xy)\big) - F\big(g(y)\big) \right|
\end{aligned}$$

and hence

$$\left| f(x + y - xy) - f(x) - f(y) + f(xy) \right| \leq 4\varepsilon \qquad (e)$$

for all $x, y \in (0, 1)$ with $x \geq y \geq 1/2$. Suppose that $x \geq 1/2 \geq y$. It then follows from (b), (c), and (d) that

$$\begin{aligned}
&\left| f(x + y - xy) - f(x) - f(y) + f(xy) \right| \\
&= \left| F\big(g(x + y - xy)\big) - F\big(g(x)\big) - F\big(g(y)\big) + F\big(g(xy)\big) \right| \\
&\leq \left| F\big(g(x + y - xy)\big) - F\big(g(x)\big) \right| + \left| F\big(g(xy)\big) - F\big(g(y)\big) \right| \\
&\leq 4\varepsilon.
\end{aligned}$$

Suppose that $y \leq x \leq 1/2$. Then, it holds true that $y^* \geq x^* \geq 1/2$, and it follows from (e) that

$$\begin{aligned}
&\left| f(x + y - xy) - f(x) - f(y) + f(xy) \right| \\
&= \left| f(x^* y^*) - f(x^*) - f(y^*) + f(x^* + y^* - x^* y^*) \right| \\
&\leq 4\varepsilon,
\end{aligned}$$

since $x + y - xy = (x^* y^*)^*$, $x = (x^*)^*$, $y = (y^*)^*$, $xy = (x^* + y^* - x^* y^*)^*$ and $f(x^*) = f(x)$. $\qquad\qquad\square$

We are now able to prove the main theorem of J. Tabor [348].

Theorem 4.12 (Tabor). *Let U be the unit interval. For every $\delta > 0$, there exists a function $f : U \to \mathbb{R}$ such that the inequality*

$$\left| f(x + y - xy) - f(x) - f(y) + f(xy) \right| \leq \delta$$

holds true for all $x, y \in U$, but the range set of $|f(x) - H(x)|$ is not bounded on U for each Hosszú function $H : U \to \mathbb{R}$.

Proof. Without loss of generality, we may assume that $\delta = 1$. Let us define a function $F : [0, \infty) \to \mathbb{R}$ by

$$F(x) = (1/8) \ln(1 + x).$$

Then, F satisfies (4.6) and (4.7) with $\varepsilon = 1/4$. Let

$$f(x) = \begin{cases} F\big(g(x)\big) & \text{(for } x \in (0, 1)), \\ 0 & \text{(for } x \in U \setminus (0, 1)), \end{cases}$$

where g is the function defined in Lemma 4.11.

We show that

$$\big| f(x + y - xy) - f(x) - f(y) + f(xy) \big| \leq 1$$

for any $x, y \in U$. If $x = 0$ or $y = 0$, then the inequality is obvious. For $x, y \in (0, 1)$ the relation holds true by Lemma 4.11. We remark that an additive function $A : \mathbb{R} \to \mathbb{R}$ is either continuous or has a dense graph in $\mathbb{R}^2$ (see Theorems 2.1 and 2.2). Since f is continuous on $(0, 1)$ and

$$\lim_{x \to 0} f(x) = \lim_{x \to 1} f(x) = \infty,$$

this implies that the range set of $|f(x) - A(x) - c|$ is unbounded on $(0, 1)$ for every additive function $A : \mathbb{R} \to \mathbb{R}$ and every constant c. Therefore, the range set of $|f(x) - H(x)|$ is unbounded on $(0, 1)$ for each solution $H : U \to \mathbb{R}$ of Hosszú's functional equation.   □

4.5 Hosszú's Functional Equation of Pexider Type

In this section, we investigate the Hyers–Ulam stability of the functional equation

$$f(x + y - \alpha xy) + g(xy) = h(x) + k(y). \tag{4.9}$$

For $\alpha = 1$, the functional equation (4.9) is a pexiderized version of Hosszú's functional equation (4.1).

In the following theorem, we prove the Hyers–Ulam stability of the functional equation (4.9) when $\alpha \neq 0$ (ref. [321]).

Theorem 4.13. *Let E be a Banach space and let α be a nonzero real number. If functions $f, g, h, k : \mathbb{R} \to E$ satisfy the inequality*

$$\| f(x + y - \alpha xy) + g(xy) - h(x) - k(y) \| \leq \delta \tag{4.10}$$

for all $x, y \in \mathbb{R}$ *and for some* $\delta > 0$, *then there exists a unique additive function* $A : \mathbb{R} \to E$ *such that*

$$\|f(x) - A(\alpha x) - a\| \leq 24\delta,$$

$$\|h(x) - A(\alpha x) - a - b_1\| \leq 25\delta,$$

$$\|k(x) - A(\alpha x) - a - b_2\| \leq 25\delta,$$

$$\|g(x) - A(\alpha^2 x) - a - b_1 - b_2\| \leq 27\delta$$

for all $x \in \mathbb{R}$, *where* $a = f(1/\alpha) - A(1)$, $b_1 = g(0) - k(0)$, *and* $b_2 = g(0) - h(0)$.

Proof. If we put $y = 0$ in (4.10), then

$$\|f(x) - h(x) + b_1\| \leq \delta, \qquad (a)$$

where $b_1 = g(0) - k(0)$. If we put $x = 0$ in (4.10), then

$$\|f(y) - k(y) + b_2\| \leq \delta, \qquad (b)$$

where $b_2 = g(0) - h(0)$. It follows from (4.10), (a), and (b) that

$$\|f(x + y - \alpha xy) + g(xy) - f(x) - f(y) - b_1 - b_2\|$$
$$\leq \|f(x + y - \alpha xy) + g(xy) - h(x) - k(y)\|$$
$$+ \|h(x) - f(x) - b_1\| + \|k(y) - f(y) - b_2\|$$
$$\leq 3\delta \qquad (c)$$

for all $x, y \in \mathbb{R}$.

Since $\alpha \neq 0$, if we put $y = 1/\alpha$ in (c), then

$$\|g(x/\alpha) - f(x) - b_1 - b_2\| \leq 3\delta \qquad (d)$$

for all $x \in \mathbb{R}$. If we replace x in (d) with αx, then we have

$$\|g(x) - f(\alpha x) - b_1 - b_2\| \leq 3\delta \qquad (e)$$

for all $x \in \mathbb{R}$. It then follows from (c) and (e) that

$$\|f(x + y - \alpha xy) + f(\alpha xy) - f(x) - f(y)\|$$
$$\leq \|f(x + y - \alpha xy) + g(xy) - f(x) - f(y) - b_1 - b_2\|$$
$$+ \|f(\alpha xy) - g(xy) + b_1 + b_2\|$$
$$\leq 6\delta \qquad (f)$$

for all $x, y \in \mathbb{R}$. If we replace x and y in (f) with x/α and y/α, respectively, then we get

$$\left\|f\big((1/\alpha)(x + y - xy)\big) + f\big((1/\alpha)xy\big) - f(x/\alpha) - f(y/\alpha)\right\| \leq 6\delta \qquad (g)$$

for any $x, y \in \mathbb{R}$.

Define a function $\psi : \mathbb{R} \to E$ by

$$\psi(x) = f(x/\alpha). \tag{h}$$

In view of (g) and (h), we obtain

$$\|\psi(x + y - xy) + \psi(xy) - \psi(x) - \psi(y)\| \leq 6\delta$$

for all $x, y \in \mathbb{R}$.

According to Theorem 4.6, there exists a unique additive function $A : \mathbb{R} \to E$ such that

$$\|\psi(x) - A(x) - a\| \leq 24\delta \tag{i}$$

for all $x \in \mathbb{R}$, where $a = \psi(1) - A(1)$. It thus follows from (h) and (i) that

$$\|f(x) - A(\alpha x) - a\| \leq 24\delta \tag{j}$$

for each $x \in \mathbb{R}$, where $a = f(1/\alpha) - A(1)$.

Now, by (a) and (j), we get

$$\|h(x) - A(\alpha x) - a - b_1\| \leq \|h(x) - f(x) - b_1\| + \|f(x) - A(\alpha x) - a\|$$
$$\leq 25\delta$$

for any $x \in \mathbb{R}$. Similarly, by (b) and (j), we obtain

$$\|k(x) - A(\alpha x) - a - b_2\| \leq \|k(x) - f(x) - b_2\| + \|f(x) - A(\alpha x) - a\|$$
$$\leq 25\delta$$

for all $x \in \mathbb{R}$.

Finally, it follows from (e) and (j) that

$$\|g(x) - A(\alpha^2 x) - a - b_1 - b_2\|$$
$$\leq \|g(x) - f(\alpha x) - b_1 - b_2\| + \|f(\alpha x) - A(\alpha^2 x) - a\|$$
$$\leq 27\delta$$

for any $x \in \mathbb{R}$.   □

Chapter 5
Homogeneous Functional Equation

The functional equation $f(yx) = y^k f(x)$ (where k is a fixed real constant) is called the *homogeneous functional equation* of degree k. In the case when $k = 1$ in the above equation, the equation is simply called the *homogeneous functional equation*. In Section 5.1, the Hyers–Ulam–Rassias stability of the homogeneous functional equation of degree k between real Banach algebras will be proved in the case when k is a positive integer. It will especially be proved that every "approximately" homogeneous function of degree k is a real homogeneous function of degree k. Section 5.2 deals with the superstability property of the homogeneous equation on a restricted domain and an asymptotic behavior of the homogeneous functions. The stability problem of the equation between vector spaces will be discussed in Section 5.3. In the last section, we will deal with the Hyers–Ulam–Rassias stability of the homogeneous functional equation of Pexider type.

5.1 Homogeneous Equation Between Banach Algebras

By $\mathbb{K}$ we denote either $\mathbb{R}$ or $\mathbb{C}$. An *algebra* over $\mathbb{K}$ is a vector space E over $\mathbb{K}$ in which a multiplication is defined such that

(A1) $x(yz) = (xy)z$ *for all* $x, y, z \in E$,
(A2) $x(y + z) = xy + xz$ *and* $(x + y)z = xz + yz$ *for any* $x, y, z \in E$,
(A3) $\alpha(xy) = (\alpha x)y = x(\alpha y)$ *for all* $x, y \in E$ *and for all* $\alpha \in \mathbb{K}$.

If an algebra E over $\mathbb{K}$ is a Banach space with a norm $\| \cdot \|$ that satisfies the multiplicative inequality

(A4) $\|xy\| \leq \|x\|\|y\|$ *for every* $x, y \in E$,

then E is called a *real* (or *complex*) *Banach algebra*. If $xy = yx$ for all $x, y \in E$, then the Banach algebra E is called *commutative*.

Throughout this section, let E be a real commutative Banach algebra with the following additional properties:

(A4') $\|xy\| = \|x\|\|y\|$ *for every* $x, y \in E$,

S.-M. Jung, *Hyers–Ulam–Rassias Stability of Functional Equations in Nonlinear Analysis*, Springer Optimization and Its Applications 48, DOI 10.1007/978-1-4419-9637-4_5, © Springer Science+Business Media, LLC 2011

(A5) E contains an identity $e \neq 0$ such that $ex = xe = x$ for each $x \in E$,
(A6) $(E, \cdot)$ is a group.

It then follows from $(A4')$ and $(A5)$ that $\|e\| = 1$. We will write $x^2, x^3, \ldots$ instead of $x \cdot x, (x \cdot x) \cdot x, \ldots$. By x^{-1} we will denote the *multiplicative inverse* of x. Analogously, we write $x^{-2}, x^{-3}, \ldots$ instead of $x^{-1} \cdot x^{-1}, (x^{-1} \cdot x^{-1}) \cdot x^{-1}, \ldots$. It is then obvious that

$$\|x^{-n}\| = \|x\|^{-n}$$

for any $n \in \mathbb{N}$.

Let k be a fixed positive integer. The equation

$$f(yx) = y^k f(x)$$

is said to be the *homogeneous functional equation* of degree k. Every solution of the homogeneous functional equation of degree k is called a *homogeneous function* of degree k. In the case of $k = 1$ in the above equation, the corresponding equation is simply called the *homogeneous functional equation* and each solution of the homogeneous functional equation is called a *homogeneous function*.

It is well-known that every homogeneous function $f : [0, \infty) \to \mathbb{R}$ of degree k is of the form $f(x) = cx^k$, where c is a real constant (see [59]).

Let $\varphi : E^2 \to [0, \infty)$ be a function such that

$$\Phi_z(x) = \sum_{j=0}^{\infty} \|z\|^{-(j+1)k} \varphi(z^j x, z) < \infty \tag{5.1}$$

or

$$\tilde{\Phi}_z(x) = \sum_{j=0}^{\infty} \|z\|^{jk} \varphi(z^{-(j+1)} x, z) < \infty \tag{5.2}$$

for some $z \in E$ with $\|z\| > 1$ and for all $x \in E$. Moreover, we assume that

$$\begin{cases} \Phi_z(w^n x) = o(\|w\|^{nk}) & \text{as } n \to \infty \quad (\text{for } \Phi_z(x) < \infty), \\ \tilde{\Phi}_z(w^n x) = o(\|w\|^{nk}) & \text{as } n \to \infty \quad (\text{for } \tilde{\Phi}_z(x) < \infty) \end{cases} \tag{5.3}$$

for some $w \in E$ and for all $x \in E$.

S.-M. Jung [161] proved the following theorem concerning the Hyers–Ulam–Rassias stability of the homogeneous functional equation of degree k.

Theorem 5.1. *Let E be a real commutative Banach algebra with properties $(A4')$, $(A5)$, and $(A6)$. If a function $f : E \to E$ satisfies $f(0) = 0$,*

$$\|f(yx) - y^k f(x)\| \leq \varphi(x, y) \tag{5.4}$$

and

$$\begin{cases} \varphi(z^n x, y) = o\big(\|f(z^n x)\|\big) & \text{as } n \to \infty \quad (\text{for } \Phi_z(x) < \infty), \\ \varphi(z^{-n} x, y) = o\big(\|f(z^{-n} x)\|\big) & \text{as } n \to \infty \quad (\text{for } \tilde{\Phi}_z(x) < \infty) \end{cases} \tag{5.5}$$

for all $x, y \in E \backslash \{0\}$, *then there exists a unique homogeneous function* $H : E \to E$ *of degree* k *such that*

$$\|f(x) - H(x)\| \le \begin{cases} \Phi_z(x) & (\text{for } \Phi_z(x) < \infty), \\ \tilde{\Phi}_z(x) & (\text{for } \tilde{\Phi}_z(x) < \infty) \end{cases} \tag{5.6}$$

for all $x \in E$.

Proof. We implement induction on n to prove

$$\left\| y^{-nk} f(y^n x) - f(x) \right\| \le \sum_{j=0}^{n-1} \|y\|^{-(j+1)k} \varphi(y^j x, y) \tag{a}$$

for any $n \in \mathbb{N}$. In view of (5.4), the inequality (a) is true for $n = 1$. If we assume the validity of (a) for some $n > 0$, then it follows from (5.4) and (a) that

$$\left\| y^{-(n+1)k} f\left(y^{n+1} x\right) - f(x) \right\|$$
$$\le \|y\|^{-(n+1)k} \left\| f(y y^n x) - y^k f(y^n x) \right\| + \left\| y^{-nk} f(y^n x) - f(x) \right\|$$
$$\le \|y\|^{-(n+1)k} \varphi(y^n x, y) + \sum_{j=0}^{n-1} \|y\|^{-(j+1)k} \varphi(y^j x, y)$$
$$= \sum_{j=0}^{n} \|y\|^{-(j+1)k} \varphi(y^j x, y),$$

which ends the proof of (a).

First, we consider the case when $\Phi_z(x) < \infty$ for some $z \in E$ with $\|z\| > 1$ and for all $x \in E$. Let $n > m > 0$. It then follows from (a) and (5.1) that

$$\left\| z^{-nk} f(z^n x) - z^{-mk} f(z^m x) \right\|$$
$$= \|z\|^{-mk} \left\| z^{-(n-m)k} f(z^{n-m} z^m x) - f(z^m x) \right\|$$
$$\le \|z\|^{-mk} \sum_{j=0}^{n-m-1} \|z\|^{-(j+1)k} \varphi(z^j z^m x, z)$$
$$= \sum_{j=m}^{n-1} \|z\|^{-(j+1)k} \varphi(z^j x, z)$$
$$\to 0 \quad \text{as } m \to \infty.$$

Therefore, $\{z^{-nk} f(z^n x)\}$ is a Cauchy sequence. Since E is a Banach space, we may define

$$H(x) = \lim_{n \to \infty} z^{-nk} f(z^n x)$$

for all $x \in E$. The validity of the first inequality in (5.6) easily follows from the definition of H, (5.1), and (a).

Suppose $x, y \in E \setminus \{0\}$ are given. It then follows from (a) that

$$\left\| y^{-k} f(yz^n x) - f(z^n x) \right\| \leq \|y\|^{-k} \varphi(z^n x, y).$$

By using the last inequality and (5.5), we have

$$\left\| f(z^n x)^{-1} y^{-k} f(yz^n x) - e \right\| \leq \|y\|^{-k} \|f(z^n x)\|^{-1} \varphi(z^n x, y)$$
$$\to 0 \quad \text{as } n \to \infty.$$

Hence,

$$\lim_{n \to \infty} f(z^n x)^{-1} y^{-k} f(yz^n x) = e. \tag{b}$$

By the definition of H and (b), we can show that

$$H(yx) = \lim_{n \to \infty} z^{-nk} f(z^n yx)$$
$$= y^k \lim_{n \to \infty} z^{-nk} f(z^n x) \lim_{n \to \infty} f(z^n x)^{-1} y^{-k} f(z^n yx)$$
$$= y^k H(x)$$

for all $x, y \in E \setminus \{0\}$. In addition, it is not difficult to show that $H(0) = 0$. Hence, we conclude that $H(yx) = y^k H(x)$ for all $x, y \in E$.

Let $H' : E \to E$ be another homogeneous function of degree k satisfying (5.6). By using (5.6) and (5.3), we get

$$\|H(x) - H'(x)\| = \|w\|^{-nk} \|H(w^n x) - H'(w^n x)\|$$
$$\leq 2\|w\|^{-nk} \Phi_z(w^n x)$$
$$\to 0 \quad \text{as } n \to \infty.$$

Hence, we conclude that $H(x) = H'(x)$ for all $x \in E$.

Now, we consider the case of $\tilde{\Phi}_z(x) < \infty$ for some $z \in E$ with $\|z\| > 1$ and for all $x \in E$. Replacing x in (a) with $y^{-n} x$ and multiplying the resulting inequality by $\|y^{nk}\|$, we get

$$\|f(x) - y^{nk} f(y^{-n} x)\| \leq \sum_{j=0}^{n-1} \|y\|^{jk} \varphi(y^{-(j+1)} x, y) \tag{c}$$

for any $n \in \mathbb{N}$. As in the first part, when $n > m > 0$, it follows from (c) and (5.2) that

$$\|z^{nk} f(z^{-n} x) - z^{mk} f(z^{-m} x)\| \leq \sum_{j=m}^{n-1} \|z\|^{jk} \varphi\left(z^{-(j+1)} x, z\right)$$
$$\to 0 \quad \text{as} \quad m \to \infty.$$

We may now define

$$H(x) = \lim_{n \to \infty} z^{nk} f(z^{-n} x)$$

for all $x \in E$. Hence, the second inequality in (5.6) is obvious in view of (c).

As in the first part, it follows from (a) and (5.5) that

$$\lim_{n \to \infty} f(z^{-n} x)^{-1} y^{-k} f(yz^{-n} x) = e \tag{d}$$

for $x, y \in E \setminus \{0\}$. By using the definition of H and (d), we get

$$\begin{aligned} H(yx) &= \lim_{n \to \infty} z^{nk} f(z^{-n} yx) \\ &= y^k \lim_{n \to \infty} z^{nk} f(z^{-n} x) \lim_{n \to \infty} f(z^{-n} x)^{-1} y^{-k} f(yz^{-n} x) \\ &= y^k H(x) \end{aligned}$$

for any $x, y \in E \setminus \{0\}$. Since $f(0) = 0$, it also holds true that $H(yx) = y^k H(x)$ for $x = 0$ or $y = 0$. The uniqueness of H can be easily proved. $\square$

Jung [161] proved the following corollary.

Corollary 5.2. *Let* $\varphi(x, y) = \delta + \theta \|x\|^a \|y\|^b$ $(\delta \geq 0, \theta \geq 0, 0 \leq a < k, b \geq 0)$ *be given in the functional inequality* (5.4). *If a function* $f : E \to E$ *satisfies* $f(0) = 0$, *the inequality* (5.4), *and the first condition in* (5.5) *for all* $x, y \in E \setminus \{0\}$, *then there exists a unique homogeneous function* $H : E \to E$ *of degree* k *such that*

$$\|f(x) - H(x)\| \leq \delta\left(\|z\|^k - 1\right)^{-1} + \theta \|z\|^b \left(\|z\|^k - \|z\|^a\right)^{-1} \|x\|^a$$

for any $x, z \in E$ *for which* $\|z\|$ *is sufficiently large. In particular, if* $\delta \geq 0$ *and* $\theta = 0$ *in the definition of* φ, *then* f *itself is a homogeneous function of degree* k.

In Corollary 5.2, it was shown that the homogeneous equation of degree k is superstable. More precisely, if E is a real commutative Banach algebra with additional properties $(A4')$, $(A5)$, and $(A6)$, if k is a positive integer, and if a function $f : E \to E$ satisfies $f(0) = 0$, the inequality

$$\|f(yx) - y^k f(x)\| \leq \delta,$$

and the first condition in (5.5) for some $\delta \geq 0$ and for all $x, y, z \in E \setminus \{0\}$ for which the value of $\|z\|$ is sufficiently large, then f is a homogeneous function of degree k. The superstability phenomenon of homogeneous functions will be summarized in the following theorem.

Theorem 5.3. *Let E be a real commutative Banach algebra with properties $(A4')$, $(A5)$, and $(A6)$ and let k be a positive integer. If a function $f : E \to E$ satisfies $f(0) = 0$ and the inequality*

$$\|f(yx) - y^k f(x)\| \leq \delta \tag{5.7}$$

for some $\delta \geq 0$ and for all $x, y \in E \setminus \{0\}$, and if f satisfies the condition

$$\|f(z^n x)\| \to \infty \quad \text{as } n \to \infty$$

for all $x, z \in E \setminus \{0\}$ for which the value of $\|z\|$ is sufficiently large, then f is homogeneous of degree k.

The following corollary was provided in [161].

Corollary 5.4. *Assume that $\varphi(x, y) = \theta \|x\|^a \|y\|^b$ $(\theta \geq 0, a > k, b \geq 0)$ is given in the functional inequality (5.4). If a function $f : E \to E$ satisfies $f(0) = 0$, the inequality (5.4), and the second condition in (5.5) for all $x, y \in E \setminus \{0\}$, then there exists a unique homogeneous function $H : E \to E$ of degree k such that*

$$\|f(x) - H(x)\| \leq \theta \|z\|^b (\|z\|^a - \|z\|^k)^{-1} \|x\|^a$$

for all $x, z \in E$ with $\|z\|$ sufficiently large.

If $\varphi(x, y) = \theta \|x\|^k g(\|y\|)$ for some function $g : [0, \infty) \to [0, \infty)$, then our method to prove the stability for the homogeneous equation of degree k cannot be applied. By modifying an example in [310] of Rassias and Šemrl, Jung introduced a function $f : \mathbb{R} \to \mathbb{R}$ which satisfies the inequality (5.4) and both conditions in (5.5) with some φ and for which $|f(x)|/|x|^k$ $(x \neq 0)$ is unbounded (see [161]):

Example. Let k be a given positive integer. Let us define $f(x) = x^k \log|x|$ for $x \neq 0$ and $f(0) = 0$. Then f satisfies the inequality (5.4) and both conditions in (5.5) with $\varphi(x, y) = |x|^k |y|^k |\log|y||$ $(y \neq 0)$ and $\varphi(x, 0) = 0$. On the other hand, φ satisfies neither (5.1) nor (5.2). In this case we can expect no analogy to the results of Corollaries 5.2 and 5.4. In fact, it holds true that

$$\lim_{x \to \infty} |f(x) - H(x)|/|x|^k = \infty$$

for every homogeneous function $H : \mathbb{R} \to \mathbb{R}$ of degree k.

5.2 Superstability on a Restricted Domain

Throughout this section, assume that E is a real commutative Banach algebra with additional properties $(A4')$, $(A5)$, and $(A6)$, and suppose k is a fixed positive integer. Moreover, let δ and ξ be given positive numbers.

In this section, we will apply Theorem 5.1 to the proof of the superstability of the homogeneous equation of degree k on a restricted domain.

Lemma 5.5. *Assume that a function* $f : E \rightarrow E$ *satisfies* $f(0) = 0$ *and the inequality* (5.7) *for all* $x, y \in E \setminus \{0\}$. *If* $f(x) \neq 0$ *for any* $x \in E \setminus \{0\}$, *then* f *satisfies the condition*

$$\| f(z^n x) \| \rightarrow \infty \quad as \ n \rightarrow \infty$$

for all $x, z \in E \setminus \{0\}$ *with* $\|z\|$ *sufficiently large.*

Proof. Let $x, y \in E \setminus \{0\}$ be given. By using induction on n, we prove that

$$\| f(y^n x) - y^{nk} f(x) \| \leq \delta \big(1 + \|y\|^k + \cdots + \|y\|^{(n-1)k} \big) \qquad (a)$$

for all $n \in \mathbb{N}$. When $n = 1$, the inequality (a) is an immediate consequence of (5.7). If we assume the validity of (a) for some $n > 0$, then it follows from (a) and (5.7) that

$$
\begin{aligned}
&\| f\big(y^{n+1}x\big) - y^{(n+1)k} f(x) \| \\
&\leq \| f\big(y^{n+1}x\big) - y^{nk} f(yx) \| + \|y\|^{nk} \| f(yx) - y^k f(x) \| \\
&\leq \delta \big(1 + \|y\|^k + \cdots + \|y\|^{nk} \big),
\end{aligned}
$$

which completes the proof of (a). It now follows from (a) that

$$
\begin{aligned}
\big\| y^{-nk} f(y^n x) - f(x) \big\| &\leq \delta \big(\|y\|^{-nk} + \|y\|^{-(n-1)k} + \cdots + \|y\|^{-k} \big) \\
&\leq \frac{\delta}{\|y\|^k - 1} \qquad (b)
\end{aligned}
$$

for any $x, y \in E \setminus \{0\}$ with $\|y\| > 1$.

Assume now that $f(x) \neq 0$ for all $x \in E \setminus \{0\}$. If we choose a $z \in E$ such that

$$\|z\| > 1 \quad \text{and} \quad \frac{\delta}{\|z\|^k - 1} < \| f(x) \|$$

(if necessary, we replace z by mz for sufficiently large $m \in \mathbb{N}$), then it follows from (b) that

$$z^{-nk} f(z^n x) \nrightarrow 0 \quad \text{as } n \rightarrow \infty,$$

i.e., since $\| z^{-nk} \| = \|z\|^{-nk} \rightarrow 0$ as $n \rightarrow \infty$, we have

$$\| f(z^n x) \| \rightarrow \infty \quad \text{as } n \rightarrow \infty,$$

which ends the proof. $\square$

Now, let us define

$$S = \{(x, y) \in E^2 \mid \|x\| < \xi \text{ and } \|y\| < \xi\}$$

for some $\xi > 0$.

Jung [166] proved the superstability of the homogeneous functional equation of degree k on a restricted domain.

Theorem 5.6. *If a function $f : E \to E$ satisfies $f(0) = 0$ and the inequality (5.7) for all $(x, y) \in E^2 \backslash S$, and if $f(x) \neq 0$ for all $x \in E \backslash \{0\}$, then f is homogeneous of degree k.*

Proof. Assume that $\|x\| < \xi$ and $0 < \|y\| < \xi$. Since $y \neq 0$, we can choose some real number η with $\eta \geq \xi$ and $\|\eta y\| \geq \xi$. Then, it is true that $(\eta^{-1}x, \eta y), (\eta^{-1}x, \eta e) \in E^2 \backslash S$. It follows from the hypothesis that

$$\begin{aligned}
&\|f(yx) - y^k f(x)\| \\
&= \|f(yx) - (\eta y)^k f(\eta^{-1}x) + (\eta y)^k f(\eta^{-1}x) - y^k f(x)\| \\
&\leq \|f(yx) - (\eta y)^k f(\eta^{-1}x)\| + \|y\|^k \|f(x) - (\eta e)^k f(\eta^{-1}x)\| \\
&\leq \delta(1 + \xi^k),
\end{aligned}$$

since $yx = (\eta y)(\eta^{-1}x)$, $x = (\eta e)(\eta^{-1}x)$, and $\|\eta e\| \geq \xi$. Therefore, f satisfies the inequality

$$\|f(yx) - y^k f(x)\| \leq \delta(1 + \xi^k) \tag{a}$$

for all $x, y \in E \backslash \{0\}$. It follows from (a), Theorem 5.3, and Lemma 5.5 that f is homogeneous of degree k. $\square$

Jung [166] also proved the following:

Theorem 5.7. *If a function $f : E \to E$ satisfies $f(0) = 0$ and the inequality (5.7) for all $x, y \in E \backslash \{0\}$ with $\|y\| \geq \xi$, and if $f(x) \neq 0$ for all $x \in E \backslash \{0\}$, then f is homogeneous of degree k.*

Proof. For any $x, y \in E \backslash \{0\}$ with $0 < \|y\| < \xi$, there exists some real number η with $\eta \geq \xi$ and $\|\eta y\| \geq \xi$. By the same way as in the proof of Theorem 5.6, we see that

$$\|f(yx) - y^k f(x)\| \leq \delta(1 + \xi^k). \tag{a}$$

Therefore, f satisfies the inequality (a) for all $x, y \in E \backslash \{0\}$. In view of Theorem 5.3 and Lemma 5.5, we conclude that f is homogeneous of degree k. $\square$

F. Skof investigated an interesting asymptotic property of the additive functions (see Theorem 2.34). In fact, she proved that a function $f : E_1 \to E_2$ is additive if and only if $\|f(x + y) - f(x) - f(y)\| \to 0$ as $\|x\| + \|y\| \to \infty$, where E_1 is a normed space and E_2 is a Banach space.

Now, we can prove the following corollary concerning an asymptotic property of homogeneous functions (ref. [166]):

Corollary 5.8. *Let* $f : E \to E$ *be a function satisfying* $f(0) = 0$ *in which* $f(x) \neq 0$ *for all* $x \in E \setminus \{0\}$. *Then* f *is a homogeneous function of degree* k *if and only if*

$$\| f(yx) - y^k f(x) \| \to 0 \quad as \quad \|x\| + \|y\| \to \infty.$$

Proof. Let $\delta > 0$ be a given number. By the hypothesis, there exists a constant $\xi > 0$ such that the inequality (5.7) holds true for $\|x\| \geq \xi$ or $\|y\| \geq \xi$. According to Theorem 5.6, f is homogeneous of degree k. The reverse assertion is trivial. $\square$

5.3 Homogeneous Equation Between Vector Spaces

In general, the "multiplication" between elements of a vector space is not defined. Therefore, the previous definition of homogeneous functions has to be modified accordingly.

Let us define the scalar homogeneity of functions between vector spaces. Suppose that $k \neq 0$ is a fixed real number. A function $f : E_1 \to E_2$ between vector spaces is called *homogeneous of degree* k if it satisfies $f(cx) = c^k f(x)$ for any $x \in E_1$ and any scalar c such that c^k is also a scalar. For the case of $k = 1$, the corresponding function is simply said to be *homogeneous*.

J. Tabor [347] proved the following superstability result of the equation for homogeneous functions.

Theorem 5.9. *Let* E_1 *be a real vector space, let* E_2 *be a real topological vector space, and let* V *be a bounded subset of* E_2. *Suppose a function* $G : \mathbb{R} \times E_1 \to \mathbb{R}$ *satisfies the inequality*

$$|G(c, x)| \leq |c|^p |G(1, x)| \tag{5.8}$$

for some $p \neq 1$ *and for all* $c \in \mathbb{R}$ *and* $x \in E_1$. *If a function* $f : E_1 \to E_2$ *satisfies*

$$f(cx) - cf(x) \in G(c, x)V \tag{5.9}$$

for any $c \in \mathbb{R}$ *and* $x \in E_1$, *then* f *is a homogeneous function.*

Proof. Assume that $p < 1$. Let $\{c_n\}$ be a sequence of nonzero real numbers such that $|c_n| \to \infty$ as $n \to \infty$. It follows from (5.9) that

$$(1/c_n) f(c_n x) \in f(x) + \big(G(c_n, x)/c_n\big) V \tag{a}$$

for all $n \in \mathbb{N}$ and for any $x \in E_1$. However, the inequality (5.8) yields

$$\big| G(c_n, x)/c_n \big| \leq |c_n|^{p-1} |G(1, x)|$$

for all $n \in \mathbb{N}$ and for any $x \in E_1$. Hence, we obtain

$$\lim_{n \to \infty} G(c_n, x)/c_n = 0$$

for all $x \in E_1$, and further by (a)

$$\lim_{n \to \infty} (1/c_n) f(c_n x) = f(x) \qquad (b)$$

for all $x \in E_1$. Thus,

$$f(cx) = \lim_{n \to \infty} (1/c_n) f(c_n cx) \qquad (c)$$

for all $c \in \mathbb{R}$ and for all $x \in E_1$.

We claim that

$$f(cx) = \lim_{n \to \infty} (1/c_n) f(c_n cx) = cf(x) \qquad (d)$$

for $c \in \mathbb{R}$ and $x \in E_1$. It is obvious that (d) holds true for $c = 0$. Now, let $c \neq 0$. Then

$$\lim_{n \to \infty} |c_n c| = \infty,$$

therefore, by (b) and (c)

$$f(cx) = \lim_{n \to \infty} (1/c_n) f(c_n cx) = c \lim_{n \to \infty} (1/(c_n c)) f(c_n cx) = cf(x),$$

i.e., f is homogeneous.

In the case $p > 1$ we consider a sequence $\{c_n\}$ of nonzero real numbers such that $c_n \to 0$ as $n \to \infty$. Then the remaining part of the proof runs analogously as in the previous case.   □

Using the last result, Tabor [347] proved the following:

Corollary 5.10. *Let E_1 and E_2 be real normed spaces, and let $p \neq 1$ and $\theta \geq 0$ be given. If a function $f : E_1 \to E_2$ satisfies the inequality*

$$\|f(cx) - cf(x)\| \leq \theta \|cx\|^p$$

for all $c \in \mathbb{R}$ and $x \in E_1$, then f is homogeneous.

Proof. Put $G(c, x) = \theta \|cx\|^p$ and $V = \{x \in E_1 \mid \|x\| \leq 1\}$ for any $c \in \mathbb{R}$ and $x \in E_1$. Applying Theorem 5.9 to this case, we end the proof.   □

S. Czerwik [88] studied the stability problems for the homogeneous functions of degree k between vector spaces. In what follows, we use the notation

$$U_k = \{c \in \mathbb{R} \mid c^k \in \mathbb{R}\}$$

for any real number $k \neq 0$.

Theorem 5.11 (Czerwik). *Let E_1 and E_2 be a real vector space and a real Banach space, respectively. Suppose $\varphi : \mathbb{R} \times E_1 \rightarrow [0, \infty)$ is a function for which there exists an $a \in U_k \setminus \{0\}$ such that*

$$\sum_{n=1}^{\infty} |a|^{-nk} \varphi(a, a^n x) < \infty \tag{5.10}$$

for each $x \in E_1$, and

$$\liminf_{n \to \infty} |a|^{-nk} \varphi(c, a^n x) = 0 \tag{5.11}$$

for any $(c, x) \in U_k \times E_1$, where $k \neq 0$ is a fixed real number. If a function $f : E_1 \rightarrow E_2$ satisfies the inequality

$$\| f(cx) - c^k f(x) \| \leq \varphi(c, x) \tag{5.12}$$

for all $(c, x) \in U_k \times E_1$, then there exists a unique function $H : E_1 \rightarrow E_2$ such that $H(cx) = c^k H(x)$ for each $(c, x) \in U_k \times E_1$ and which satisfies

$$\| f(x) - H(x) \| \leq \sum_{n=1}^{\infty} |a|^{-nk} \varphi(a, a^{n-1} x) \tag{5.13}$$

for any $x \in E_1$.

Proof. First, we claim that

$$\| f(a^n x) - a^{nk} f(x) \| \leq \sum_{j=0}^{n-1} |a|^{jk} \varphi(a, a^{n-j-1} x) \tag{a}$$

for all $n \in \mathbb{N}$ and $(a, x) \in U_k \times E_1$. By putting $c = a$ in (5.12), we immediately see the validity of (a) for $n = 1$. Assume that the inequality (a) is true for some $n > 0$. Replacing x in (a) with ax yields

$$\| f(a^{n+1} x) - a^{nk} f(ax) \| \leq \sum_{j=0}^{n-1} |a|^{jk} \varphi(a, a^{n-j} x).$$

Replacing c in (5.12) with a and multiplying the resulting inequality by $|a|^{nk}$, we obtain

$$\| a^{nk} f(ax) - a^{(n+1)k} f(x) \| \leq |a|^{nk} \varphi(a, x).$$

Combining the last two inequalities yields

$$\| f(a^{n+1} x) - a^{(n+1)k} f(x) \| \leq \sum_{j=0}^{n} |a|^{jk} \varphi(a, a^{n-j} x),$$

which completes the proof of (a).

Put

$$h_n(x) = a^{-nk} f(a^n x)$$

for all $n \in \mathbb{N}$ and any $x \in E_1$. From (a) we get

$$\|h_n(x) - f(x)\| \leq \sum_{j=1}^{n} |a|^{-jk} \varphi(a, a^{j-1} x) \qquad (b)$$

for $n \in \mathbb{N}$ and $x \in E_1$. If we replace x in (a) with $a^m x$ and n by $n - m$ $(n > m)$, then we have

$$
\begin{aligned}
\|h_n(x) - h_m(x)\| &= |a|^{-nk} \|f(a^n x) - a^{(n-m)k} f(a^m x)\| \\
&= |a|^{-nk} \|f(a^{n-m} a^m x) - a^{(n-m)k} f(a^m x)\| \\
&\leq |a|^{-nk} \sum_{j=0}^{n-m-1} |a|^{jk} \varphi(a, a^{n-j-1} x) \\
&= \sum_{i=m+1}^{n} |a|^{-ik} \varphi(a, a^{i-1} x)
\end{aligned}
$$

and it follows from (5.10) that $\{h_n(x)\}$ is a Cauchy sequence for each $x \in E_1$. Since E_2 is complete, we can define

$$H(x) = \lim_{n \to \infty} h_n(x) = \lim_{n \to \infty} a^{-nk} f(a^n x)$$

for any $x \in E_1$.

By (5.12) and (5.11), we get

$$
\begin{aligned}
\|H(cx) - c^k H(x)\| &= \lim_{n \to \infty} \|a^{-nk} (f(a^n cx) - c^k f(a^n x))\| \\
&\leq \lim_{n \to \infty} |a|^{-nk} \varphi(c, a^n x) \\
&= 0.
\end{aligned}
$$

Thus, H is a homogeneous function of degree k when $c \in U_k$. Also, from (b) we get (5.13).

Suppose that $H' : E_1 \to E_2$ is another homogeneous function of degree k when $c \in U_k$ satisfying (5.13). Then, by (5.13) and (5.10), we have

$$
\begin{aligned}
&\|H(x) - H'(x)\| \\
&= |a|^{-mk} \|H(a^m x) - H'(a^m x)\| \\
&\leq |a|^{-mk} (\|H(a^m x) - f(a^m x)\| + \|f(a^m x) - H'(a^m x)\|)
\end{aligned}
$$

$$\leq |a|^{-mk} 2 \sum_{n=1}^{\infty} |a|^{-nk} \varphi(a, a^{n+m-1}x)$$

$$\leq 2 \sum_{i=m+1}^{\infty} |a|^{-ik} \varphi(a, a^{i-1}x)$$

$$\to 0 \text{ as } m \to \infty,$$

which implies $H = H'$. $\qquad\qquad\square$

On account of Theorem 5.11, Czerwik could also prove the following corollary (see [88]).

Corollary 5.12. *Let E_1 and E_2 be a real vector space and a real Banach space, respectively. Suppose $k \neq 0$ is a given real constant. If a function $f : E_1 \to E_2$ satisfies the inequality*

$$\| f(cx) - c^k f(x) \| \leq \delta + \theta |c|^k$$

for some δ, $\theta \geq 0$ and for all $(c, x) \in U_k \times E_1$, then there exists a unique function $H : E_1 \to E_2$ such that $H(cx) = c^k H(x)$ for each $(c, x) \in U_k \times E_1$ and which satisfies

$$\| f(x) - H(x) \| \leq \theta$$

for any $x \in E_1$.

Proof. Assume that $k > 0$. By Theorem 5.11, for each integer $m \geq 2$, there exists a homogeneous function of degree k

$$H_m(x) = \lim_{n \to \infty} m^{-nk} f(m^n x)$$

such that

$$\| f(x) - H_m(x) \| \leq \sum_{n=1}^{\infty} m^{-nk}(\delta + \theta m^k) = \frac{\delta + \theta m^k}{m^k - 1} \qquad (a)$$

for all $x \in E_1$. By (a) we have

$$\| H_m(x) - H_\ell(x) \| = 2^{-nk} \| H_m(2^n x) - H_\ell(2^n x) \|$$

$$\leq 2^{-nk} \left(\frac{\delta + \theta m^k}{m^k - 1} + \frac{\delta + \theta \ell^k}{\ell^k - 1} \right)$$

$$\to 0 \quad \text{as } n \to \infty,$$

and hence $H_m = H_\ell$ for all integers ℓ and m larger than 1. We put $H(x) = H_2(x)$ for all $x \in E_1$. By (a) we have

$$\| f(x) - H(x) \| \leq \frac{\delta + \theta m^k}{m^k - 1}$$

for all $x \in E_1$ and letting $m \to \infty$, we find that $\|f(x) - H(x)\| \le \theta$.

In the case when $k < 0$, the proof runs analogously. $\square$

Czerwik [88] has remarked that the function $f(x) = \sin x$ $(x \in \mathbb{R})$ is not a homogeneous function of degree k, but it satisfies

$$|\sin(cx) - c^k \sin x| \le 1 + |c|^k$$

for each $(c, x) \in U_k \times \mathbb{R}$. From this example we see that not all cases under Corollary 5.12 are superstable (see [88]).

Corollary 5.13. *Assume that the assumptions in Corollary 5.12 are satisfied. If either δ or θ is zero, then $f(cx) = c^k f(x)$ for all $(c, x) \in (U_k \setminus \{0\}) \times E_1$.*

Proof. Suppose that $\delta = 0$. Then,

$$\|f(cx) - c^k f(x)\| \le \theta |c|^k$$

for all $(c, x) \in U_k \times E_1$. Putting $x = (1/c)y$ with $c \in U_k \setminus \{0\}$, we get

$$\left\| f(y) - c^k f\big((1/c)y\big) \right\| \le \theta |c|^k. \qquad (a)$$

Assume that $k > 0$. It then follows from the last inequality that

$$f(y) = \lim_{c \to 0} c^k f\big((1/c)y\big)$$

for any $y \in E_1$. Therefore, for $(a, x) \in (U_k \setminus \{0\}) \times E_1$, we have

$$f(ax) = \lim_{c \to 0} c^k f\big((1/c)ax\big) = \lim_{c \to 0} a^k (c/a)^k f\big((a/c)x\big) = a^k f(x).$$

For $k < 0$, it follows from (a) that

$$\lim_{|c| \to \infty} c^k f\big((1/c)y\big) = f(y),$$

and, as before, we find that the corollary is true for $\delta = 0$ and $k < 0$.

Now, suppose that $\theta = 0$. Then, we have

$$\left\| c^{-k} f(cx) - f(x) \right\| \le \delta |c|^{-k}$$

for any $(c, x) \in (U_k \setminus \{0\}) \times E_1$. Hence, we get

$$f(x) = \begin{cases} \displaystyle\lim_{|c| \to \infty} c^{-k} f(cx) & \text{(for } k > 0), \\ \displaystyle\lim_{c \to 0} c^{-k} f(cx) & \text{(for } k < 0). \end{cases}$$

As before, the assertion is valid in these cases as well. $\square$

5.4 Homogeneous Equation of Pexider Type

In this section, the Hyers–Ulam–Rassias stability of the *homogeneous equation of Pexider type*

$$f(\alpha x) = \psi(\alpha)g(x)$$

will be investigated. In 2005, S. Czerwik [91] introduced the following lemma.

Lemma 5.14. *Let E_1 and E_2 be a real vector space and a real normed space, respectively. Assume that $f : E_1 \to E_2$, $g : E_1 \to E_2$, $\psi : \mathbb{R} \to \mathbb{R}$, and $\varphi : \mathbb{R} \times E_1 \to [0, \infty)$ are given functions with $\psi(1) = 1$. If f satisfies the inequality*

$$\|f(\alpha x) - \psi(\alpha)g(x)\| \le \varphi(\alpha, x) \tag{5.14}$$

for all $\alpha \in \mathbb{R}$ and $x \in E_1$, then it holds true for all $\alpha \in \mathbb{R}$, $x \in E_1$, and $n \in \mathbb{N}$ that

$$\|f(\alpha^n x) - \psi(\alpha)^n f(x)\| \le \sum_{i=1}^{n} |\psi(\alpha)|^{i-1} \Phi_1(\alpha, \alpha^{n-i} x) \tag{5.15}$$

and

$$\|g(\alpha^n x) - \psi(\alpha)^n g(x)\| \le \sum_{i=1}^{n} |\psi(\alpha)|^{i-1} \Phi_2(\alpha, \alpha^{n-i} x), \tag{5.16}$$

where $\Phi_1(\alpha, x) = \varphi(\alpha, x) + |\psi(\alpha)|\varphi(1, x)$ and $\Phi_2(\alpha, x) = \varphi(\alpha, x) + \varphi(1, \alpha x)$.

Proof. It follows from (5.14) that

$$
\begin{aligned}
\|f(\alpha x) - \psi(\alpha) f(x)\| &\le \|f(\alpha x) - \psi(\alpha)g(x)\| + \|\psi(\alpha)g(x) - \psi(\alpha)f(x)\| \\
&\le \varphi(\alpha, x) + |\psi(\alpha)|\varphi(1, x) \\
&= \Phi_1(\alpha, x)
\end{aligned}
$$

for any $\alpha \in \mathbb{R}$ and $x \in E_1$. Hence, (5.15) is true for $n = 1$. Assume that (5.15) holds true for some $n > 0$. It then follows from (5.15) that

$$
\begin{aligned}
\|f(\alpha^{n+1} x) &- \psi(\alpha)^{n+1} f(x)\| \\
&\le \|f(\alpha^{n+1} x) - \psi(\alpha) f(\alpha^n x)\| + \|\psi(\alpha) f(\alpha^n x) - \psi(\alpha)^{n+1} f(x)\| \\
&\le \Phi_1(\alpha, \alpha^n x) + |\psi(\alpha)| \|f(\alpha^n x) - \psi(\alpha)^n f(x)\| \\
&\le \Phi_1(\alpha, \alpha^n x) + \sum_{i=1}^{n} |\psi(\alpha)|^i \Phi_1(\alpha, \alpha^{n-i} x) \\
&= \sum_{i=1}^{n+1} |\psi(\alpha)|^{i-1} \Phi_1(\alpha, \alpha^{n+1-i} x),
\end{aligned}
$$

which proves the inequality (5.15) for all $n \in \mathbb{N}$. Similarly, we can prove the inequality (5.16). $\square$

Using Lemma 5.14, Czerwik proved the following theorem concerning the Hyers–Ulam–Rassias stability of the homogeneous functional equation of Pexider type (see [91, Theorem 1]).

Theorem 5.15. *Let E_1 and E_2 be a real vector space and a real Banach space, respectively. Assume that $f : E_1 \to E_2$, $g : E_1 \to E_2$, $\psi : \mathbb{R} \to \mathbb{R}$, and $\varphi : \mathbb{R} \times E_1 \to [0, \infty)$ are given functions with $\psi(1) = 1$. Moreover, assume that there exists a $\beta \in \mathbb{R}$ such that $\psi(\beta) \neq 0$ and the series*

$$\sum_{n=1}^{\infty} |\psi(\beta)|^{-n} \Phi_1(\beta, \beta^{n-1}x) < \infty \tag{5.17}$$

for all $x \in E_1$ and

$$\lim_{n \to \infty} |\psi(\beta)|^{-n} \Phi_1(\alpha, \beta^{n-1}x) = 0 \tag{5.18}$$

for all $\alpha \in \mathbb{R}$ and $x \in E_1$. If f satisfies the inequality (5.14) for all $\alpha \in \mathbb{R}$ and $x \in E_1$, then there exists a unique ψ-homogeneous function $H : E_1 \to E_2$ such that

$$\|H(x) - f(x)\| \leq \sum_{n=1}^{\infty} |\psi(\beta)|^{-n} \Phi_1(\beta, \beta^{n-1}x) \tag{5.19}$$

and

$$\|H(x) - g(x)\| \leq \sum_{n=1}^{\infty} |\psi(\beta)|^{-n} \Phi_2(\beta, \beta^{n-1}x) \tag{5.20}$$

for all $x \in E_1$. (See Lemma 5.14 for the definitions of Φ_1 and Φ_2.)

Proof. For each $x \in E_1$ and $n \in \mathbb{N}$, define

$$H_n(x) = \psi(\beta)^{-n} f(\beta^n x). \tag{a}$$

It then follows from (5.15) that

$$\|H_n(x) - f(x)\| \leq \sum_{i=1}^{n} |\psi(\beta)|^{-(n-i)-1} \Phi_1(\beta, \beta^{n-i}x),$$

which implies that

$$\|H_n(x) - f(x)\| \leq \sum_{i=1}^{n} |\psi(\beta)|^{-i} \Phi_1(\beta, \beta^{i-1}x) \tag{b}$$

for any $x \in E_1$ and $n \in \mathbb{N}$.

Now, we will verify that $\{H_n(x)\}$ is a Cauchy sequence for each $x \in E_1$. The inequality (5.15) yields

$$\|H_n(x) - H_m(x)\| \leq |\psi(\beta)|^{-n} \|f(\beta^n x) - \psi(\beta)^{n-m} f(\beta^m x)\|$$
$$\leq \sum_{i=m+1}^{n} |\psi(\beta)|^{-i} \Phi_1(\beta, \beta^{i-1} x)$$

for all $x \in E_1$ and $m, n \in \mathbb{N}$ with $n > m$. In view of (5.17), we see that $\{H_n(x)\}$ is a Cauchy sequence for any $x \in E_1$. Therefore, we can define a function $H : E_1 \to E_2$ by

$$H(x) = \lim_{n \to \infty} H_n(x).$$

It follows from (5.14) and (5.18) that

$$\|H(\alpha x) - \psi(\alpha) H(x)\| = \lim_{n \to \infty} |\psi(\beta)|^{-n} \|f(\alpha \beta^n x) - \psi(\alpha) f(\beta^n x)\|$$
$$\leq \lim_{n \to \infty} |\psi(\beta)|^{-n} \Phi_1(\alpha, \beta^n x)$$
$$= 0$$

for all $\alpha \in \mathbb{R}$ and $x \in E_1$, which implies that H is ψ-homogeneous. Due to (b) and the definition of H, the inequality (5.19) is true.

By a similar way as for the $\{H_n(x)\}$, if we define

$$H_n'(x) = \psi(\beta)^{-n} g(\beta^n x)$$

for every $x \in E_1$, then $\{H_n'(x)\}$ is a Cauchy sequence. Hence, we may define

$$H'(x) = \lim_{n \to \infty} H_n'(x).$$

Then, by (5.16), we get

$$\|H_n'(x) - g(x)\| \leq \sum_{i=1}^{n} |\psi(\beta)|^{-i} \Phi_2(\beta, \beta^{i-1} x) \qquad (c)$$

for all $x \in E_1$. Furthermore, by (5.14), we have

$$\|H_n(x) - H_n'(x)\| = |\psi(\beta)|^{-n} \|f(\beta^n x) - g(\beta^n x)\|$$
$$\leq |\psi(\beta)|^{-n} \varphi(1, \beta^n x)$$
$$= (1/2)|\psi(\beta)|^{-n} \Phi_1(1, \beta^n x).$$

In view of (5.18), it follows that $H(x) = H'(x)$ for all $x \in E_1$ and hence, the inequality (5.20) follows from (c).

Finally, it remains to prove the uniqueness of the ψ-homogeneous function H. Assume that $H_0 : E_1 \to E_2$ is another ψ-homogeneous function which satisfies the inequalities (5.19) and (5.20). Without loss of generality, we suppose that there exists an $x_0 \in E_1$ with $H_0(x_0) \neq 0$. Then, by using the ψ-homogeneity of H_0, we have

$$H_0(\alpha\beta x_0) = \psi(\alpha\beta)H_0(x_0) = \psi(\alpha)\psi(\beta)H_0(x_0)$$

for all $\alpha, \beta \in \mathbb{R}$, and thus

$$\psi(\alpha\beta) = \psi(\alpha)\psi(\beta) \tag{d}$$

for any $\alpha, \beta \in \mathbb{R}$.

Consequently, it follows from (5.17), (5.19), and (d) that

$$
\begin{aligned}
\|H(x) &- H_0(x)\| \\
&= |\psi(\beta^m)|^{-1}\|H(\beta^m x) - H_0(\beta^m x)\| \\
&\leq |\psi(\beta)|^{-m}\left(\|H(\beta^m x) - f(\beta^m x)\| + \|f(\beta^m x) - H_0(\beta^m x)\|\right) \\
&\leq 2\sum_{n=m+1}^{\infty} |\psi(\beta)|^{-n}\Phi_1\left(\beta, \beta^{n-1}x\right) \\
&\to 0 \quad \text{as } m \to \infty
\end{aligned}
$$

for all $x \in E_1$, which proves the uniqueness of H. $\square$

From the proof of Theorem 5.15, we may guess that if H does not identically equal the zero function, then the function ψ has to be multiplicative.

Corollary 5.16. *Let E_1 and E_2 be a real vector space and a real Banach space, respectively. Assume that functions $f : E_1 \to E_2$ and $g : E_1 \to E_2$ satisfy the inequality*

$$\|f(\alpha x) - |\alpha|^v g(x)\| \leq \delta + |\alpha|^v \varepsilon$$

for all $x \in E_1$ and $\alpha \in \mathbb{R}$, where $v > 0$, $\delta \geq 0$, and $\varepsilon \geq 0$ are constants. Then, there exists a unique function $H : E_1 \to E_2$ such that

$$
\begin{aligned}
H(\alpha x) &= |\alpha|^v H(x), \\
\|H(x) - f(x)\| &\leq \delta + 2\varepsilon, \\
\|H(x) - g(x)\| &\leq \varepsilon
\end{aligned}
$$

for all $x \in E_1$.

Corollary 5.17. *Let E_1 and E_2 be a real vector space and a real Banach space, respectively. Assume that functions $f : E_1 \to E_2$ and $g : E_1 \to E_2$ satisfy the inequality*

$$\|f(\alpha x) - |\alpha|^v g(x)\| \leq \delta + |\alpha|^v \varepsilon$$

for all $x \in E_1$ and $\alpha \in \mathbb{R}$, where $v > 0$, $\delta \geq 0$, and $\varepsilon \geq 0$ are constants.
  (i) If $\delta = 0$, then f is v-homogeneous and

$$\|f(x) - g(x)\| \leq \varepsilon$$

for any $x \in E_1$;
  (ii) If $\varepsilon = 0$, then g is v-homogeneous and

$$\|f(x) - g(x)\| \leq \delta$$

for any $x \in E_1$.

A function $f : E_1 \to E_2$ is called a quadratic function if and only if

$$f(x + y) + f(x - y) = 2f(x) + 2f(y)$$

for all $x, y \in E_1$.

Corollary 5.18. *Let E_1 and E_2 be a real vector space and a real Banach space, respectively. Assume that functions $f, g, h : E_1 \to E_2$ are given and a constant $\delta > 0$ is given. Then, $f = g + h$, where g is a quadratic 2-homogeneous function and $\|h(x)\| \leq \delta$ for all $x \in E_1$, if and only if*

$$\|f(x + y) + f(x - y) - 2f(x) - 2f(y)\| \leq 6\delta \tag{5.21}$$

and

$$\|f(\alpha x) - \alpha^2 f(x)\| \leq \delta + \alpha^2 \delta \tag{5.22}$$

for all $x, y \in E_1$ and $\alpha \in \mathbb{R}$.

Proof. Assume that $f = g + h$, where g is a quadratic 2-homogeneous function and $\|h(x)\| \leq \delta$ for all $x \in E_1$. Obviously, we have

$$\begin{aligned}
\|f(x + y) &+ f(x - y) - 2f(x) - 2f(y)\| \\
&= \|h(x + y) + h(x - y) - 2h(x) - 2h(y)\| \\
&\leq 6\delta
\end{aligned}$$

and

$$\|f(\alpha x) - \alpha^2 f(x)\| = \|h(\alpha x) - \alpha^2 h(x)\| \leq \delta + \alpha^2 \delta$$

for all $x, y \in E_1$ and $\alpha \in \mathbb{R}$.

We now assume that the inequalities (5.21) and (5.22) are true. According to Corollary 5.16, there exists a 2-homogeneous function $g : E_1 \to E_2$ such that

$$\|f(x) - g(x)\| \leq \delta$$

for each $x \in E_1$. If we define $h = f - g$, then $f = g + h$ and $\|h(x)\| \le \delta$ for all $x \in E_1$. By (5.21), we get

$$\|g(\alpha(x + y)) + g(\alpha(x - y)) - 2g(\alpha x) - 2g(\alpha y) \\ + h(\alpha(x + y)) + h(\alpha(x - y)) - 2h(\alpha x) - 2h(\alpha y)\| \le 6\delta,$$

and hence

$$\|g(x + y) + g(x - y) - 2g(x) - 2g(y) \\ + (1/\alpha^2)(h(\alpha(x + y)) + h(\alpha(x - y)) - 2h(\alpha x) - 2h(\alpha y))\| \le 6\delta/\alpha^2$$

for any $\alpha \ne 0$. Finally, if we let $\alpha \to \infty$, then

$$\|g(x + y) + g(x - y) - 2g(x) - 2g(y)\| = 0$$

for all $x, y \in E_1$, which implies that g is a quadratic function.   □

Chapter 6
Linear Functional Equations

A function is called a *linear function* if it is homogeneous as well as additive. The homogeneity of a function, however, is a consequence of additivity if the function is assumed to be continuous. There are a number of (systems of) functional equations which include all the linear functions as their solutions. In this chapter, only a few (systems of) functional equations among them will be introduced. In Section 6.1, the superstability property of the "intuitive" system (6.1) of functional equations $f(x + y) = f(x) + f(y)$ and $f(cx) = cf(x)$ which stands for the linear functions is introduced. The stability problem for the functional equation $f(x + cy) = f(x) + cf(y)$ is proved in the second section and the result is applied to the proof of the Hyers–Ulam stability of the "intuitive" system (6.1). In the final section, stability problems of other systems, which describe linear functions, are discussed.

6.1 A System for Linear Functions

It is natural for one to expect that the additive Cauchy equation, together with the homogeneous equation, may determine all linear functions. Thus, the following system of the functional equations

$$\begin{cases} f(x + y) = f(x) + f(y), \\ f(cx) = cf(x) \end{cases} \tag{6.1}$$

may be introduced as a system of equations for linear functions. This section is devoted to the study of superstability problems of the system (6.1) of linear functions.

J. Tabor [347] proved a theorem concerning the superstability of the system (6.1).

Theorem 6.1 (Tabor). *Let E_1 be a real vector space, E_2 a locally convex topological vector space, V a bounded subset of E_2, and let $F : E_1^2 \to \mathbb{R}$ and $G : \mathbb{R} \times E_1 \to \mathbb{R}$ be such functions that there exists a sequence $\{c_n\}$ of real numbers satisfying*

$$\lim_{n \to \infty} F(c_n x, c_n y)/c_n = 0 \tag{6.2}$$

S.-M. Jung, *Hyers–Ulam–Rassias Stability of Functional Equations in Nonlinear Analysis*, Springer Optimization and Its Applications 48,
DOI 10.1007/978-1-4419-9637-4_6, © Springer Science+Business Media, LLC 2011

and

$$\lim_{n\to\infty} G(c_n, x)/c_n = 0 \tag{6.3}$$

for all $x, y \in E_1$. If a function $f : E_1 \to E_2$ satisfies

$$\begin{cases} f(x + y) - f(x) - f(y) \in F(x, y)V, \\ f(cx) - cf(x) \in G(c, x)V \end{cases} \tag{6.4}$$

for all $c \in \mathbb{R}$ and $x, y \in E_1$, then f is additive. Moreover, if for each $x \in E_1$ the function $G(\cdot, x)$ is bounded on a set of positive inner Lebesgue measure or on a set of the second category with the Baire property, then f is a linear function.

Proof. It follows from the second relation in (6.4) that

$$f(c_n x)/c_n \in f(x) + \big(G(c_n, x)/c_n\big)V$$

for $n \in \mathbb{N}$ and $x \in E_1$. Since V is bounded, by (6.3) we obtain

$$\lim_{n\to\infty} f(c_n x)/c_n = f(x) \tag{a}$$

for each $x \in E_1$. From the first relation in (6.4) we get

$$f\big(c_n(x + y)\big)/c_n - f(c_n x)/c_n - f(c_n y)/c_n \in \big(F(c_n x, c_n y)/c_n\big)V$$

for all $n \in \mathbb{N}$ and $x, y \in E_1$, where by (6.2) and (a) we obtain

$$f(x + y) - f(x) - f(y) = 0$$

for all $x, y \in E_1$, which means that f is additive.

For the proof of the second part we arbitrarily fix an $x \in E_1$ and a $\mu \in E_2^*$ (the dual space of E_2). We put

$$f^*(c) = \mu\big(f(cx)\big)$$

for all real numbers c. Obviously, f^* is additive. From the second relation in (6.4) it follows that

$$f^*(c) \in \mu\big(cf(x) + G(c, x)V\big) \tag{b}$$

for any $c \in \mathbb{R}$. By the assumption, $G(\cdot, x)$ is bounded on a set A of positive inner Lebesgue measure or of the second category with the Baire property. Without loss of generality, we may assume that A is bounded. Since μ is a continuous linear function, it maps a bounded set into a bounded set. Thus, by (b), f^* is bounded on A, and hence linear (cf. Theorem 2.1, [1, 228]). Now, for an arbitrarily fixed $c \in \mathbb{R}$, we get

$$\mu\big(f(cx)\big) = f^*(c) = cf^*(1) = c\mu\big(f(x)\big) = \mu\big(cf(x)\big).$$

Since this equality holds true for each $\mu \in E_2^*$, we have

$$f(cx) = cf(x)$$

for any $c \in \mathbb{R}$ and $x \in E_1$, which ends the proof.   □

Tabor [347] noted that the boundedness of the function $G(\cdot, x)$ on a respective set is an essential assumption of Theorem 6.1: Consider an additive discontinuous function $f : \mathbb{R} \to \mathbb{R}$ and put

$$\begin{cases} F(x, y) = 0 & \text{(for } x, y \in \mathbb{R}\text{)}, \\ G(c, x) = f(cx) - cf(x) & \text{(for } c, x \in \mathbb{R}\text{)}, \\ V = \{1\}. \end{cases}$$

Clearly, both relations in (6.4) hold true. We also have $G(c, x) = 0$ for $c \in \mathbb{Q}$ and $x \in \mathbb{R}$. Therefore, for an arbitrary sequence $\{c_n\}$ of nonzero rational numbers, (6.2) and (6.3) are valid. However, f is not linear.

Using the last theorem, Tabor [347] obtained a more familiar result:

Corollary 6.2. *Let E_1 and E_2 be normed spaces, let $p \neq 1$ be given, and let a function $H : [0, \infty)^2 \to [0, \infty)$ be homogeneous of degree p and a function $K : [0, \infty)^2 \to [0, \infty)$ be homogeneous of degree p with respect to the first variable. If a function $f : E_1 \to E_2$ satisfies*

$$\begin{cases} \|f(x + y) - f(x) - f(y)\| \leq H\big(\|x\|, \|y\|\big), \\ \|f(cx) - cf(x)\| \leq K\big(|c|, \|x\|\big) \end{cases} \tag{6.5}$$

for all $c \in \mathbb{R}$ and all $x, y \in E_1$, then f is linear.

Proof. Put

$$\begin{cases} F(x, y) = H\big(\|x\|, \|y\|\big) & \text{(for } x, y \in E_1\text{)}, \\ G(c, x) = K\big(|c|, \|x\|\big) & \text{(for } c \in \mathbb{R}, \ x \in E_1\text{)}, \\ V = \{x \in E_1 \mid \|x\| \leq 1\}. \end{cases}$$

Then, in view of (6.5), both conditions in (6.4) hold true. We now claim that there exists a sequence $\{c_n\}$ satisfying (6.2) and (6.3). In the case of $p < 1$ we take $c_n = 2^n$. Then we obtain

$$\lim_{n \to \infty} 2^{-n} F(2^n x, 2^n y) = \lim_{n \to \infty} 2^{n(p-1)} H\big(\|x\|, \|y\|\big) = 0$$

and

$$\lim_{n \to \infty} 2^{-n} G(2^n, x) = \lim_{n \to \infty} 2^{n(p-1)} K\big(1, \|x\|\big) = 0$$

for all $x, y \in E_1$.

In the case $p > 1$ we take $c_n = 2^{-n}$. Then,

$$\lim_{n \to \infty} 2^n F(2^{-n}x, 2^{-n}y) = \lim_{n \to \infty} 2^{n(1-p)} H(\|x\|, \|y\|) = 0$$

and

$$\lim_{n \to \infty} 2^n G(2^{-n}, x) = \lim_{n \to \infty} 2^{n(1-p)} K(1, \|x\|) = 0$$

for all $x, y \in E_1$. We have

$$G(c, x) = |c|^p K(1, \|x\|)$$

for $c \in \mathbb{R}$ and $x \in E_1$. This equality shows that the function $G(\cdot, x)$ is bounded on every bounded interval, isolated from zero. Hence, we can apply Theorem 6.1, which ends the proof. □

Taking in Corollary 6.2

$$\begin{cases} H(\|x\|, \|y\|) = \theta(\|x\|^p + \|y\|^p) & \text{(for } x, y \in E_1), \\ K(|c|, \|x\|) = \theta \|cx\|^p & \text{(for } c \in \mathbb{R}, \ x \in E_1), \end{cases}$$

Tabor [347] obtained the following corollary.

Corollary 6.3. *Let E_1 and E_2 be normed spaces, and let $p \neq 1$ and $\theta \geq 0$ be given. If a function $f : E_1 \to E_2$ satisfies*

$$\begin{cases} \|f(x + y) - f(x) - f(y)\| \leq \theta(\|x\|^p + \|y\|^p), \\ \|f(cx) - cf(x)\| \leq \theta \|cx\|^p \end{cases}$$

for all $c \in \mathbb{R}$ and for all $x, y \in E_1$, then f is linear.

6.2 Functional Equation $f(x + cy) = f(x) + cf(y)$

In the previous section, it was shown that the "intuitive" *linear functional equations* (6.1) is superstable. J. Schwaiger [325] introduced the functional equation

$$f(x + cy) = f(x) + cf(y)$$

which is equivalent to the system (6.1) if the related domain and range are assumed to be vector spaces and proved the stability of the given equation.

Theorem 6.4 (Schwaiger). *Let E_1 and E_2 be a real vector space and a real Banach space, respectively, and let $\varphi : \mathbb{R} \to [0, \infty)$ be a function. If a function $f : E_1 \to E_2$ satisfies the inequality*

$$\|f(x + cy) - f(x) - cf(y)\| \leq \varphi(c) \tag{6.6}$$

for all $x, y \in E_1$ and for any $c \in \mathbb{R}$, then there exists a unique linear function $L : E_1 \to E_2$ such that

$$\|f(x) - L(x)\| \leq \varphi(1)$$

for all $x \in E_1$.

Proof. For $c = 1$ we see that the conditions in Theorem 2.3 (see also the comment just below the theorem) are satisfied with $\delta = \varphi(1)$, where $L(x) = \lim_{n \to \infty} 2^{-n} f(2^n x)$ is the unique additive function such that

$$\|f(x) - L(x)\| \leq \varphi(1)$$

for all $x \in E_1$. It remains to show the homogeneity of L. Putting $x = 0$ and replacing y in (6.6) with $2^n y$ yield

$$\|f(2^n cy) - cf(2^n y) - f(0)\| \leq \varphi(c).$$

Dividing this inequality by 2^n and letting $n \to \infty$, we find that

$$L(cy) = cL(y)$$

for all $y \in E_1$ and $c \in \mathbb{R}$. $\qquad\qquad\qquad\qquad\qquad\qquad\qquad\qquad$ □

Using the above theorem we can prove the Hyers–Ulam stability of the system (6.1) which cannot be deduced from Theorem 6.1.

Corollary 6.5. *Let E_1 and E_2 be a real vector space and a real Banach space, respectively, and let $\delta_1, \delta_2 \geq 0$ be given. If a function $f : E_1 \to E_2$ satisfies the inequalities*

$$\begin{cases} \|f(x + y) - f(x) - f(y)\| \leq \delta_1, \\ \|f(cx) - cf(x)\| \leq \delta_2 \end{cases}$$

for all $x, y \in E_1$ and for any $c \in \mathbb{R}$, then there exists a unique linear function $L : E_1 \to E_2$ such that

$$\|f(x) - L(x)\| \leq \delta_1 + \delta_2$$

for any $x \in E_1$.

Proof. From the hypothesis we get

$$\begin{aligned} \|f(x &+ cy) - f(x) - cf(y)\| \\ &\leq \|f(x + cy) - f(x) - f(cy)\| + \|f(cy) - cf(y)\| \\ &\leq \delta_1 + \delta_2 \end{aligned}$$

for all $x, y \in E_1$ and for any $c \in \mathbb{R}$. In view of Theorem 6.4 with $\varphi(c) = \delta_1 + \delta_2$ ($c \in \mathbb{R}$), there exists a unique linear function $L : E_1 \to E_2$ such that

$$\| f(x) - L(x) \| \leq \delta_1 + \delta_2$$

for any $x \in E_1$. $\square$

6.3 Stability for Other Equations

We start this section with an old theorem which is often called the Eidelheit's separation theorem (see [93] for the proof of this theorem). It plays an important role in the proof of Theorem 6.7.

Theorem 6.6 (Eidelheit). *Suppose E is a topological vector space. Let K_1 and K_2 be convex subsets of E such that K_1 has an interior point and K_2 contains no interior point of K_1. Then there exists a $\mu \in E^*$ such that*

$$\mu(x) \leq \mu(y)$$

for all $x \in K_1$ and for all $y \in K_2$.

It was shown in Theorem 2.6 that the statement of Theorem 2.5 is no more valid for the case when $p = 1$ is assumed in the functional inequality (2.5). Such counterexamples have stimulated many authors to attempt to surpass such awkwardness of the inequality (2.5) for the case $p = 1$. B. E. Johnson was one of those mathematicians who worked for this purpose. Johnson [147] provided the following theorem.

Theorem 6.7 (Johnson). *Let E be a real Banach space. If a continuous functional $f : E \to \mathbb{R}$ satisfies the functional inequalities*

$$\left| f(c_1 x_1 + \cdots + c_n x_n) - c_1 f(x_1) - \cdots - c_n f(x_n) \right|$$
$$\leq \theta \left(\| c_1 x_1 \| + \cdots + \| c_n x_n \| \right)$$

for some $\theta > 0$ and for all $n \in \mathbb{N}$, all $c_1, \ldots, c_n \in \mathbb{R}$, and all $x_1, \ldots, x_n \in E$, then there exists a linear functional $L : E \to \mathbb{R}$ such that

$$| f(x) - L(x) | \leq 3\theta \| x \|$$

for all $x \in E$.

Proof. For the special case of $n = 1$, the above inequality may be expressed as

$$| f(cx) - cf(x) | \leq \theta \| cx \| \tag{a}$$

for all $c \in \mathbb{R}$ and for all $x \in E$, and putting $c = 0$ in (a) yields $f(0) = 0$. Since f is assumed to be continuous, there exists an $\eta > 0$ such that $|f(x)| = |f(x) - f(0)| < 1$ when $\|x\| < \eta$. Given any $x' \neq 0$ in E, put $c = (2/\eta)\|x'\|$ and $x = (1/c)x'$. Thus, we get

$$|f(x')| = |f(cx)| \leq |cf(x)| + \theta\|cx\| < (2/\eta)\|x'\| + \theta\|x'\|$$

for all $x' \in E \setminus \{0\}$.

Put

$$\Gamma = \{(x, y) \in E \times \mathbb{R} \mid \|x\| \leq 1, \ y = f(x)\}$$

and let Γ' be the closed convex hull of Γ. Given $(x, y) \in \Gamma'$, let $\{(x^{(m)}, y^{(m)})\}$ be a sequence of convex combinations of elements of Γ converging to (x, y). Thus, we have

$$x^{(m)} = \sum_i c_i^{(m)} x_i^{(m)} \quad \text{and} \quad y^{(m)} = \sum_i c_i^{(m)} y_i^{(m)} = \sum_i c_i^{(m)} f(x_i^{(m)}),$$

where

$$c_i^{(m)} \geq 0, \quad \sum_i c_i^{(m)} = 1, \quad \text{and} \quad \|x_i^{(m)}\| \leq 1.$$

By hypothesis, we obtain

$$\begin{aligned} \left| f(x^{(m)}) - y^{(m)} \right| &= \left| f\left(\sum_i c_i^{(m)} x_i^{(m)} \right) - \sum_i c_i^{(m)} f(x_i^{(m)}) \right| \\ &\leq \theta \sum_i \|c_i^{(m)} x_i^{(m)}\| \\ &= \theta \sum_i c_i^{(m)} \|x_i^{(m)}\| \\ &\leq \theta. \end{aligned}$$

Taking the limit as $m \to \infty$, we have

$$|f(x) - y| \leq \theta \quad \text{and} \quad \|x\| \leq 1 \tag{b}$$

for $(x, y) \in \Gamma'$.

Now, put

$$\Gamma^+ = \{(x, y) \in E \times \mathbb{R} \mid \|x\| \leq 1 \ \text{and} \ \exists z \leq y \ \text{with} \ (x, z) \in \Gamma'\}.$$

Let Γ'' be the translate $\Gamma' - (0, 2\theta)$ of Γ'. Since Γ is a subset of $B \times [-2/\eta - \theta, 2/\eta + \theta]$ and Γ' is also a subset of $B \times [-2/\eta - \theta, 2/\eta + \theta]$, where B is the closed unit ball in E, we see that Γ^+ contains $B \times (2/\eta + \theta, \infty)$ and hence is a convex body.

Since the closed interval $[-2/\eta - \theta, 2/\eta + \theta]$ is compact, it is easily shown that Γ^+ is closed and hence that it is the closure of its interior. If $(x, y) \in \Gamma^+$, then there exists a z with $z \leq y$ and $(x, z) \in \Gamma'$, so by (b), we have $z \geq f(x) - \theta$, and thus $y \geq f(x) - \theta$. Hence, if (x, y) belongs to the interior of Γ^+, then $y > f(x) - \theta$. If $(x, y) \in \Gamma''$, then $(x, y + 2\theta)$ is in Γ', so by (b), we get $y + 2\theta \leq f(x) + \theta$, i.e., $y \leq f(x) - \theta$. Therefore,

$$(\Gamma^+)^\circ \cap \Gamma'' = \emptyset.$$

Thus, by Eidelheit's separation theorem (see above), there exist a nonzero linear functional G on the Banach space $E \times \mathbb{R}$ and a number k with $G(w) \geq k$ for $w \in (\Gamma^+)^\circ$ and hence for w in Γ^+, and $G(w) \leq k$ for $w \in \Gamma''$. There is an $L \in E^*$ (where E^* is the dual space of E) and $\mu \in \mathbb{R}$ such that $G(x, y) = \mu y - L(x)$ for $x \in E$ and $y \in \mathbb{R}$. If $y > f(x)$ and $\|x\| \leq 1$, then $(x, y) \in \Gamma^+$, so $\mu y - L(x) = G(x, y) \geq k$. By letting $y \to \infty$, we see that $\mu \geq 0$. Since $(x, f(x)) \in \Gamma^+$ and $(x, f(x) - 2\theta) \in \Gamma''$ for $\|x\| \leq 1$, we have $\mu f(x) - L(x) \geq k$ while $\mu(f(x) - 2\theta) - L(x) \leq k$ for $\|x\| \leq 1$. If μ were 0, then we would have $L(x) = -k$ for $\|x\| \leq 1$ and hence we would obtain $k = 0$, $L = 0$, and $G = 0$ which would lead to a contradiction. Thus, we conclude that $\mu > 0$. Dividing by μ, we may assume that $\mu = 1$. Then, we get $k \leq f(x) - L(x) \leq k + 2\theta$ when $\|x\| \leq 1$. As $f(0) = 0 = L(0)$, this shows that $0 \geq k \geq -2\theta$, and hence

$$|f(x) - L(x)| \leq 2\theta \qquad\qquad (c)$$

when $\|x\| \leq 1$.

For any $x' \neq 0$ in E, put $x = x'/\|x'\|$, then $\|x\| = 1$ holds true. Then, by (a) and (c), we have

$$
\begin{aligned}
|f(x') - L(x')| &\leq \left| f(\|x'\|x) - \|x'\|f(x) \right| + \|x'\| \, |f(x) - L(x)| \\
&\leq \theta\|x'\| + 2\theta\|x'\|
\end{aligned}
$$

for all $x' \in E$. $\square$

P. Šemrl simplified the functional inequalities appearing in the last theorem and proved the stability result. First, we introduce a lemma provided by Šemrl [326].

Lemma 6.8. *Given normed spaces E_1 and E_2, let $f : E_1 \to E_2$ be a continuous function satisfying the functional inequality*

$$\|f(x_1 + \cdots + x_n) - f(x_1) - \cdots - f(x_n)\| \leq \theta(\|x_1\| + \cdots + \|x_n\|) \qquad (6.7)$$

for some $\theta > 0$ and for all $n \in \mathbb{N}$, $x_1, \ldots, x_n \in E_1$. Then

$$\|f(tx) - tf(x)\| \leq \begin{cases} 2\theta\|tx\| & (\text{for } t \geq 0), \\ 4\theta\|tx\| & (\text{for } t < 0) \end{cases} \qquad (6.8)$$

for all $t \in \mathbb{R}$ and all $x \in E_1$.

Proof. Putting each $x_i = (1/m)x$ in (6.7) yields

$$\|f((n/m)x) - nf((1/m)x)\| \le \theta\|(n/m)x\|,$$

where n and m are any positive integers. It follows that

$$\begin{aligned}
&\|f((n/m)x) - (n/m)f(x)\| \\
&\le \|f((n/m)x) - nf((1/m)x)\| + (n/m)\|mf((1/m)x) - f(x)\| \\
&\le \theta\|(n/m)x\| + (n/m)\theta\|x\|
\end{aligned}$$

and hence

$$\|f((n/m)x) - (n/m)f(x)\| \le 2\theta\|(n/m)x\| \tag{a}$$

for all $n, m \in \mathbb{N}$ and all $x \in E_1$. By (6.7), we get $f(0) = 0$.

Now, put $x_1 = -(n/m)x$, $x_2 = (n/m)x$, and $x_3 = \cdots = x_n = 0$ in (6.7) to obtain

$$\|f(-(n/m)x) + f((n/m)x)\| \le 2\theta\|(n/m)x\|,$$

while, by (a), we have

$$\|-f((n/m)x) + (n/m)f(x)\| \le 2\theta\|(n/m)x\|.$$

Combining the last two inequalities yields

$$\|f(-(n/m)x) - (-n/m)f(x)\| \le 4\theta\|(n/m)x\|.$$

Thus, we have shown that

$$\|f(tx) - tf(x)\| \le \begin{cases} 2\theta\|tx\| & (\text{for } t \ge 0), \\ 4\theta\|tx\| & (\text{for } t < 0) \end{cases}$$

for all $t \in \mathbb{Q}$ and for any $x \in E_1$. Since f is continuous, (6.8) is satisfied for all $t \in \mathbb{R}$ and $x \in E_1$. $\qquad\square$

If E_1 is a real Banach space and if a continuous function $f : E_1 \to \mathbb{R}$ satisfies the system (6.7), then

$$\begin{aligned}
&\|f(c_1x_1 + \cdots + c_nx_n) - c_1f(x_1) - \cdots - c_nf(x_n)\| \\
&\le \|f(c_1x_1 + \cdots + c_nx_n) - f(c_1x_1) - \cdots - f(c_nx_n)\| \\
&\quad + \|f(c_1x_1) - c_1f(x_1)\| + \cdots + \|f(c_nx_n) - c_nf(x_n)\| \\
&\le \theta(\|c_1x_1\| + \cdots + \|c_nx_n\|) + 4\theta\|c_1x_1\| + \cdots + 4\theta\|c_nx_n\| \\
&= 5\theta(\|c_1x_1\| + \cdots + \|c_nx_n\|)
\end{aligned}$$

for all $n \in \mathbb{N}$, all $c_1, \ldots, c_n \in \mathbb{R}$, and for all $x_1, \ldots, x_n \in E_1$. Theorem 6.7, together with this inequality, implies that there exists a linear functional $L : E_1 \to \mathbb{R}$ such that $|f(x) - L(x)| \le 15\theta \|x\|$ for any $x \in E_1$.

In the special case where $E_1 = E_2 = \mathbb{R}$, Šemrl [326] proved the following theorem.

Theorem 6.9 (Šemrl). *If a continuous function $f : \mathbb{R} \to \mathbb{R}$ satisfies the inequality (6.7) for all $n \in \mathbb{N}$ and all $x_1, \ldots, x_n \in \mathbb{R}$, then there exists a linear function $L : \mathbb{R} \to \mathbb{R}$ such that*

$$|f(x) - L(x)| \le \theta|x| \tag{6.9}$$

for all $x \in \mathbb{R}$.

Proof. Applying Lemma 6.8 to the present case, we have

$$|f(ts) - tf(s)| \le \begin{cases} 2\theta|ts| & \text{(for } t \ge 0), \\ 4\theta|ts| & \text{(for } t < 0) \end{cases}$$

for all $s, t \in \mathbb{R}$. Thus, for any two nonzero real numbers s and t, we find that

$$|f(t)/t - f(s)/s| = (1/|t|)|f((t/s)s) - (t/s)f(s)|$$
$$\le \begin{cases} 2\theta & \text{(for } ts \ge 0), \\ 4\theta & \text{(for } ts < 0). \end{cases}$$

Let us define

$$a = \sup\{f(t)/t \mid t \ne 0\} \quad \text{and} \quad b = \inf\{f(t)/t \mid t \ne 0\}.$$

First, consider the case that

$$a = \sup\{f(t)/t \mid t > 0\} \quad \text{and} \quad b = \inf\{f(t)/t \mid t > 0\}.$$

By the last inequality, we have $a - b \le 2\theta$, and by putting

$$L(x) = \frac{a+b}{2}x,$$

we obtain the inequality (6.9) for all $x \in \mathbb{R}$. The same argument can also be applied if $a = \sup\{f(t)/t \mid t < 0\}$ and $b = \inf\{f(t)/t \mid t < 0\}$.

It remains to consider the case that

$$a = \sup\{f(t)/t \mid t > 0\} \quad \text{and} \quad b = \inf\{f(t)/t \mid t < 0\}.$$

(We omit the proof of the case

$$a = \sup\{f(t)/t \mid t < 0\} \quad \text{and} \quad b = \inf\{f(t)/t \mid t > 0\}$$

because it goes through in the same way.) As mentioned before, we have to show that $a - b \le 2\theta$. Replacing $f(t)$ with $f(t) - (1/2)(a + b)t$ we may assume that $a = -b$. Now, we must prove that $a \le \theta$.

Assume on the contrary that $a > \theta$. Since f is continuous, there exist real numbers r_1, r_2 $(0 < r_1 < r_2)$ such that

$$f(t) > \theta t \qquad (a)$$

for all $t \in [r_1, r_2)$. We claim that there exists $c_1 > 0$ such that for every $t > c_1$ we can find a positive integer m and a real number $u \in [r_1, r_2)$ satisfying $t = mu$. Put $n = 1 + [r_1/(r_2 - r_1)]$, where the square brackets denote the integer part of $r_1/(r_2 - r_1)$. Then $c_1 = nr_1$ has the above property. For, if $t > nr_1$, put $m = [t/r_1]$. Then $m \ge n$ and $mr_1 \le t < (m + 1)r_1$, so $t \in [mr_1, (m + 1)r_1)$. Also, since $m \ge n$, we have $m > r_1/(r_2 - r_1)$ which implies that $(m + 1)r_1 < mr_2$, so that $t \in [mr_1, mr_2)$, i.e., there exists a $u \in [r_1, r_2)$ with $t = mu$. Similarly, there exist negative numbers p_1, p_2, c_2 with $p_1 < p_2$ such that $t \in (p_1, p_2]$ implies

$$f(t) > -\theta t, \qquad (b)$$

while for each $t < c_2$ there exist a positive integer k and a real number $w \in (p_1, p_2]$ with $t = kw$.

Now, choose $t > \max\{c_1, |c_2|\}$. Then, we may take $t = mu = -kw$ with $u \in [r_1, r_2)$ and $w \in (p_1, p_2]$. Since $f(w) > -\theta w$, $f(w) > 0$. By (a) and (b), we get

$$
\begin{aligned}
|f(0) - mf(u) - kf(w)| &= |mf(u) + kf(w)| \\
&= mf(u) + kf(w) \\
&> \theta(mu - kw) \\
&= \theta(m|u| + k|w|),
\end{aligned}
$$

since $u \in [r_1, r_2)$ and $w \in (p_1, p_2]$. On the other hand, by (6.7), we obtain

$$
\begin{aligned}
|f(0) - mf(u) - kf(w)| &= |f(mu + kw) - mf(u) - kf(w)| \\
&\le \theta(m|u| + k|w|),
\end{aligned}
$$

since $mu = -kw$ and m, k are positive integers. This contradiction ends the proof of the theorem. $\qquad\qquad\square$

Chapter 7
Jensen's Functional Equation

There are a number of variations of the additive Cauchy functional equation, for example, generalized additive Cauchy equations appearing in Chapter 3, Hosszú's equation, homogeneous equation, linear functional equation, etc. However, *Jensen's functional equation* is the simplest and the most important one among them. The Hyers–Ulam–Rassias stability problems of Jensen's equation are proved in Section 7.1, and the Hyers–Ulam stability problems of that equation on restricted domains will be discussed in Section 7.2. Moreover, the stability result on a restricted domain will be applied to the study of an asymptotic property of additive functions. In Section 7.3, another approach to prove the stability will be introduced. This approach is called the fixed point method. The superstability and Ger type stability of the Lobačevskiĭ functional equation will be surveyed in the last section.

7.1 Hyers–Ulam–Rassias Stability

The simplest and most elegant variation of the additive Cauchy equation is *Jensen's functional equation* which may be expressed in the form

$$2f\left(\frac{x+y}{2}\right) = f(x) + f(y).$$

Every solution of Jensen's functional equation is called a *Jensen function*. It is well-known that a function f between real vector spaces with $f(0) = 0$ is a Jensen function if and only if it is an additive function (see [258, 278]). We may refer to the paper [129] of H. Haruki and Th. M. Rassias for the entire solutions of a generalized Jensen's functional equation.

Using the ideas from Theorems 2.3 and 2.5, S.-M. Jung [163] proved the Hyers–Ulam–Rassias stability of Jensen's functional equation.

Theorem 7.1 (Jung). *Let E_1 and E_2 be a real normed space and a real Banach space, respectively. Assume that δ, $\theta \geq 0$ are fixed, and let $p > 0$ be given with $p \neq 1$. Suppose a function $f : E_1 \to E_2$ satisfies the functional inequality*

$$\left\| 2f\left(\frac{x+y}{2}\right) - f(x) - f(y) \right\| \leq \delta + \theta\left(\|x\|^p + \|y\|^p\right) \qquad (7.1)$$

for all $x, y \in E_1$. Furthermore, assume $f(0) = 0$ and $\delta = 0$ in (7.1) for the case of $p > 1$. Then there exists a unique additive function $A : E_1 \to E_2$ such that

$$\|f(x) - A(x)\| \leq \begin{cases} \delta + \|f(0)\| + (2^{1-p} - 1)^{-1}\theta\|x\|^p & (for\ p < 1), \\ 2^{p-1}(2^{p-1} - 1)^{-1}\theta\|x\|^p & (for\ p > 1) \end{cases} \qquad (7.2)$$

for all $x \in E_1$.

Proof. If we put $y = 0$ in (7.1), then we have

$$\|2f(x/2) - f(x)\| \leq \delta + \|f(0)\| + \theta\|x\|^p \qquad (a)$$

for all $x \in E_1$. By induction on n, we prove

$$\|2^{-n}f(2^n x) - f(x)\| \leq (\delta + \|f(0)\|)\sum_{k=1}^{n} 2^{-k} + \theta\|x\|^p \sum_{k=1}^{n} 2^{-(1-p)k} \qquad (b)$$

for the case $0 < p < 1$. By substituting $2x$ for x in (a) and dividing the resulting inequality by 2, we see the validity of (b) for $n = 1$. Assume now that the inequality (b) holds true for some $n \in \mathbb{N}$. If we replace x in (a) with $2^{n+1}x$ and divide by 2 the resulting inequality, then it follows from (b) that

$$\|2^{-(n+1)}f(2^{n+1}x) - f(x)\|$$
$$\leq 2^{-n}\|2^{-1}f(2^{n+1}x) - f(2^n x)\| + \|2^{-n}f(2^n x) - f(x)\|$$
$$\leq (\delta + \|f(0)\|)\sum_{k=1}^{n+1} 2^{-k} + \theta\|x\|^p \sum_{k=1}^{n+1} 2^{-(1-p)k}.$$

This ends the proof of the inequality (b).

Let us define

$$A(x) = \lim_{n\to\infty} 2^{-n}f(2^n x) \qquad (c)$$

for all $x \in E_1$. The function A is well defined because E_2 is a Banach space and the sequence $\{2^{-n}f(2^n x)\}$ IS a Cauchy sequence for all $x \in E_1$: For $n > m$ we use (b) to obtain

$$\|2^{-n}f(2^n x) - 2^{-m}f(2^m x)\|$$
$$= 2^{-m}\|2^{-(n-m)}f(2^{n-m} \cdot 2^m x) - f(2^m x)\|$$
$$\leq 2^{-m}\left(\delta + \|f(0)\| + \frac{2^{mp}}{2^{1-p} - 1}\theta\|x\|^p\right)$$
$$\to 0 \quad \text{as } m \to \infty.$$

Let $x, y \in E_1$ be arbitrary. It then follows from (c) and (7.1) that

$$
\begin{aligned}
\|A(x+y) &- A(x) - A(y)\| \\
&= \lim_{n\to\infty} 2^{-(n+1)} \left\| 2f\left(\frac{2^{n+1}(x+y)}{2}\right) - f\left(2^{n+1}x\right) - f\left(2^{n+1}y\right) \right\| \\
&\leq \lim_{n\to\infty} 2^{-(n+1)} \left(\delta + \theta 2^{(n+1)p}\left(\|x\|^p + \|y\|^p\right)\right) \\
&= 0.
\end{aligned}
$$

Hence, A is an additive function, and the inequality (b) and the definition (c) imply the validity of the first inequality in (7.2).

Now, let $A' : E_1 \to E_2$ be another additive function which satisfies the first inequality in (7.2). It then follows

$$
\begin{aligned}
\|A(x) - A'(x)\| &= 2^{-n}\|A(2^n x) - A'(2^n x)\| \\
&\leq 2^{-n}\left(\|A(2^n x) - f(2^n x)\| + \|f(2^n x) - A'(2^n x)\|\right) \\
&\leq 2^{-n}\left(2\delta + 2\|f(0)\| + \frac{2\theta}{2^{1-p}-1}2^{np}\|x\|^p\right)
\end{aligned}
$$

for all $x \in E_1$ and for any $n \in \mathbb{N}$. Since the right-hand side of the last inequality tends to zero as $n \to \infty$, we conclude that $A(x) = A'(x)$ for all $x \in E_1$, which proves the uniqueness of A.

For the case when $p > 1$ and $\delta = 0$ in the functional inequality (7.1) we can analogously prove the inequality

$$
\|2^n f(2^{-n}x) - f(x)\| \leq \theta\|x\|^p \sum_{k=0}^{n-1} 2^{-(p-1)k}
$$

instead of (b). The remainder of the proof for this case continues in an analogous way. $\square$

As mentioned in [163], the proof of the Hyers–Ulam–Rassias stability of Jensen's functional equation for the case of $p = 0$ can be achieved similarly as in that of Theorem 7.1.

Corollary 7.2. *Let E_1 and E_2 be a real normed space and a real Banach space, respectively. Assume $\delta \geq 0$ is fixed. Suppose a function $f : E_1 \to E_2$ satisfies the inequality (7.1) with $\theta = 0$ for all $x, y \in E_1$. Then there exists a unique additive function $A : E_1 \to E_2$ satisfying the first inequality in (7.2) with $\theta = 0$.*

Let $p \in [0, 1)$ be given. By substituting $x + y$ for x and putting $y = 0$ in (7.1) we get

$$
\left\| 2f\left(\frac{x+y}{2}\right) - f(x+y) \right\| \leq \delta + \|f(0)\| + \theta\left(\|x\|^p + \|y\|^p\right).
$$

This inequality, together with (7.1), yields

$$\|f(x + y) - f(x) - f(y)\| \le 2\delta + \|f(0)\| + 2\theta(\|x\|^p + \|y\|^p)$$

for all $x, y \in E_1$. According to Theorems 2.3 and 2.5, there exists a unique additive function $A : E_1 \to E_2$ such that

$$\|f(x) - A(x)\| \le 2\delta + \|f(0)\| + \frac{2\theta}{1 - 2^{p-1}}\|x\|^p,$$

for any $x \in E_1$, which is by no means attractive in comparison with the first inequality in (7.2).

We also remark that the ideas from the proof of Theorem 7.1 cannot be applied to the proof of the stability of (7.1) for the case $p < 0$. An essential process in the proof of Theorem 7.1 was to put $y = 0$ in the inequality (7.1) which is impossible in the case $p < 0$. The Hyers–Ulam–Rassias stability problem for the case of $p < 0$ still remains as an open problem.

As discussed at the end of Theorem 2.6, Th. M. Rassias and P. Šemrl have constructed a continuous real-valued function in their paper [310] to prove that the functional inequality

$$\|f(x + y) - f(x) - f(y)\| \le \theta(\|x\| + \|y\|)$$

is not stable in the sense of Hyers, Ulam, and Rassias. By using this result, S.-M. Jung [163] proved that the function constructed by Rassias and Šemrl serves as a counterexample to Theorem 7.1 for the case $p = 1$ as follows:

Theorem 7.3. *The continuous real-valued function defined by*

$$f(x) = \begin{cases} x \log_2(x + 1) & \text{(for } x \ge 0), \\ x \log_2 |x - 1| & \text{(for } x < 0) \end{cases}$$

satisfies the inequality

$$\left|2f\left(\frac{x + y}{2}\right) - f(x) - f(y)\right| \le 2(|x| + |y|), \tag{7.3}$$

for all $x, y \in \mathbb{R}$, and the image set of $|f(x) - A(x)|/|x|$ for $x \ne 0$ is unbounded for each additive function $A : \mathbb{R} \to \mathbb{R}$.

Proof. The given function f is continuous, odd, and convex on $(0, \infty)$. Let x and y be positive numbers. Since f is convex on $(0, \infty)$, it follows from the fact

$$|f(x + y) - f(x) - f(y)| \le f(x + y) - 2f\left(\frac{x + y}{2}\right) \tag{a}$$

that

$$|f(x+y) - f(x) - f(y)| \leq (x+y) \log_2 \frac{2+2x+2y}{2+x+y} < |x| + |y| \quad (b)$$

for all $x, y > 0$. Since f is an odd function, (b) holds true for $x, y < 0$ as well. Since (b) holds true for $x = 0$, $y = 0$, or $x + y = 0$, it only remains to consider the case when $x > 0$ and $y < 0$. Without loss of generality, assume $|x| > |y|$. By oddness and convexity of f and by (a), we get

$$
\begin{aligned}
|f(x+y) - f(x) - f(y)| &= |f(x) - f(x+y) - f(-y)| \\
&\leq f(x) - 2f(x/2) \\
&= x \log_2 \frac{2x+2}{x+2} \\
&< |x| + |y|,
\end{aligned}
$$

since $x + y$ and $-y$ are positive numbers. Thus, the inequality (b) holds true for all $x, y \in \mathbb{R}$.

By substituting $x/2$ and $y/2$ for x and y in (b), respectively, and multiplying by 2 both sides, we have

$$\left| 2f\left(\frac{x+y}{2}\right) - 2f(x/2) - 2f(y/2) \right| \leq |x| + |y| \quad (c)$$

for any $x, y \in \mathbb{R}$. Putting $x = y$ and dividing by 2 both sides in (c) yield

$$|f(x) - 2f(x/2)| \leq |x| \quad (d)$$

for $x \in \mathbb{R}$. By using (c) we obtain

$$
\begin{aligned}
&\left| 2f\left(\frac{x+y}{2}\right) - 2f(x/2) - 2f(y/2) \right| \\
&= \left| 2f\left(\frac{x+y}{2}\right) - f(x) - f(y) + f(x) - 2f(x/2) + f(y) - 2f(y/2) \right| \\
&\leq |x| + |y|
\end{aligned}
$$

for $x, y \in \mathbb{R}$. The validity of (7.3) follows immediately from (d) and the last inequality. It is well-known that if an additive function $A : \mathbb{R} \rightarrow \mathbb{R}$ is continuous at a point, then $A(x) = cx$ where c is a real number (see Theorem 2.1). It is trivial that $|f(x) - cx|/|x| \rightarrow \infty$ as $x \rightarrow \infty$ for any real number c, and that the image set of $|f(x) - A(x)|/|x|$ for $x \neq 0$ is also unbounded for every additive function $A : \mathbb{R} \rightarrow \mathbb{R}$ which is not continuous because the graph of the function A is everywhere dense in $\mathbb{R}^2$ (see Theorem 2.2). $\qquad \square$

It should be remarked that K.-W. Jun, D.-S. Shin, and B.-D. Kim proved the Hyers–Ulam–Rassias stability of Jensen's functional equation of Pexider type (see [155, Corollary 2.4]).

7.2 Stability on a Restricted Domain

In the following lemma, Z. Kominek [224] proved the stability of Jensen's functional equation on a restricted domain. This lemma is a version for Jensen's equation of Lemma 2.29.

Lemma 7.4. *Let E be a real Banach space and let N be a given positive integer. Given $c > 0$, let $f : (-c, c)^N \to E$ be a function satisfying*

$$\left\| 2f\left(\frac{x+y}{2}\right) - f(x) - f(y) \right\| \leq \delta$$

for some $\delta \geq 0$ and for all $x, y \in (-c, c)^N$ with $(1/2)(x + y) \in (-c, c)^N$. Then there exists a Jensen function $J : \mathbb{R}^N \to E$ such that

$$\| f(x) - J(x) \| \leq (25N - 4)\delta$$

for any $x \in (-c, c)^N$.

Proof. If we define a function $f_1 : (-c, c)^N \to E$ by $f_1(x) = f(x) - f(0)$, then f_1 satisfies the inequality

$$\left\| 2f_1\left(\frac{x+y}{2}\right) - f_1(x) - f_1(y) \right\| \leq \delta \qquad (a)$$

for all $x, y \in (-c, c)^N$ with $(1/2)(x + y) \in (-c, c)^N$. Put

$$A_n = \left(-2^{-n+1}c, \, 2^{-n+1}c \right)^N \setminus (-2^{-n}c, \, 2^{-n}c)^N$$

for each $n \in \mathbb{N}$. We define a function $g : (-c, c)^N \to E$ by

$$g(x) = 2^{-n+1} f_1\left(2^{n-1}x\right)$$

for all $x \in A_n$ and any $n \in \mathbb{N}$.

Since $f_1(0) = 0$, putting $y = 0$ in (a) yields

$$\| 2f_1(x/2) - f_1(x) \| \leq \delta.$$

Replacing x with $x/2$ in the last inequality and multiplying the resulting inequality by 2 yield

$$\left\| 2^2 f_1(2^{-2}x) - 2f_1(x/2) \right\| \leq 2\delta.$$

Similarly, we obtain

$$\left\| 2^k f_1(2^{-k}x) - 2^{k-1} f_1(2^{-k+1}x) \right\| \leq 2^{k-1}\delta \qquad (b)$$

for all $x \in (-c, c)^N$ and any $k \in \mathbb{N}$. Hence, using the triangle inequality and summing the inequalities in (b) corresponding to $k \in \{1, \ldots, n-1\}$, we have

$$\left\| 2^{n-1} f_1(2^{-n+1}x) - f_1(x) \right\| \leq (2^{n-1} - 1)\delta.$$

Replacing x with $2^{n-1}x$ and dividing the resulting inequality by 2^{n-1} yield

$$\| f_1(x) - g(x) \| \leq \delta \qquad (c)$$

for all $x \in A_n$. Moreover, we have $g(x) = 2g(x/2)$ for any $x \in (-c, c)^N$. This fact, together with (a) and (c), implies that

$$
\begin{aligned}
\| g(x & + y) - g(x) - g(y) \| \\
&= \left\| 2g\left(\frac{x+y}{2}\right) - g(x) - g(y) \right\| \\
&\leq 2\left\| g\left(\frac{x+y}{2}\right) - f_1\left(\frac{x+y}{2}\right) \right\| + \| f_1(x) - g(x) \| \\
&\quad + \| f_1(y) - g(y) \| + \left\| 2f_1\left(\frac{x+y}{2}\right) - f_1(x) - f_1(y) \right\| \\
&\leq 5\delta
\end{aligned}
$$

for all $x, y \in (-c, c)^N$ such that $(1/2)(x+y) \in (-c, c)^N$.

Due to Lemma 2.29, there exists an additive function $A : \mathbb{R}^N \to E$ such that

$$\| g(x) - A(x) \| \leq (5N - 1)5\delta \qquad (d)$$

for $x \in (-c, c)^N$. Define a function $J : \mathbb{R}^N \to E$ by $J(x) = A(x) + f(0)$. Then, J is a Jensen function. By (c) and (d), we have

$$
\begin{aligned}
\| f(x) - J(x) \| &= \| f_1(x) - A(x) \| \\
&\leq \| f_1(x) - g(x) \| + \| g(x) - A(x) \| \\
&\leq (25N - 4)\delta
\end{aligned}
$$

for each $x \in (-c, c)^N$. $\qquad \square$

Using the last lemma, Kominek [224] proved a more generalized result on the Hyers–Ulam stability of Jensen's equation on a restricted domain.

Theorem 7.5 (Kominek). *Let E be a real Banach space and let N be a given positive integer. Let D_1 be a bounded subset of $\mathbb{R}^N$. Assume that there exists an x_0 in the interior of D_1 such that the set $D = D_1 - x_0$ satisfies the following conditions:*

(i) $(1/2)D \subset D$,
(ii) $(-c, c)^N \subset D$ for some $c > 0$,
(iii) $D \subset (-2^n c, 2^n c)^N$ for some nonnegative integer n.

If a function $f : D_1 \to E$ satisfies the functional inequality

$$\left\| 2f\left(\frac{x+y}{2}\right) - f(x) - f(y) \right\| \leq \delta$$

for some $\delta \geq 0$ and for all $x, y \in D_1$ with $(1/2)(x + y) \in D_1$, then there exists a Jensen function $J : \mathbb{R}^N \to E$ such that

$$\| f(x) - J(x) \| \leq \big(2^n(25N - 3) - 1\big)\delta$$

for each $x \in D_1$.

Proof. If we define a function $f_0 : D \to E$ by $f_0(x) = f(x + x_0)$ for all $x \in D$, then f_0 satisfies the inequality

$$\left\| 2f_0\left(\frac{x+y}{2}\right) - f_0(x) - f_0(y) \right\| \leq \delta$$

for all $x, y \in D$ with $(1/2)(x + y) \in D$.

Similarly, as in the proof of Lemma 7.4, we define the functions f_1 and g, namely,

$$f_1(x) = f_0(x) - f_0(0)$$

for $x \in D$ and

$$g(x) = 2^{-k+1} f_1\big(2^{k-1}x\big)$$

for $x \in A_k$ ($k \in \mathbb{N}$), where

$$A_k = \big(-2^{-k+1}c,\ 2^{-k+1}c\big)^N \backslash \big(-2^{-k}c,\ 2^{-k}c\big)^N.$$

We note that

$$\| f_1(x) - 2^n f_1(2^{-n}x) \| \leq (2^n - 1)\delta \tag{a}$$

for each $x \in D$, and

$$\| f_1(x) - g(x) \| \leq \delta \tag{b}$$

for all $x \in (-c, c)^N$. Let $A : \mathbb{R}^N \to E$ be an additive function such that

$$\| g(x) - A(x) \| \leq (5N - 1)5\delta \tag{c}$$

for $x \in (-c, c)^N$ (see (d) in the proof of Lemma 7.4). Taking any $x \in D$, by (a), (b), and (c), we obtain

$$\begin{aligned}
\|f_1(x) - A(x)\| &\leq \|f_1(x) - 2^n f_1(2^{-n}x)\| + 2^n \|f_1(2^{-n}x) - A(2^{-n}x)\| \\
&\leq \|f_1(x) - 2^n f_1(2^{-n}x)\| + 2^n \|f_1(2^{-n}x) - g(2^{-n}x)\| \\
&\quad + 2^n \|g(2^{-n}x) - A(2^{-n}x)\| \\
&\leq \left(2^n(25N - 3) - 1\right)\delta.
\end{aligned}$$

Now, we put

$$J(x) = A(x - x_0) + f_0(0)$$

for $x \in \mathbb{R}^N$. Then, J is a Jensen function. By the last inequality, we get

$$\begin{aligned}
\|f(x) - J(x)\| &= \|f_0(x - x_0) - A(x - x_0) - f_0(0)\| \\
&= \|f_1(x - x_0) - A(x - x_0)\| \\
&\leq \left(2^n(25N - 3) - 1\right)\delta
\end{aligned}$$

for each $x \in D_1$. $\qquad\square$

Let D be an open and convex subset of $\mathbb{R}^N$. A function $f : D \to \mathbb{R}$ is said to be J-convex (convex in the sense of Jensen) if the inequality

$$2f\left(\frac{x + y}{2}\right) \leq f(x) + f(y)$$

holds true for all $x, y \in D$. If the inequality sign "$\leq$" is replaced with "$\geq$" in the above inequality, f is said to be a J-concave function.

We say that a subset T of $\mathbb{R}^N$ *belongs to the class* $\mathcal{A}$ if and only if every J-convex function defined on a convex open domain $D \supset T$ bounded above on T is continuous on D.

The following theorem was presented by Kominek [224].

Theorem 7.6. *Let D be an open convex subset of $\mathbb{R}^N$ and let $T \subset D$ be a fixed set belonging to the class $\mathcal{A}$. If $f : D \to \mathbb{R}$ is a J-convex function and $g : D \to \mathbb{R}$ is a J-concave function and, moreover,*

$$f(x) \leq g(x)$$

for all $x \in T$, then there exist an additive function $A : \mathbb{R}^N \to \mathbb{R}$, a convex function $F : D \to \mathbb{R}$, and a concave function $G : D \to \mathbb{R}$ such that

$$\begin{aligned}
f(x) &= A(x) + F(x), \\
g(x) &= A(x) + G(x)
\end{aligned}$$

for every $x \in D$.

Proof. Put

$$\varphi(x) = f(x) - g(x)$$

for all $x \in D$. We note that φ is a J-convex function bounded above on T. Thus, φ is continuous on D. Let D_1 be an open convex and bounded subset of D for which there exists a constant $M > 0$ such that

$$|\varphi(x)| \leq M \qquad\qquad (a)$$

for any $x \in D_1$. From the definition of φ, J-concavity of g, J-convexity of f, and (a) it follows that

$$
\begin{aligned}
0 \leq{} & 2g\left(\frac{x+y}{2}\right) - g(x) - g(y) \\
={} & 2f\left(\frac{x+y}{2}\right) - f(x) - f(y) - \left(2\varphi\left(\frac{x+y}{2}\right) - \varphi(x) - \varphi(y)\right) \\
\leq{} & 4M
\end{aligned}
$$

for all $x, y \in D_1$. In particular,

$$\left|2g\left(\frac{x+y}{2}\right) - g(x) - g(y)\right| \leq 4M$$

for all $x, y \in D_1$.

On account of Theorem 7.5 there exist a Jensen function $J : \mathbb{R}^N \to \mathbb{R}$ and a nonnegative integer n such that

$$|g(x) - J(x)| \leq \left(2^n(25N - 3) - 1\right)4M \qquad\qquad (b)$$

for each $x \in D_1$.

Now, we define functions A, F, and G by

$$
\begin{aligned}
A(x) &= J(x) - J(0) \quad (x \in \mathbb{R}^N), \\
G(x) &= g(x) - A(x) \quad (x \in D), \\
F(x) &= \varphi(x) + G(x) \quad (x \in D).
\end{aligned}
$$

Then, A is an additive function. On account of (b), the function G is J-concave bounded below on D_1, and hence it is concave on D. The function F is convex since it is continuous and J-convex. Moreover, it is easily indicated that $f(x) = A(x) + F(x)$ and $g(x) = A(x) + G(x)$ for any $x \in D$. $\qquad\square$

S.-M. Jung [163] proved the stability of Jensen's functional equation on a restricted and unbounded domain, and applied the result to the study of an asymptotic behavior of additive functions.

Theorem 7.7 (Jung). *Let E_1 and E_2 be a real normed space and a real Banach space, respectively. Assume that $d > 0$ and $\delta \geq 0$ are given. If a function f :*

$E_1 \to E_2$ *satisfies the functional inequality*

$$\left\| 2f\left(\frac{x+y}{2}\right) - f(x) - f(y) \right\| \leq \delta \tag{7.4}$$

for all $x, y \in E_1$ *with* $\|x\| + \|y\| \geq d$, *then there exists a unique additive function* $A : E_1 \to E_2$ *such that*

$$\|f(x) - A(x)\| \leq 5\delta + \|f(0)\| \tag{7.5}$$

for all $x \in E_1$.

Proof. Suppose $\|x\| + \|y\| < d$. If $x = y = 0$, we can choose a $z \in E_1$ such that $\|z\| = d$. Otherwise, let $z = \left(1 + d/\|x\|\right)x$ for $\|x\| \geq \|y\|$ or $z = \left(1 + d/\|y\|\right)y$ for $\|x\| < \|y\|$. It is then obvious that

$$\|x - z\| + \|y + z\| \geq d,$$
$$\|2z\| + \|x - z\| \geq d,$$
$$\|y\| + \|2z\| \geq d, \tag{a}$$
$$\|y + z\| + \|z\| \geq d,$$
$$\|x\| + \|z\| \geq d.$$

From (7.4), (a), and the relation

$$2f\left(\frac{x+y}{2}\right) - f(x) - f(y)$$
$$= 2f\left(\frac{x+y}{2}\right) - f(x-z) - f(y+z)$$
$$- \left(2f\left(\frac{x+z}{2}\right) - f(2z) - f(x-z)\right) + 2f\left(\frac{y+2z}{2}\right) - f(y) - f(2z)$$
$$- \left(2f\left(\frac{y+2z}{2}\right) - f(y+z) - f(z)\right) + 2f\left(\frac{x+z}{2}\right) - f(x) - f(z),$$

we get

$$\left\| 2f\left(\frac{x+y}{2}\right) - f(x) - f(y) \right\| \leq 5\delta. \tag{b}$$

In view of (7.4) and (b), the function f satisfies the inequality (b) for all $x, y \in E_1$. Therefore, it follows from (b) and Theorem 7.1 that there exists a unique additive function $A : E_1 \to E_2$ satisfying the inequality (7.5) for all $x \in E_1$. $\qquad \square$

Using the result of Theorem 7.7, Jung [163] proved an asymptotic behavior of additive functions.

Corollary 7.8. *Suppose a function* $f : E_1 \to E_2$ *satisfies the condition* $f(0) = 0$, *where* E_1 *and* E_2 *are a real normed space and a real Banach space, respectively. The function* f *is additive if and only if*

$$\left\| 2f\left(\frac{x+y}{2}\right) - f(x) - f(y) \right\| \to 0 \quad as \;\; \|x\| + \|y\| \to \infty. \tag{7.6}$$

Proof. On account of (7.6), there exists a sequence $\{\delta_n\}$ monotonically decreasing to 0 such that

$$\left\| 2f\left(\frac{x+y}{2}\right) - f(x) - f(y) \right\| \le \delta_n \tag{a}$$

for all $x, y \in E_1$ with $\|x\| + \|y\| \ge n$. It then follows from (a) and Theorem 7.7 that there exists a unique additive function $A_n : E_1 \to E_2$ such that

$$\| f(x) - A_n(x) \| \le 5\delta_n \tag{b}$$

for all $x \in E_1$. Let $\ell, m \in \mathbb{N}$ satisfy $m \ge \ell$. Obviously, it follows from (b) that

$$\| f(x) - A_m(x) \| \le 5\delta_m \le 5\delta_\ell$$

for all $x \in E_1$ since $\{\delta_n\}$ is a monotonically decreasing sequence. The uniqueness of A_n implies $A_m = A_\ell$. Hence, by letting $n \to \infty$ in (b), we conclude that f is additive. The reverse assertion is trivial. □

7.3 Fixed Point Method

Using the fixed point method (Theorem 2.43), L. Cădariu and V. Radu [55] proved the Hyers–Ulam–Rassias stability of the Jensen's functional equation.

Theorem 7.9 (Cădariu and Radu). *Let E_1 and E_2 be a (real or complex) vector space and a Banach space, respectively. Assume that a function $f : E_1 \to E_2$ satisfies $f(0) = 0$ and the inequality*

$$\left\| 2f\left(\frac{x+y}{2}\right) - f(x) - f(y) \right\| \le \varphi(x, y) \tag{7.7}$$

for all $x, y \in E_1$, where $\varphi : E_1^2 \to [0, \infty)$ is a given function. Moreover, assume that there exists a positive constant $L < 1$ such that

$$\varphi(x, 0) \le Lq_i\varphi(x/q_i, 0) \tag{7.8}$$

for any $x \in E_1$, where $q_0 = 2$ and $q_1 = 1/2$. If φ satisfies

$$\lim_{n\to\infty} q_i^{-n}\varphi(q_i^n x, q_i^n y) = 0 \tag{7.9}$$

for all $x, y \in E_1$, then there exists a unique additive function $A : E_1 \to E_2$ such that

$$\| f(x) - A(x) \| \le \frac{L^{1-i}}{1-L}\varphi(x, 0) \tag{7.10}$$

for any $x \in E_1$.

Proof. Let us define

$$X = \{g : E_1 \to E_2 \mid g(0) = 0\}$$

and introduce the generalized metric d on X:

$$d(g,h) = \inf \{C \in [0,\infty] \mid \|g(x) - h(x)\| \le C\varphi(x,0) \text{ for all } x \in E_1\}.$$

Then, (X,d) is complete (ref. [181, Theorem 2.1]).

We now define an operator $\Lambda : X \to X$ by

$$(\Lambda g)(x) = (1/q_i)g(q_i x).$$

For any $g, h \in X$, $d(g,h) \le C$ implies that

$$\|g(x) - h(x)\| \le C\varphi(x,0)$$

or

$$\left\| (1/q_i)g(q_i x) - (1/q_i)h(q_i x) \right\| \le (1/q_i)C\varphi(q_i x, 0)$$

for all $x \in E_1$. It follows from (7.8) that

$$\left\| (1/q_i)g(q_i x) - (1/q_i)h(q_i x) \right\| \le LC\varphi(x,0)$$

for any $x \in E_1$. That is, if $d(g,h) \le C$, then $d(\Lambda g, \Lambda h) \le LC$. Therefore, we conclude that $d(\Lambda g, \Lambda h) \le Ld(g,h)$ for all $g, h \in X$. Indeed, Λ is strictly contractive on X with the Lipschitz constant L.

Assume that $i = 0$. If we set $x = 2t$ and $y = 0$ in (7.7), then it follows from (7.8) that

$$\|f(t) - (1/2)f(2t)\| \le (1/2)\varphi(2t,0) \le L\varphi(t,0)$$

for each $t \in E_1$, i.e., $d(f, \Lambda f) \le L = L^1 < \infty$.

For $i = 1$, we put $y = 0$ in (7.7) to obtain

$$\|2f(x/2) - f(x)\| \le \varphi(x,0)$$

for all $x \in E_1$. Thus, $d(f, \Lambda f) \le 1 = L^0 < \infty$.

In both cases we can apply Theorem 2.43 and conclude that there exists a function $A : E_1 \to E_2$ with $A(0) = 0$ such that

$$A(2x) = 2A(x) \tag{a}$$

for any $x \in E_1$ and A is the unique function satisfying (a) in the set

$$X^* = \{g \in X \mid d(f, g) < \infty\},$$

i.e., there exists a constant $C > 0$ such that

$$\|A(x) - f(x)\| \leq C\varphi(x, 0) \qquad (b)$$

for all $x \in E_1$.

Moreover, according to Theorem 2.43 (i), $d(\Lambda^n f, A) \to 0$ as $n \to \infty$, which implies that

$$A(x) = \lim_{n \to \infty} q_i^{-n} f(q_i^n x) \qquad (c)$$

for each $x \in E_1$. Due to Theorem 2.43 (iii), we have

$$d(f, A) \leq \frac{1}{1 - L} d(f, \Lambda f) \quad \text{or} \quad d(f, A) \leq \frac{L^{1-i}}{1 - L},$$

which implies the validity of (7.10).

If we replace x and y in (7.7) with $2q_i^n x$ and $2q_i^n y$, respectively, then we get

$$\left\| q_i^{-n} f\left(q_i^n(x + y)\right) - (1/2)q_i^{-n} f(2q_i^n x) - (1/2)q_i^{-n} f(2q_i^n y) \right\|$$
$$\leq (1/2)q_i^{-n}\varphi(2q_i^n x, 2q_i^n y)$$

for all $x, y \in E_1$. In view of (7.9), if we let $n \to \infty$ in the last inequality, then we obtain

$$A(x + y) = A(x) + A(y)$$

for any $x, y \in E_1$.   □

7.4 Lobačevskiĭ's Functional Equation

In this section, the superstability of Lobačevskiĭ's functional equation

$$f\left(\frac{x + y}{2}\right)^2 = f(x)f(y) \qquad (7.11)$$

will be investigated. P. Găvruta contributed to the following theorem (ref. [114]).

Theorem 7.10. *Let G be a 2-divisible abelian group. If a function $f : G \to \mathbb{C}$ satisfies the inequality*

$$\left| f\left(\frac{x + y}{2}\right)^2 - f(x)f(y) \right| \leq \delta \qquad (7.12)$$

for all $x, y \in G$ *and for some* $\delta > 0$, *then either*

$$|f(x)| \le (1/2)\big(|f(0)| + (|f(0)|^2 + 4\delta)^{1/2}\big)$$

for all $x \in G$, *or* f *is a solution of Lobačevskiǐ's functional equation* (7.11).

Proof. Suppose there exists an $x_0 \in G$ with

$$|f(x_0)| > (1/2)\big(|f(0)| + (|f(0)|^2 + 4\delta)^{1/2}\big). \tag{a}$$

If we replace x and y in (7.12) with $2x$ and 0, respectively, then we have

$$\big|f(x)^2 - f(0)f(2x)\big| \le \delta \tag{b}$$

for all $x \in G$. If we had $f(0) = 0$, then it would follow from (b) that $|f(x)| \le \sqrt{\delta}$ for any $x \in G$. However, by (a), we get $|f(x_0)| > \sqrt{\delta}$, which leads to a contradiction. Hence, we conclude that $f(0) \ne 0$.

In view of (b), we obtain

$$\begin{aligned}
|f(0)||f(2x)| &= \big|f(x)^2 + \big(f(0)f(2x) - f(x)^2\big)\big| \\
&\ge |f(x)|^2 - \big|f(0)f(2x) - f(x)^2\big| \\
&\ge |f(x)|^2 - \delta
\end{aligned}$$

for each $x \in G$. Thus, we have

$$|f(2x)| \ge \frac{|f(x)|^2 - \delta}{|f(0)|} \tag{c}$$

for all $x \in G$. We now set

$$\alpha = (1/2)\big(|f(0)| + (|f(0)|^2 + 4\delta)^{1/2}\big) \quad \text{and} \quad \beta = |f(x_0)| - \alpha. \tag{d}$$

It then follows from (a) and (d) that

$$|f(x_0)| = \alpha + \beta \quad \text{and} \quad \beta > 0.$$

Now, we will prove that

$$|f(2^n x_0)| \ge \alpha + 2^n \beta \tag{e}$$

for all $n \in \mathbb{N}$. Putting $x = x_0$ in (c) yields

$$|f(2x_0)| \geq \frac{(\alpha + \beta)^2 - \delta}{|f(0)|} = \alpha + \frac{2\alpha\beta + \beta^2}{|f(0)|} \geq \alpha + 2\beta$$

since

$$\alpha^2 = |f(0)|\alpha + \delta, \quad \alpha \geq |f(0)|, \quad \text{and} \quad \beta > 0.$$

That is, the inequality (e) is true for $n = 1$. Assume that (e) is true for some integer $n > 0$. If we put $x = 2^n x_0$ in (c), then

$$\left| f\left(2^{n+1} x_0\right) \right| \geq \frac{(\alpha + 2^n \beta)^2 - \delta}{|f(0)|} = \alpha + \frac{2^{n+1}\alpha\beta + 2^{2n}\beta^2}{|f(0)|} \geq \alpha + 2^{n+1}\beta,$$

which ends the proof of (e).

We set $x_n = 2^n x_0$ for any $n \in \mathbb{N}$. Then, by (7.12), we get

$$\left| f(x_n)f(x) - f\left(\frac{x + x_n}{2}\right)^2 \right| \leq \delta$$

for all $x \in G$ and $n \in \mathbb{N}$. Hence, it follows from (e) that

$$f(x) = \lim_{n \to \infty} f(x_n)^{-1} f\left(\frac{x + x_n}{2}\right)^2$$

for all $x \in G$. Hence, we obtain

$$f(x)f(y) = \lim_{n \to \infty} \left(f(x_n)^{-1} f\left(\frac{x + x_n}{2}\right) f\left(\frac{y + x_n}{2}\right) \right)^2,$$
$$f\left(\frac{x + y}{2}\right)^2 = \left(\lim_{n \to \infty} f(x_n)^{-1} f\left(\frac{x + y + 2x_n}{4}\right)^2 \right)^2. \qquad (f)$$

On the other hand, it follows from (7.12) that

$$\left| f(x_n)^{-1} \left(f\left(\frac{x + x_n}{2}\right) f\left(\frac{y + x_n}{2}\right) - f\left(\frac{x + y + 2x_n}{4}\right)^2 \right) \right| \leq \delta/|f(x_n)|$$

for all $x, y \in G$ and $n \in \mathbb{N}$. If we let $n \to \infty$ in the last inequality, then (f) implies that

$$f\left(\frac{x + y}{2}\right)^2 = f(x)f(y)$$

for any $x, y \in G$. $\qquad \square$

We introduce a theorem concerning the Ger type stability of Lobačevskiĭ's functional equation (7.11) presented by S.-M. Jung [170].

Lemma 7.11. *Let $(G, +)$ be a cancellative abelian group which is uniquely 2-divisible, and let $\varepsilon \in (0, 1/8)$. If a function $f : G \to \mathbb{R}$ satisfies the congruence*

$$2f\left(\frac{x+y}{2}\right) - f(x) - f(y) \in (-\varepsilon, \varepsilon) + \mathbb{Z} \tag{7.13}$$

for all $x, y \in G$, then there exists a function $p : G \to \mathbb{R}$ such that

$$p(x + y) - p(x) - p(y) \in \mathbb{Z}$$

for any $x, y \in G$, and

$$|f(x) - p(x) - f(0)| \le 2\varepsilon$$

for all $x \in G$.

Proof. If we put $y = 0$ in (7.13), then we have

$$2f(x/2) - f(x) - f(0) \in (-\varepsilon, \varepsilon) + \mathbb{Z} \tag{a}$$

for every $x \in G$. Let us define a function $g : G \to \mathbb{R}$ by $g(x) = f(x) - f(0)$. It then follows from (7.13) and (a) that

$$
\begin{aligned}
&g(x + y) - g(x) - g(y) \\
&= 2g\left(\frac{x+y}{2}\right) - g(x) - g(y) + g(x + y) - 2g\left(\frac{x+y}{2}\right) \\
&\in (-2\varepsilon, 2\varepsilon) + \mathbb{Z}
\end{aligned}
$$

for any $x, y \in G$.

According to Corollary 2.48 (or [170, Theorem 3]), there exists a function $p : G \to \mathbb{R}$ such that

$$p(x + y) - p(x) - p(y) \in \mathbb{Z}$$

for any $x, y \in G$, and

$$|g(x) - p(x)| \le 2\varepsilon$$

for all $x \in G$, which ends the proof of our theorem. $\qquad \square$

Let $\delta_0 \approx 0.344446...$ be one of the real solutions of the equation

$$x\sqrt{5 - 4x} = 1 - x.$$

More precisely, we define δ_0 by

$$\delta_0 = \frac{1}{3} + \frac{1}{3}\left(\frac{5\sqrt{10}}{4}\right)^{1/3}$$
$$\cdot \left(\cos\frac{\theta + 4\pi}{3} + i\sin\frac{\theta + 4\pi}{3} + \cos\frac{2\pi - \theta}{3} + i\sin\frac{2\pi - \theta}{3}\right),$$

where $\theta \approx 91.812153\ldots^\circ$ satisfying $\sin\theta = (9/10)\sqrt{37/30}$.

Theorem 7.12. *Let E be a real normed space and let $\delta \in [0, \delta_0)$ be a constant. If a function $f : E \to \mathbb{C}\setminus\{0\}$ satisfies the inequality*

$$\left| f\left(\frac{x+y}{2}\right)^2 f(x)^{-1} f(y)^{-1} - 1 \right| \leq \delta \tag{7.14}$$

for all $x, y \in E$, then there exists a unique exponential function $F : E \to \mathbb{C}\setminus\{0\}$ such that

$$\max\{|f(x)f(0)^{-1}F(x)^{-1} - 1|, |f(0)F(x)f(x)^{-1} - 1|\}$$
$$\leq \frac{\delta}{1-\delta}(5 - 4\delta)^{1/2} \tag{7.15}$$

for all $x \in E$.

Proof. It follows from (7.14) that

$$\left| \frac{\left|f\left(\frac{x+y}{2}\right)\right|^2}{|f(x)||f(y)|} \exp\left(i\left(2\arg f\left(\frac{x+y}{2}\right) - \arg f(x) - \arg f(y)\right)\right) - 1 \right| \leq \delta$$

for all $x, y \in E$. Therefore, we obtain

$$1 - \delta \leq \left| f\left(\frac{x+y}{2}\right) \right|^2 |f(x)|^{-1}|f(y)|^{-1} \leq 1 + \delta \tag{a}$$

and

$$2\arg f\left(\frac{x+y}{2}\right) - \arg f(x) - \arg f(y) \in \left[-\sin^{-1}\delta, \ \sin^{-1}\delta\right] + 2\pi\mathbb{Z}$$

for any $x, y \in E$.

We define a function $g : E \to \mathbb{R}$ by $g(x) = (1/2\pi)\arg f(x)$. Since $\delta < 1/\sqrt{2}$ means $(1/2\pi)\sin^{-1}\delta < 1/8$, we can apply Lemma 7.11 to this case. Thus, there exists a function $p : E \to \mathbb{R}$ such that

$$p(x + y) - p(x) - p(y) \in 2\pi\mathbb{Z} \tag{b}$$

and

$$| \arg f(x) - p(x) - \arg f(0)| \leq 2 \sin^{-1} \delta \qquad (c)$$

for any $x, y \in E$.

If we define a function $h : E \to \mathbb{R}$ by $h(x) = \ln |f(x)|$, it then follows from (a) that

$$\left| 2h\left(\frac{x+y}{2}\right) - h(x) - h(y) \right| \leq -\ln(1-\delta)$$

for $x, y \in E$. According to Theorem 7.1, there exists a unique additive function $a : E \to \mathbb{R}$ such that

$$|h(x) - a(x) - h(0)| \leq -\ln(1-\delta) \qquad (d)$$

for each $x \in E$.

We now define a function $F : E \to \mathbb{C} \setminus \{0\}$ by

$$F(x) = \exp\left(a(x) + i p(x)\right). \qquad (e)$$

From the additivity of a, (b), and (e), we can easily show that F is an exponential function.

We observe

$$\left| f(x) f(0)^{-1} F(x)^{-1} - 1 \right|$$
$$= \left| \exp\left(h(x) - a(x) - h(0)\right) \exp\left(i\left(\arg f(x) - p(x) - \arg f(0)\right)\right) - 1 \right|.$$

In view of (c) and (d), the complex number $f(x) f(0)^{-1} F(x)^{-1}$ belongs to the set

$$\Lambda = \{\lambda \in \mathbb{C} \mid 1 - \delta \leq |\lambda| \leq (1-\delta)^{-1} \text{ and } -2 \sin^{-1} \delta \leq \arg \lambda \leq 2 \sin^{-1} \delta\}.$$

From this fact, it is not difficult to see

$$\left| f(x) f(0)^{-1} F(x)^{-1} - 1 \right| \leq \frac{\delta}{1-\delta}(5 - 4\delta)^{1/2}$$

for every $x \in E$. The inequality for $|f(0) F(x) f(x)^{-1} - 1|$ in (7.15) may be proved in a similar way.

The hypothesis for δ means that there is a constant $0 < \beta < 1$ with

$$\frac{\delta}{1-\delta}(5 - 4\delta)^{1/2} < \beta < 1. \qquad (f)$$

Let $F' : E \to \mathbb{C} \setminus \{0\}$ be another exponential function satisfying the inequality (7.15) instead of F. Since

$$F(2^n x) = F(x)^{2^n} \quad \text{and} \quad F'(2^n x) = F'(x)^{2^n}$$

for all $x \in E$ and for any $n \in \mathbb{N}$, it follows from (7.15) and (f) that

$$
\begin{aligned}
\frac{F(x)}{F'(x)} &= \left(\frac{F(2^n x)}{F'(2^n x)} \right)^{2^{-n}} \\
&= \left(\frac{f(0)F(2^n x)}{f(2^n x)} \right)^{2^{-n}} \left(\frac{f(2^n x)}{f(0)F'(2^n x)} \right)^{2^{-n}} \\
&\to 1 \quad \text{as} \quad n \to \infty,
\end{aligned}
$$

implying the uniqueness of F. $\qquad\qquad\qquad\qquad\qquad\qquad\qquad\qquad\qquad\qquad\qquad$ $\square$

The stability problem of Lobačevskiǐ's functional equation on restricted domains and an asymptotic property of the exponential functions were investigated in [170].

Chapter 8
Quadratic Functional Equations

So far, we have discussed the stability problems of functional equations in connection with additive or linear functions. In this chapter, the Hyers–Ulam–Rassias stability of quadratic functional equations will be proved. Most mathematicians may be interested in the study of the quadratic functional equation since the quadratic functions are applied to almost every field of mathematics. In Section 8.1, the Hyers–Ulam–Rassias stability of the quadratic equation is surveyed. The stability problems for that equation on a restricted domain are discussed in Section 8.2, and the Hyers–Ulam–Rassias stability of the quadratic functional equation will be proved by using the fixed point method in Section 8.3. In Section 8.4, the Hyers–Ulam stability of an interesting quadratic functional equation different from the "original" quadratic functional equation is proved. Finally, the stability problem of the quadratic equation of Pexider type is discussed in Section 8.5.

8.1 Hyers–Ulam–Rassias Stability

The quadratic function $f(x) = cx^2$ ($x \in \mathbb{R}$), where c is a real constant, clearly satisfies the equation

$$f(x + y) + f(x - y) = 2f(x) + 2f(y). \tag{8.1}$$

Hence, the equation (8.1) is called the *quadratic functional equation*.

There are a number of functional equations considered as quadratic and one of them will be introduced in Section 8.4. A *quadratic function* implies a solution of the quadratic functional equation (8.1).

A function $f : E_1 \to E_2$ between real vector spaces is a quadratic function if and only if there exists a symmetric biadditive function $B : E_1^2 \to E_2$ such that $f(x) = B(x, x)$. (A function $B : E_1^2 \to E_2$ is called *biadditive* if and only if B is additive in each variable.) If f is a quadratic function, then the biadditive function B is sometimes called the *polar* of f and given by

$$B(x, y) = (1/4)\big(f(x + y) - f(x - y)\big).$$

S.-M. Jung, *Hyers–Ulam–Rassias Stability of Functional Equations in Nonlinear Analysis*, Springer Optimization and Its Applications 48, DOI 10.1007/978-1-4419-9637-4_8, © Springer Science+Business Media, LLC 2011

F. Skof [331] was the first person to prove the Hyers–Ulam stability of the quadratic functional equation (8.1) for functions $f : E_1 \to E_2$ where E_1 and E_2 are a normed space and a Banach space, respectively. P. W. Cholewa [70] demonstrated that Skof's theorem is also valid if E_1 is replaced with an abelian group G.

Theorem 8.1 (Skof). *Let G be an abelian group and let E be a Banach space. If a function $f : G \to E$ satisfies the inequality*

$$\| f(x + y) + f(x - y) - 2f(x) - 2f(y) \| \le \delta$$

for some $\delta \ge 0$ and for all $x, y \in G$, then there exists a unique quadratic function $Q : G \to E$ such that

$$\| f(x) - Q(x) \| \le (1/2)\delta$$

for any $x \in G$.

I. Fenyö [102] improved Theorem 8.1 by replacing the bound $(1/2)\delta$ with the best possible one, $(1/3)(\delta + \| f(0) \|)$. The proof of Theorem 8.1 is a special case of that of Theorem 8.3 by S. Czerwik [87], and hence, it will be omitted. Before starting the theorem of Czerwik, we need a lemma provided by the same author.

Lemma 8.2. *Let E_1 and E_2 be normed spaces. Assume that there exist $\delta, \theta \ge 0$ and $p \in \mathbb{R}$ such that a function $f : E_1 \to E_2$ satisfies the inequality*

$$\| f(x + y) + f(x - y) - 2f(x) - 2f(y) \| \le \delta + \theta(\|x\|^p + \|y\|^p) \qquad (8.2)$$

for all $x, y \in E_1 \setminus \{0\}$. Then for $x \in E_1 \setminus \{0\}$ and $n \in \mathbb{N}$

$$\| f(2^n x) - 4^n f(x) \|$$
$$\le (1/3)(4^n - 1)(\delta + c) + 2 \cdot 4^{n-1} \theta \|x\|^p (1 + a + \cdots + a^{n-1}) \qquad (8.3)$$

and

$$\| f(x) - 4^n f(2^{-n} x) \|$$
$$\le (1/3)(4^n - 1)(\delta + c) + 2^{1-p} \theta \|x\|^p (1 + b + \cdots + b^{n-1}), \qquad (8.4)$$

where $a = 2^{p-2}$, $b = 2^{2-p}$, and $c = \| f(0) \|$.

Proof. Putting $x = y \ne 0$ in (8.2) yields

$$\| f(2x) - 4f(x) \| \le \| f(0) \| + \delta + 2\theta \|x\|^p$$

which proves (8.3) for $n = 1$. Assume now that (8.3) holds true for each $k \le n$ and $x \in E_1 \setminus \{0\}$. Then, for $n + 1$, we have

$$\|f(2^{n+1}x) - 4^{n+1}f(x)\|$$

$$\leq \|f(2 \cdot 2^n x) - 4f(2^n x)\| + 4\|f(2^n x) - 4^n f(x)\|$$

$$\leq \delta + c + 2\theta\|2^n x\|^p + (4/3)(\delta + c)(4^n - 1)$$

$$+ 2 \cdot 4^n \theta\|x\|^p (1 + a + \cdots + a^{n-1})$$

$$= (1/3)(4^{n+1} - 1)(\delta + c) + 2 \cdot 4^n \theta\|x\|^p (1 + a + \cdots + a^n),$$

which proves the validity of the inequality (8.3).

Similarly, taking $x = y = t/2$, we can verify the inequality (8.4) for $n = 1$. Applying the induction principle we get the result for all $n \in \mathbb{N}$, which ends the proof. $\qquad\square$

S. Czerwik [87] has proved the Hyers–Ulam–Rassias stability of the quadratic functional equation (8.1).

Theorem 8.3 (Czerwik). *Let E_1 and E_2 be a normed space and a Banach space, respectively. If a function $f : E_1 \rightarrow E_2$ satisfies the inequality (8.2) for some $\delta, \theta \geq 0$, $p < 2$ and for all $x, y \in E_1 \setminus \{0\}$, then there exists a unique quadratic function $Q : E_1 \rightarrow E_2$ such that*

$$\|f(x) - Q(x)\| \leq (1/3)(\delta + c) + 2(4 - 2^p)^{-1}\theta\|x\|^p \qquad (8.5)$$

for any $x \in E_1 \setminus \{0\}$, where $c = \|f(0)\|$.

Proof. Let us define

$$Q_n(x) = 4^{-n} f(2^n x) \qquad (a)$$

for $x \in E_1$ and $n \in \mathbb{N}$. Then $\{Q_n(x)\}$ is a Cauchy sequence for every $x \in E_1$. Really, for $x = 0$ it is trivial. Let $x \in E_1 \setminus \{0\}$. For $n > m$ we obtain by (8.3)

$$\|Q_n(x) - Q_m(x)\| = 4^{-n}\|f(2^{n-m} \cdot 2^m x) - 4^{n-m} f(2^m x)\|$$

$$\leq 4^{-n} 3^{-1}(4^{n-m} - 1)(\delta + c)$$

$$+ 2 \cdot 4^{-m-1}\theta\|2^m x\|^p (1 + a + \cdots + a^{n-m-1})$$

and hence

$$\|Q_n(x) - Q_m(x)\| \leq 3^{-1}4^{-m}(\delta + c) + 2^{m(p-2)-1}(1 - a)^{-1}\theta\|x\|^p. \qquad (b)$$

The assumption $p < 2$ implies that $\{Q_n(x)\}$ is a Cauchy sequence. Since E_2 is complete, we can define

$$Q(x) = \lim_{n \to \infty} Q_n(x)$$

for any $x \in E_1$.

We will check that Q is a quadratic function. It is clear for $x = y = 0$, since $Q(0) = 0$. For $y = 0$ and $x \neq 0$ we have

$$Q(x + 0) + Q(x - 0) - 2Q(x) - 2Q(0) = 0.$$

Let us now consider the case $x, y \in E_1 \setminus \{0\}$. Then,

$$
\begin{aligned}
\| Q_n(x + y) &+ Q_n(x - y) - 2Q_n(x) - 2Q_n(y) \| \\
&= 4^{-n} \| f\left(2^n(x + y)\right) + f\left(2^n(x - y)\right) - 2f(2^n x) - 2f(2^n y) \| \\
&\leq 4^{-n} \left(\delta + \theta\left(\|2^n x\|^p + \|2^n y\|^p\right)\right) \\
&= 4^{-n} \delta + 2^{n(p-2)} \theta\left(\|x\|^p + \|y\|^p\right).
\end{aligned}
$$

By letting $n \to \infty$ we get the equality

$$Q(x + y) + Q(x - y) - 2Q(x) - 2Q(y) = 0.$$

If we put $x = y$ in the last equality, then we get $Q(2x) = 4Q(x)$ for each $x \in E_1$. Moreover, putting $y = -x$ yields $Q(-x) = Q(x)$ for $x \in E_1$. Therefore, if $x = 0$ and $y \neq 0$, then

$$Q(y) + Q(-y) - 2Q(0) - 2Q(y) = 0,$$

i.e., Q is a quadratic function.

The inequality (8.5) immediately follows from the inequality (8.3).

To prove the uniqueness, assume that there exist two quadratic functions $q_i :$ $E_1 \to E_2$ $(i \in \{1, 2\})$ such that

$$\| f(x) - q_i(x) \| \leq c_i + b_i \|x\|^p$$

for any $x \in E_1 \setminus \{0\}$ and $i \in \{1, 2\}$, where c_i, b_i $(i \in \{1, 2\})$ are given nonnegative constants. Then,

$$q_i(2^n x) = 4^n q_i(x)$$

for any $x \in E_1$, $n \in \mathbb{N}$, and $i \in \{1, 2\}$. Now, we have

$$
\begin{aligned}
\| q_1(x) - q_2(x) \| &\leq 4^{-n} \left(\|q_1(2^n x) - f(2^n x)\| + \|f(2^n x) - q_2(2^n x)\|\right) \\
&\leq 4^{-n}(c_1 + c_2) + 2^{n(p-2)}(b_1 + b_2)\|x\|^p.
\end{aligned}
$$

If we let $n \to \infty$, we get $q_1(x) = q_2(x)$ for all $x \in E_1$. $\square$

Czerwik [87] also proved the following theorem.

Theorem 8.4 (Czerwik). *Let E_1 and E_2 be a normed space and a Banach space, respectively. If a function $f : E_1 \to E_2$ satisfies the inequality*

$$\| f(x + y) + f(x - y) - 2f(x) - 2f(y) \| \leq \theta\left(\|x\|^p + \|y\|^p\right) \tag{8.6}$$

for some $\theta \geq 0$, $p > 2$ *and for all* $x, y \in E_1$, *then there exists a unique quadratic function* $Q : E_1 \rightarrow E_2$ *such that*

$$\|f(x) - Q(x)\| \leq 2(2^p - 4)^{-1}\theta\|x\|^p \qquad (8.7)$$

for all $x \in E_1$.

Proof. Define the sequence

$$Q_n(x) = 4^n f(2^{-n}x) \qquad (a)$$

for all $x \in E_1$ and all $n \in \mathbb{N}$. Since $f(0) = 0$, applying (8.4) yields

$$\|Q_n(x) - Q_m(x)\| \leq 2^{1-p}2^{m(2-p)}(1-b)^{-1}\theta\|x\|^p \qquad (b)$$

for any $x \in E_1$ and $n > m$. Hence, the assumption $p > 2$ implies that $\{Q_n(x)\}$ is a Cauchy sequence for every $x \in E_1$. Define

$$Q(x) = \lim_{n \to \infty} Q_n(x)$$

for each $x \in E_1$. Then, in a similar way as in the proof of Theorem 8.3, we may verify that Q is a quadratic function. Using (8.4) again, we obtain (8.7).

It is not difficult to prove that

$$Q(rx) = r^2 Q(x) \qquad (c)$$

for all $x \in E_1$ and for all $r \in \mathbb{Q}$. Assume now that there exist two quadratic functions $q_i : E_1 \rightarrow E_2$ ($i \in \{1, 2\}$) such that

$$\|f(x) - q_i(x)\| \leq d_i\|x\|^p$$

for all $x \in E_1$, where d_i, $i \in \{1, 2\}$, are nonnegative constants. By (c) we get

$$\begin{aligned} \|q_1(x) - q_2(x)\| &= 4^n\|q_1(2^{-n}x) - q_2(2^{-n}x)\| \\ &\leq 4^n(d_1 + d_2)\|2^{-n}x\|^p \\ &= 2^{n(2-p)}(d_1 + d_2)\|x\|^p, \end{aligned}$$

from which we conclude that $q_1 = q_2$. $\qquad \square$

The following corollary is derived from the book [137] written by D. H. Hyers, G. Isac, and Th. M. Rassias.

Corollary 8.5. *If in Theorems 8.3 and 8.4 the function* f *is continuous on* E_1, *then the quadratic function* Q *is also continuous at each point* $x \in E_1 \backslash \{0\}$. *When* $p > 0$, *this restriction is unnecessary.*

Proof. Suppose that f is continuous at each point $x \in E_1$ and that $x_0 \in E_1 \setminus \{0\}$. Put $s = \|x_0\|/2$ and define an open ball by

$$B(x_0, s) = \{x \in E_1 \mid \|x - x_0\| < s\}.$$

For $x \in B(x_0, s)$ we have $s < \|x\| < 3s$. By letting $n \to \infty$ in inequalities (b)'s in the proofs of Theorems 8.3 and 8.4, we get

$$\|Q(x) - Q_m(x)\|$$
$$\leq \begin{cases} (1/3)4^{-m}(\delta + c) + 2^{m(p-2)-1}(1 - a)^{-1}\theta\|x\|^p & \text{(for } p < 2\text{)}, \\ 2^{1-p}2^{m(2-p)}(1 - b)^{-1}\theta\|x\|^p & \text{(for } p > 2\text{)}. \end{cases}$$

For $x \in B(x_0, s)$ we have $s^p > \|x\|^p > (3s)^p$ when $p < 0$, while the inequalities are reversed when $p > 0$. Consequently, Q_m converges uniformly to Q on $B(x_0, s)$ as $m \to \infty$. Since each function Q_m is continuous on $B(x_0, s)$, it follows that the limit Q is also continuous on $B(x_0, s)$. Thus, the quadratic function Q is continuous at any point $x_0 \neq 0$ in E_1. Obviously, the restriction $x_0 \neq 0$ is not needed when $p > 0$.   □

It should be noted that Czerwik [87] proved the following corollary under the weaker assumption that $f(tx)$ is Borel measurable in t for each fixed $x \in E_1$.

Corollary 8.6. *Let E_1 and E_2 be a normed space and a Banach space, respectively. If a function $f : E_1 \to E_2$ satisfies either the inequality (8.2) for $x, y \neq 0$ or the inequality (8.6) for $x, y \in E_1$, according to $p < 2$ or $p > 2$, and if moreover $f(tx)$ is continuous in t for each fixed $x \in E_1$, then the unique quadratic function $Q : E_1 \to E_2$ defined by*

$$Q(x) = \begin{cases} \lim_{n \to \infty} 4^{-n} f(2^n x) & \text{(for } p < 2\text{)}, \\ \lim_{n \to \infty} 4^n f(2^{-n} x) & \text{(for } p > 2\text{)} \end{cases}$$

satisfies $Q(tx) = t^2 Q(x)$ for all $t \in \mathbb{R}$ and $x \in E_1$.

Proof. It is obvious that $Q(rx) = r^2 Q(x)$ holds true for all $x \in E_1$ and for all $r \in \mathbb{Q}$ (ref. (c) in the proof of Theorem 8.4). To prove that Q is homogeneous of degree 2 for all real numbers as well, it suffices to prove that $Q(tx)$ is continuous in t for each fixed $x \in E_1$. By hypothesis, $f(tx)$ is continuous in t for each fixed x in E_1. Apply Corollary 8.5 to the case where $E_1 = \mathbb{R}$ to show that if $x \neq 0$ and $t_0 \neq 0$, then $Q(tx)$ is continuous at $t = t_0$. Thus, $Q(t_0 x) = t_0^2 Q(x)$ for all $x \neq 0$ and $t_0 \neq 0$. But this equality is also valid for $x = 0$ or $t_0 = 0$. Therefore, we conclude that $Q(tx) = t^2 Q(x)$ holds true for all $t \in \mathbb{R}$ and $x \in E_1$.   □

Czerwik [87] presented an example concerning the special case $p = 2$. (This case was excluded in Theorems 8.3 and 8.4.) This is a modification of the example contained in [112] (or see [137]).

Theorem 8.7. *Let us define a function $f : \mathbb{R} \to \mathbb{R}$ by*

$$f(x) = \sum_{n=0}^{\infty} 4^{-n} \varphi(2^n x),$$

where the function $\varphi : \mathbb{R} \to \mathbb{R}$ is given by

$$\varphi(x) = \begin{cases} a & \text{(for } |x| \geq 1), \\ ax^2 & \text{(for } |x| < 1) \end{cases}$$

with a positive number a. The function f satisfies the inequality

$$\left| f(x+y) + f(x-y) - 2f(x) - 2f(y) \right| \leq 32a(x^2 + y^2) \qquad (8.8)$$

for all $x, y \in \mathbb{R}$. Moreover, there exists no quadratic function $Q : \mathbb{R} \to \mathbb{R}$ such that the image set of $|f(x) - Q(x)|/x^2$ ($x \neq 0$) is bounded.

Proof. For $x = y = 0$ or for $x, y \in \mathbb{R}$ such that $x^2 + y^2 \geq 1/4$, it is clear that the inequality (8.8) holds true because f is bounded by $(4/3)a$. Consider the case $0 < x^2 + y^2 < 1/4$. Then there exists a $k \in \mathbb{N}$ such that

$$4^{-k-1} \leq x^2 + y^2 < 4^{-k}, \qquad (a)$$

where $4^{k-1}x^2 < 1/4$ and $4^{k-1}y^2 < 1/4$ and consequently

$$2^{k-1}x, \ 2^{k-1}y, \ 2^{k-1}(x+y), \ 2^{k-1}(x-y) \in (-1, 1).$$

Therefore, for each $n \in \{0, 1, \dots, k-1\}$, we have

$$2^n x, \ 2^n y, \ 2^n(x+y), \ 2^n(x-y) \in (-1, 1)$$

and

$$\varphi\big(2^n(x+y)\big) + \varphi\big(2^n(x-y)\big) - 2\varphi(2^n x) - 2\varphi(2^n y) = 0$$

for $n \in \{0, 1, \dots, k-1\}$. Using (a) we obtain

$$\left| f(x+y) + f(x-y) - 2f(x) - 2f(y) \right|$$

$$\leq \sum_{n=0}^{\infty} 4^{-n} \left| \varphi\big(2^n(x+y)\big) + \varphi\big(2^n(x-y)\big) - 2\varphi(2^n x) - 2\varphi(2^n y) \right|$$

$$\leq \sum_{n=k}^{\infty} 4^{-n} 6a \ = \ 2 \cdot 4^{1-k} a \ \leq \ 32a(x^2 + y^2),$$

i.e., the inequality (8.8) holds true.

Assume that there exist a quadratic function $Q : \mathbb{R} \to \mathbb{R}$ and a constant $b > 0$ such that

$$|f(x) - Q(x)| \leq bx^2$$

for all $x \in \mathbb{R}$. Since Q is locally bounded, it is of the form $Q(x) = cx^2$ $(x \in \mathbb{R})$, where c is a constant (see [230]). Therefore, we have

$$|f(x)| \leq (b + |c|)x^2 \qquad (b)$$

for all $x \in \mathbb{R}$. Let $k \in \mathbb{N}$ satisfy $ka > b + |c|$. If $x \in (0, 2^{1-k})$, then $2^n x \in (0, 1)$ for $n \leq k - 1$ and we have

$$f(x) = \sum_{n=0}^{\infty} 4^{-n} \varphi(2^n x) \geq \sum_{n=0}^{k-1} a 4^{-n} (2^n x)^2 = kax^2 > (b + |c|)x^2,$$

which in comparison with (b) is a contradiction. $\qquad \square$

The results of Theorems 8.3 and 8.4 are immediate consequences of the following stability result, which was presented by C. Borelli and G. L. Forti [24] for a wide class of functional equations which contains the quadratic functional equation as a particular case:

Let G be an abelian group, E a Banach space, and let $f : G \to E$ be a function with $f(0) = 0$ and satisfy the inequality

$$\|f(x + y) + f(x - y) - 2f(x) - 2f(y)\| \leq \varphi(x, y)$$

for all $x, y \in G$. Assume that one of the series

$$\sum_{i=1}^{\infty} 2^{-2i} \varphi(2^{i-1}x, 2^{i-1}x) \quad and \quad \sum_{i=1}^{\infty} 2^{2(i-1)} \varphi(2^{-i}x, 2^{-i}x)$$

converges for each $x \in G$ and denote by $\Phi(x)$ its sum. If

$$2^{-2i} \varphi(2^{i-1}x, 2^{i-1}y) \to 0 \quad or \quad 2^{2(i-1)} \varphi(2^{-i}x, 2^{-i}y) \to 0,$$

as $i \to \infty$, then there exists a unique quadratic function $Q : G \to E$ such that

$$\|f(x) - Q(x)\| \leq \Phi(x)$$

for any $x \in G$.

8.2 Stability on a Restricted Domain

Before initially proving the stability of the quadratic functional equation for functions defined on a restricted domain, we discuss the stability problem for biadditive functions on a restricted domain.

F. Skof and S. Terracini [336] proved the following stability theorem for symmetric biadditive functions based on results from F. Skof [330].

Theorem 8.8. *Let E be a Banach space and let c, $\delta > 0$ be given. If a symmetric function $f : [0, c)^2 \to E$ satisfies the inequality*

$$\| f(x_1 + x_2, y) - f(x_1, y) - f(x_2, y) \| \leq \delta$$

for all $x_1, x_2, y \in [0, c)$ with $x_1 + x_2 < c$, then there exists a symmetric and biadditive function $B : [0, c)^2 \to E$ such that

$$\| f(x, y) - B(x, y) \| \leq 9\delta$$

for all $x, y \in [0, c)$.

Proof. By hypothesis, for each fixed $y \in [0, c)$, the function $f_y(x) = f(x, y)$ satisfies the inequality

$$\| f_y(x_1 + x_2) - f_y(x_1) - f_y(x_2) \| \leq \delta$$

for all $x_1, x_2 \in [0, c)$ with $x_1 + x_2 < c$. Following the proof of Lemma 2.28, we define the function $f_y^* : [0, \infty) \to E$ for fixed y by

$$f_y^*(x) = f_y(\mu) + n f_y(c/2)$$

for $x = \mu + (c/2)n, n \in \mathbb{N}$ and $0 \leq \mu < c/2$. Thus, we have

$$\| f_y(x) - f_y^*(x) \| \leq \delta \qquad\qquad (a)$$

for any $x \in [0, c)$. This function is extended to $\mathbb{R}$ by putting $f_y^*(x) = -f_y^*(-x)$ when $x < 0$. It follows that f_y^* $(0 \leq y < c)$ satisfies

$$\| f_y^*(x_1 + x_2) - f_y^*(x_1) - f_y^*(x_2) \| \leq 2\delta$$

for all $x_1, x_2 \in \mathbb{R}$ (cf. (b) in the proof of Lemma 2.28), and hence, by Theorem 2.3, there exists a unique additive function

$$A_y^*(x) = \lim_{n \to \infty} 2^{-n} f_y^*(2^n x),$$

for all $x \in \mathbb{R}$ and $y \in [0, c)$, such that

$$\|f_y^*(x) - A_y^*(x)\| \leq 2\delta \tag{b}$$

for any $x \in \mathbb{R}$ and $y \in [0, c)$.

Now, let us define $A(x, y) = A_y^*(x)$ for $x, y \in [0, c)$. $A(x, y)$ is additive in the first variable.

Fix $x, y, z \in [0, c)$ with $y + z < c$. Put $2^n x = \mu_n + (c/2)k_n$, where $k_n \in \mathbb{N}$ and $0 \leq \mu_n < c/2$, so that $k_n = (2/c)(2^n x - \mu_n)$. Then,

$$
\begin{aligned}
A(x, & y + z) - A(x, y) - A(x, z) \\
&= \lim_{n \to \infty} 2^{-n} \left(f_{y+z}^*(2^n x) - f_y^*(2^n x) - f_z^*(2^n x) \right) \\
&= \lim_{n \to \infty} 2^{-n} \left(f_{y+z}(\mu_n) - f_y(\mu_n) - f_z(\mu_n) \right) \\
&\quad + \lim_{n \to \infty} 2^{-n} k_n \left(f_{y+z}(c/2) - f_y(c/2) - f_z(c/2) \right) \\
&= 2(x/c) \left(f_{y+z}(c/2) - f_y(c/2) - f_z(c/2) \right).
\end{aligned}
$$

Hence,

$$\|A(x, y + z) - A(x, y) - A(x, z)\| \leq 2\delta$$

for all $x, y, z \in [0, c)$ with $y + z < c$.

Next, we extend A to a function $A' : [0, c) \times \mathbb{R} \to E$. With $x \in [0, c)$ and $y \geq 0$, let $y = \mu + (c/2)n$, where $n \in \mathbb{N}$ and $0 \leq \mu < c/2$, and define A' by

$$A'(x, y) = A(x, \mu) + nA(x, c/2)$$

and put

$$A'(x, y) = -A'(x, -y)$$

for $y < 0$. In view of (b) and (a) in the proof of Lemma 2.28 and the first part of the proof of Lemma 2.27, we obtain

$$\|A'(x, y + z) - A'(x, y) - A'(x, z)\| \leq 4\delta$$

for all $y, z \in \mathbb{R}$ and

$$\|A(x, y) - A'(x, y)\| \leq 2\delta \tag{c}$$

for all $x, y \in [0, c)$. Also, for each $y \in \mathbb{R}$, A' is additive in the first variable, i.e.,

$$A'(x_1 + x_2, y) = A'(x_1, y) + A'(x_2, y) \tag{d}$$

for $x_1, x_2 \in [0, c)$ with $x_1 + x_2 < c$, since A has this property.

For each fixed $x \in [0, c)$, it follows from Theorem 2.3 that the function

$$B(x, y) = \lim_{n \to \infty} 2^{-n} A'(x, 2^n y) \tag{e}$$

is additive in y and satisfies

$$\|A'(x, y) - B(x, y)\| \le 4\delta \qquad (f)$$

for all $x \in [0, c)$ and $y \in \mathbb{R}$. By (d) and (e) we have

$$B(x_1 + x_2, y) = B(x_1, y) + B(x_2, y)$$

for $x_1, x_2 \in [0, c)$ with $x_1 + x_2 < c$. By (a), (b), (c), and (f), we obtain

$$\|f(x, y) - B(x, y)\| \le 9\delta \qquad (g)$$

for $x, y \in [0, c)$.

Since f is symmetric, it follows from (g) that

$$\|B(x, y) - B(y, x)\| \le 18\delta$$

for any $x, y \in [0, c)$. If $y = 0$, we have $B(x, 0) = 0 = B(0, x)$ for $x \in [0, c)$. For a given $y \in (0, c)$, put

$$b_y(x) = B(y, x) - B(x, y)$$

for any $x \in [0, c)$. Now, $b_y(x)$ is bounded and additive, more precisely,

$$b_y(x_1 + x_2) = b_y(x_1) + b_y(x_2)$$

for $x_1, x_2 \in [0, c)$ with $x_1 + x_2 < c$, so it is the restriction to $[0, c)$ of a function of x of the form $b_y(x) = a(y)x$. Since $b_y(y) = 0$ for all $y \in (0, c)$, $a(y) \equiv 0$. Hence, B is symmetric on $[0, c)^2$. □

Using the result of the last theorem, Skof and Terracini [336] have proved the following theorem.

Theorem 8.9. *Let E be a Banach space and let $c, \delta > 0$ be given. If a function $f : [0, c) \to E$ satisfies the inequality*

$$\|f(x + y) + f(x - y) - 2f(x) - 2f(y)\| \le \delta \qquad (8.9)$$

for all $x \ge y \ge 0$ with $x + y < c$, then there exists a quadratic function $Q : \mathbb{R} \to E$ such that

$$\|f(x) - Q(x)\| \le (79/2)\delta \qquad (8.10)$$

for any $x \in [0, c)$.

Proof. Putting $x = y (= 0)$ in (8.9) yields

$$\|f(2x) + f(0) - 4f(x)\| \le \delta \quad \text{and} \quad \|f(0)\| \le \delta/2,$$

and hence

$$\| f(2x) - 4f(x) \| \leq (3/2)\delta \tag{a}$$

for $x \in [0, c/2)$. Define

$$g(x) = \begin{cases} f(x) & (\text{for } x \in [0, c)), \\ f(-x) & (\text{for } x \in (-c, 0)) \end{cases}$$

and put $\mu(x, y) = \| g(x + y) + g(x - y) - 2g(x) - 2g(y) \|$. It follows from (8.9) that

$$\mu(x, y) \leq \delta \quad \text{for } x \geq y \geq 0 \text{ with } x + y < c. \tag{b}$$

For $y \geq x \geq 0$ with $y + x < c$ we have

$$\mu(x, y) = \mu(y, x) = \| f(x + y) + f(-x + y) - 2f(x) - 2f(y) \|.$$

Hence, from (8.9) again, we get

$$\mu(x, y) \leq \delta \quad \text{for } y \geq x \geq 0 \text{ with } y + x < c. \tag{c}$$

If $x < 0$, $y \geq 0$, and $y - x < c$, then $\mu(x, y) = \mu(-x, y)$, since g is even. For this case it holds true that either $-x \geq y \geq 0$ with $-x + y < c$ or $y \geq -x > 0$ with $y - x < c$. Hence, from (b) or (c) we obtain

$$\mu(x, y) \leq \delta \quad \text{for } x < 0, \ y \geq 0, \text{ and } \ y - x < c. \tag{d}$$

Finally, assume that $y < 0$, $|x + y| < c$, and $|x - y| < c$. It then holds true that either $x \geq -y > 0$ with $x - y < c$ or $-y \geq x \geq 0$ with $x - y < c$ or $x < 0$, $-y > 0$, and $-y - x < c$. Therefore, it follows from (b), (c), or (d) that

$$\mu(x, y) \leq \delta \quad \text{for } y < 0, \ |x + y| < c, \text{ and } \ |x - y| < c \tag{e}$$

since $\mu(x, -y) = \mu(x, y)$. According to (b), (c), (d), and (e), we conclude that

$$\| g(x + y) + g(x - y) - 2g(x) - 2g(y) \| \leq \delta \tag{f}$$

for all $(x, y) \in D(c)$, where we set

$$D(c) = \{ (x, y) \in \mathbb{R}^2 \mid |x + y| < c, \ |x - y| < c \}.$$

Let us define the auxiliary function $h : D(c) \to E$ by

$$h(x, y) = (1/4)\big(g(x + y) - g(x - y)\big).$$

Clearly, $h(x, y) = h(y, x)$ for all $(x, y) \in D(c)$. When $y \in [0, c/2)$, h satisfies the inequality

$$\|h(x_1 + x_2, y) - h(x_1, y) - h(x_2, y)\| \leq \delta$$

for any $x_1, x_2 \in [0, c/2)$ with $x_1 + x_2 < c/2$ (see below), and by the interchange of x and y, we also have, when $x \in [0, c/2)$,

$$\|h(x, y + z) - h(x, y) - h(x, z)\| \leq \delta$$

for all $y, z \in [0, c/2)$ with $y + z < c/2$. From the definition of h it follows that

$$
\begin{aligned}
4\big(h(x_1 + x_2, y) &- h(x_1, y) - h(x_2, y)\big) \\
= &\big(g(x_1 + x_2 + y) + g(x_1 - x_2 - y) - 2g(x_1) - 2g(x_2 + y)\big) \\
&- \big(g(x_1 - y - x_2) + g(x_1 - y + x_2) - 2g(x_1 - y) - 2g(x_2)\big) \\
&- \big(g(x_1 + y) + g(x_1 - y) - 2g(x_1) - 2g(y)\big) \\
&+ \big(g(x_2 + y) + g(x_2 - y) - 2g(x_2) - 2g(y)\big)
\end{aligned}
$$

for any $(x_1, x_2 + y), (x_1 - y, x_2), (x_1, y), (x_2, y) \in D(c)$. Hence, by (f),

$$\|h(x_1 + x_2, y) - h(x_1, y) - h(x_2, y)\| \leq \delta.$$

Thus, h satisfies the hypothesis of Theorem 8.8 with $c/2$ in place of c. Hence, there exists a symmetric and biadditive function $B : [0, c/2)^2 \to E$ such that

$$\|h(x, y) - B(x, y)\| \leq 9\delta \qquad (g)$$

for $(x, y) \in [0, c/2)^2$. When $x \in [0, c/2)$, we have

$$
\begin{aligned}
\|f(x) - h(x, x)\| &= (1/4)\|4f(x) - f(2x) + f(0)\| \\
&\leq (1/4)\|4f(x) - f(2x) - f(0)\| + (1/2)\|f(0)\|,
\end{aligned}
$$

and it follows from (8.9) that

$$\|f(x) - h(x, x)\| \leq \delta/2 \qquad (h)$$

for $x \in [0, c/2)$.

The function $B(x, x)$ is quadratic for $x \geq y \geq 0$ with $x + y < c/2$. According to a theorem by F. Skof [331], it may be extended to a quadratic function $Q : \mathbb{R} \to E$ such that $Q(x) = B(x, x)$ for all $x \in [0, c/2)$. Thus, from (g) and (h), we get

$$
\begin{aligned}
\|f(x) - Q(x)\| &= \|f(x) - B(x, x)\| \\
&\leq \|f(x) - h(x, x)\| + \|h(x, x) - B(x, x)\| \\
&\leq (19/2)\delta
\end{aligned}
$$

for $x \in [0, c/2)$. Now, let $x \in [c/2, c)$. Taking account of (a) and the last inequality yields

$$\|f(x) - Q(x)\| \leq \|f(x) - 4f(x/2)\| + \|4f(x/2) - 4Q(x/2)\| \leq (79/2)\delta,$$

which ends the proof.   □

With the help of Theorem 8.9, Skof and Terracini [336] proved the following theorem.

Theorem 8.10 (Skof and Terracini). *Let E be a Banach space and let c, $\delta > 0$ be given. If a function $f : (-c, c) \to E$ satisfies the inequality (8.9) for all $x, y \in \mathbb{R}$ with $|x + y| < c$ and $|x - y| < c$, then there exists a quadratic function $Q : \mathbb{R} \to E$ such that*

$$\|f(x) - Q(x)\| \leq (81/2)\delta$$

for any $x \in (-c, c)$.

Proof. Obviously we have

$$2\|f(y) - f(-y)\| \leq \|f(x + y) + f(x - y) - 2f(x) - 2f(y)\|$$
$$+ \| - f(x - y) - f(x + y) + 2f(x) + 2f(-y)\|$$

and hence, it follows from the hypothesis that

$$\|f(y) - f(-y)\| \leq \delta \tag{a}$$

for any $y \in (-c, c)$ because for each $y \in (-c, c)$ there exists an $x \in (-c, c)$ such that $|x + y| < c$ and $|x - y| < c$. Let us denote by f_0 the restriction of f to $[0, c)$. According to Theorem 8.9, there exists a quadratic function $Q : \mathbb{R} \to E$ such that

$$\|f_0(x) - Q(x)\| \leq (79/2)\delta \tag{b}$$

for $x \in [0, c)$. For $x \in (-c, 0)$, the inequalities (a) and (b) imply

$$\|f(x) - Q(x)\| \leq \|f(x) - f(-x)\| + \|f(-x) - Q(-x)\|$$
$$\leq \delta + \|f_0(-x) - Q(-x)\|$$
$$\leq (81/2)\delta,$$

which ends the proof.   □

The last theorem was generalized by extending the domains of the relevant functions f to bounded subsets of $\mathbb{R}^n$. Indeed, S.-M. Jung and B. Kim proved the following result in the paper [187]:

Let E be a Banach space and let $c, \delta > 0$ be given constants. If a function $f : [-c, c]^n \to E$ satisfies the inequality (8.9) for all $x, y \in [-c, c]^n$ with $x + y, x - y \in [-c, c]^n$, then there exists a quadratic function $Q : \mathbb{R}^n \to E$ such that

$$\|f(x) - Q(x)\| < (2912n^2 + 1872n + 334)\delta$$

for any $x \in [-c, c]^n$.

In 1998, Jung investigated the Hyers–Ulam stability of the quadratic functional equation (8.1) on the unbounded restricted domains (see [167, Theorem 2]).

Theorem 8.11 (Jung). *Let E_1 and E_2 be a real normed space and a real Banach space, respectively, and let $d > 0$ and $\delta \geq 0$ be given. Assume that a function $f : E_1 \to E_2$ satisfies the inequality (8.9) for all $x, y \in E_1$ with $\|x\| + \|y\| \geq d$. Then there exists a unique quadratic function $Q : E_1 \to E_2$ such that*

$$\|f(x) - Q(x)\| \leq (7/2)\delta \tag{8.11}$$

for all $x \in E_1$.

Proof. Assume $\|x\| + \|y\| < d$. If $x = y = 0$, then we choose a $z \in E_1$ with $\|z\| = d$. Otherwise, let $z = (1 + d/\|x\|)x$ for $\|x\| \geq \|y\|$ or $z = (1 + d/\|y\|)y$ for $\|x\| < \|y\|$. Clearly, we see

$$\|x - z\| + \|y + z\| \geq d, \quad \|x + z\| + \|y + z\| \geq d,$$
$$\|y + z\| + \|z\| \geq d, \quad \|x\| + \|y + 2z\| \geq d, \quad \|x\| + \|z\| \geq d. \tag{a}$$

From (8.9), (a), and the relation

$$
\begin{aligned}
&f(x + y) + f(x - y) - 2f(x) - 2f(y) \\
&= f(x + y) + f(x - y - 2z) - 2f(x - z) - 2f(y + z) \\
&\quad + f(x + y + 2z) + f(x - y) - 2f(x + z) - 2f(y + z) \\
&\quad - 2f(y + 2z) - 2f(y) + 4f(y + z) + 4f(z) \\
&\quad - f(x + y + 2z) - f(x - y - 2z) + 2f(x) + 2f(y + 2z) \\
&\quad + 2f(x + z) + 2f(x - z) - 4f(x) - 4f(z),
\end{aligned}
$$

we get

$$\|f(x + y) + f(x - y) - 2f(x) - 2f(y)\| \leq 7\delta. \tag{b}$$

Obviously, the inequality (b) holds true for all $x, y \in E_1$. According to (b) and Theorem 8.1, there exists a unique quadratic function $Q : E_1 \to E_2$ which satisfies the inequality (8.11) for all $x \in E_1$. $\qquad\square$

For more detailed information on the stability of the quadratic equation (8.1), we can refer to a paper [298] of Th. M. Rassias.

8.3 Fixed Point Method

In this section, we fix a real number β with $0 < \beta \leq 1$. Suppose E is a vector space over $\mathbb{K}$. A function $\| \cdot \|_\beta : E \to [0, \infty)$ is called a β-*norm* if and only if it satisfies

(N1) $\|x\|_\beta = 0$ *if and only if* $x = 0$;
(N2) $\|\lambda x\|_\beta = |\lambda|^\beta \|x\|_\beta$ *for all* $\lambda \in \mathbb{K}$ *and all* $x \in E$;
(N3) $\|x + y\|_\beta \leq \|x\|_\beta + \|y\|_\beta$ *for all* $x, y \in E$.

By using the idea of L. Cădariu and V. Radu [55, 57], S.-M. Jung, T.-S. Kim, and K.-S. Lee [190] proved the Hyers–Ulam–Rassias stability of the quadratic functional equation in a more general setting. (Before the paper [190], Cădariu and Radu [56] applied the fixed point method to the proof of the stability of quadratic functional equations.) In the following theorem, we will introduce some results obtained by Jung, Kim, and Lee.

Theorem 8.12. *Let E_1 and E_2 be vector spaces over $\mathbb{K}$. In particular, let E_2 be a complete β-normed space, where $0 < \beta \leq 1$. Suppose $\varphi : E_1^2 \to [0, \infty)$ is a given function and there exists a constant L, $0 < L < 1$, such that*

$$\varphi(2x, 2x) \leq 4^\beta L \varphi(x, x) \tag{8.12}$$

for all $x \in E_1$. Furthermore, let $f : E_1 \to E_2$ be a function with $f(0) = 0$ which satisfies

$$\| f(x + y) + f(x - y) - 2f(x) - 2f(y) \|_\beta \leq \varphi(x, y) \tag{8.13}$$

for all $x, y \in E_1$. If φ satisfies

$$\lim_{n \to \infty} 4^{-n\beta} \varphi(2^n x, 2^n y) = 0 \tag{8.14}$$

for any $x, y \in E_1$, then there exists a unique quadratic function $Q : E_1 \to E_2$ such that

$$\| f(x) - Q(x) \|_\beta \leq \frac{1}{4^\beta} \frac{1}{1 - L} \varphi(x, x) \tag{8.15}$$

for all $x \in E_1$.

Proof. If we define

$$X = \{ g : E_1 \to E_2 \mid g(0) = 0 \}$$

and introduce a generalized metric on X as follows:

$$d(g, h) = \inf \{ C \in [0, \infty] \mid \|g(x) - h(x)\|_\beta \leq C \varphi(x, x) \text{ for all } x \in E_1 \},$$

then (X, d) is complete (ref. [181, Theorem 2.1]).

We define an operator $\Lambda : X \to X$ by

$$(\Lambda g)(x) = (1/4) g(2x)$$

for all $x \in E_1$. First, we assert that Λ is strictly contractive on X. Given $g, h \in X$, let $C \in [0, \infty]$ be an arbitrary constant with $d(g, h) \leq C$, i.e.,

$$\|g(x) - h(x)\|_\beta \leq C \varphi(x, x)$$

for all $x \in E_1$. If we replace x in the last inequality with $2x$ and make use of (8.12), then we have

$$\|(\Lambda g)(x) - (\Lambda h)(x)\|_\beta \leq LC \varphi(x, x)$$

for every $x \in E_1$, i.e., $d(\Lambda g, \Lambda h) \leq LC$. Hence, we conclude that $d(\Lambda g, \Lambda h) \leq Ld(g, h)$ for any $g, h \in X$.

Next, we assert that $d(\Lambda f, f) < \infty$. If we substitute x for y in (8.13) and we divide both sides by 4^β, then (8.12) establishes

$$\|(\Lambda f)(x) - f(x)\|_\beta \leq 4^{-\beta} \varphi(x, x)$$

for any $x \in E_1$, i.e.,

$$d(\Lambda f, f) \leq 4^{-\beta} < \infty. \tag{a}$$

Now, it follows from Theorem 2.43 (i) that there exists a function $Q : E_1 \to E_2$ with $Q(0) = 0$, which is a fixed point of Λ, such that $\Lambda^n f \to Q$, i.e.,

$$Q(x) = \lim_{n \to \infty} 4^{-n} f(2^n x) \tag{b}$$

for all $x \in E_1$.

Since the integer n_0 of Theorem 2.43 is 0 and $f \in X^*$ (see Theorem 2.43 for the definition of X^*), by Theorem 2.43 (iii) and (a), we obtain

$$d(f, Q) \leq \frac{1}{1 - L} d(\Lambda f, f) \leq \frac{1}{4^\beta} \frac{1}{1 - L}, \tag{c}$$

i.e., the inequality (8.15) is true for all $x \in E_1$.

Now, substitute $2^n x$ and $2^n y$ for x and y in (8.13), respectively. If we divide both sides of the resulting inequality by $4^{n\beta}$ and let n go to infinity, it follows from (8.14) and (b) that Q is a quadratic function.

Assume that the inequality (8.15) is also satisfied with another quadratic function $Q' : E_1 \to E_2$ besides Q. (As Q' is a quadratic function, Q' satisfies that $Q'(x) = (1/4)Q'(2x) = (\Lambda Q')(x)$ for all $x \in E_1$. That is, Q' is a fixed point of Λ.) In view of (8.15) and the definition of d, we know that

$$d(f, Q') \leq \frac{1}{4^\beta} \frac{1}{1 - L} < \infty,$$

i.e., $Q' \in X^* = \{y \in X \mid d(\Lambda f, y) < \infty\}$. (In view of (a), the integer n_0 of Theorem 2.43 is 0.) Thus, Theorem 2.43 (ii) implies that $Q = Q'$. This proves the uniqueness of Q. $\qquad \square$

We will now generalize the above theorem by removing the hypothesis $f(0) = 0$ and get the following theorem.

Theorem 8.13. *Let E_1 and E_2 be a vector space over $\mathbb{K}$ and a complete β-normed space over $\mathbb{K}$, respectively, where $0 < \beta \leq 1$. Suppose a function $\varphi : E_1^2 \to [0, \infty)$ satisfies the condition (8.14) for all $x, y \in E_1$ and there exists a constant L, $4^{-\beta} \leq L < 1$, for which the inequality (8.12) holds true for any $x \in E_1$. If a function $f : E_1 \to E_2$ satisfies the inequality (8.13) for all $x, y \in E_1$, then there exists a unique quadratic function $Q : E_1 \to E_2$ such that*

$$\|f(x) - f(0) - Q(x)\|_\beta \leq \frac{1}{4^\beta} \frac{1}{1 - L} \left(\inf \{ \varphi(z, 0) \mid z \in E_1 \} + \varphi(x, x) \right)$$

for all $x \in E_1$.

Proof. Putting $y = 0$ in (8.13) yields

$$\|2 f(0)\|_\beta \leq \varphi(x, 0)$$

for any $x \in E_1$. We define a function $g : E_1 \to E_2$ by $g(x) = f(x) - f(0)$. If we set

$$\psi(x, y) = \varphi_0 + \varphi(x, y)$$

for each $x, y \in E_1$, where $\varphi_0 = \inf \{ \varphi(x, 0) \mid x \in E_1 \}$, it then follows from (8.13) that

$$\|g(x + y) + g(x - y) - 2g(x) - 2g(y)\|_\beta \leq \psi(x, y)$$

for all $x, y \in E_1$.

Considering (8.12) and $L \geq 4^{-\beta}$, we see that

$$\psi(2x, 2x) = \varphi_0 + \varphi(2x, 2x) \leq \varphi_0 + 4^\beta L \varphi(x, x) \leq 4^\beta L \psi(x, x)$$

for any $x \in E_1$. Moreover, we make use of (8.14) to verify that

$$\lim_{n \to \infty} 4^{-n\beta} \psi(2^n x, 2^n y) = \lim_{n \to \infty} 4^{-n\beta} \left(\varphi_0 + \varphi(2^n x, 2^n y) \right) = 0$$

for every $x, y \in E_1$.

According to Theorem 8.12, there exists a unique quadratic function $Q : E_1 \to E_2$ satisfying the inequality (8.15) with g instead of f. $\square$

By a similar way as in the proof of Theorem 8.12, we also apply Theorem 2.43 and prove the following theorem.

Theorem 8.14. *Let E_1 and E_2 be a vector space over $\mathbb{K}$ and a complete β-normed space over $\mathbb{K}$, respectively. Assume that $\varphi : E_1^2 \to [0, \infty)$ is a given function and there exists a constant L, $0 < L < 1$, such that*

$$\varphi(x, x) \leq 4^{-\beta} L \varphi(2x, 2x) \tag{8.16}$$

for all $x \in E_1$. Furthermore, assume that $f : E_1 \to E_2$ is a given function with $f(0) = 0$ and satisfies the inequality (8.13) for all $x, y \in E_1$. If φ satisfies

$$\lim_{n \to \infty} 4^{n\beta} \varphi(2^{-n}x, 2^{-n}y) = 0$$

for every $x, y \in E_1$, then there exists a unique quadratic function $Q : E_1 \to E_2$ such that

$$\|f(x) - Q(x)\|_\beta \leq \frac{1}{4^\beta} \frac{L}{1-L} \varphi(x, x) \tag{8.17}$$

for any $x \in E_1$.

Proof. We use the definitions for X and d, the generalized metric on X, as in the proof of Theorem 8.12. Then, (X, d) is complete. We define an operator $\Lambda : X \to X$ by

$$(\Lambda g)(x) = 4g(x/2)$$

for all $x \in E_1$. We apply the same argument as in the proof of Theorem 8.12 and prove that Λ is a strictly contractive operator. Moreover, we prove that

$$d(\Lambda f, f) \leq 4^{-\beta} L \tag{a}$$

instead of (a) in the proof of Theorem 8.12.

According to Theorem 2.43 (i), there exists a function $Q : E_1 \to E_2$ with $Q(0) = 0$, which is a fixed point of Λ, such that

$$Q(x) = \lim_{n \to \infty} 4^n f(2^{-n}x)$$

for each $x \in E_1$.

Since the integer n_0 of Theorem 2.43 is 0 and $f \in X^*$ (see Theorem 2.43 for the definition of X^*), using Theorem 2.43 (iii) and (a) yields

$$d(f, Q) \leq \frac{1}{1-L} d(\Lambda f, f) \leq \frac{1}{4^\beta} \frac{L}{1-L},$$

which implies the validity of the inequality (8.17).

In the last part of the proof of Theorem 8.12, if we replace $2^n x$, $2^n y$, and $4^{n\beta}$ with $2^{-n}x$, $2^{-n}y$, and $4^{-n\beta}$, respectively, then we can prove that Q is a unique quadratic function satisfying inequality (8.17) for all $x \in E_1$. $\qquad \square$

Theorem 8.14 cannot be generalized to the case without the condition $f(0) = 0$. For example, if φ is continuous at $(0, 0)$, then the condition (8.16) implies that

$$\varphi(x, x) \geq (4^\beta / L)^n \varphi(2^{-n}x, 2^{-n}x)$$

for any $n \in \mathbb{N}$. By letting $n \to \infty$, we conclude that $\varphi(0, 0) = 0$. If we put $x = y = 0$ in (8.13), then we get $f(0) = 0$.

Let E_1 and E_2 be real vector spaces. If an additive function $\sigma : E_1 \to E_1$ satisfies $\sigma(\sigma(x)) = x$ for all $x \in E_1$, then σ is called an *involution* of E_1. For a given involution $\sigma : E_1 \to E_1$, the functional equation

$$f(x + y) + f(x + \sigma(y)) = 2f(x) + 2f(y)$$

is called the *quadratic functional equation with involution*.

In 2008, S.-M. Jung and Z.-H. Lee proved the Hyers–Ulam–Rassias stability of the quadratic functional equation with involution by applying the fixed point method while B. Belaid, E. Elhoucien, and Th. M. Rassias proved it by using the direct method (see [18, 192]).

8.4 Quadratic Functional Equation of Other Type

A quadratic functional equation different from the "original" quadratic functional equation (8.1) is introduced:

$$f(x + y + z) + f(x) + f(y) + f(z) = f(x + y) + f(y + z) + f(z + x) \quad (8.18)$$

This equation is sometimes called the *functional equation of Deeba*. In 1995, Pl. Kannappan [211] investigated the general solution of the functional equation of Deeba:

The general solution $f : \mathbb{R} \to \mathbb{R}$ of the functional equation (8.18) is given by $f(x) = B(x, x) + A(x)$, where $B : \mathbb{R}^2 \to \mathbb{R}$ is a symmetric biadditive function and $A : \mathbb{R} \to \mathbb{R}$ is an additive function.

Throughout this section, assume that E_1 and E_2 are a real normed space and a real Banach space, respectively.

We now prove the Hyers–Ulam stability of the functional equation (8.18) under suitable condition by using a direct method. We first introduce a lemma presented in [167]:

Lemma 8.15. *If a function $f : E_1 \to E_2$ satisfies the inequality*

$$\| f(x+y+z) + f(x) + f(y) + f(z) - f(x+y) - f(y+z) - f(z+x) \| \le \delta \quad (8.19)$$

for some $\delta \ge 0$ and for all $x, y, z \in E_1$, then

$$\left\| f(x) - \frac{2^n + 1}{2^{2n+1}} f(2^n x) + \frac{2^n - 1}{2^{2n+1}} f(-2^n x) \right\| \le 3\delta \sum_{k=1}^{n} 2^{-k} \quad (8.20)$$

for all $x \in E_1$ and $n \in \mathbb{N}$.

Proof. If we replace x, y, and z in (8.19) with 0, we get $\|f(0)\| \leq \delta$. Putting $x = y = -z$ in (8.19) yields

$$\|3f(x) + f(-x) - f(2x)\| \leq 3\delta. \tag{a}$$

By substituting $-x$ for x in (a), we obtain

$$\|3f(-x) + f(x) - f(-2x)\| \leq 3\delta. \tag{b}$$

We use induction on n to prove our lemma. By (a) and (b), we have

$$\|f(x) - (3/8)f(2x) + (1/8)f(-2x)\|$$
$$\leq (3/8)\|3f(x) + f(-x) - f(2x)\|$$
$$+ (1/8)\| - 3f(-x) - f(x) + f(-2x)\|$$
$$\leq (3/2)\delta,$$

which proves the validity of the inequality (8.20) for $n = 1$. Assume now that the inequality (8.20) holds true for some $n \in \mathbb{N}$. By using (a), (b), and the relation

$$f(x) - \frac{2^{n+1} + 1}{2^{2n+3}} f\left(2^{n+1}x\right) + \frac{2^{n+1} - 1}{2^{2n+3}} f\left(-2^{n+1}x\right)$$

$$= f(x) - \frac{2^n + 1}{2^{2n+1}} f(2^n x) + \frac{2^n - 1}{2^{2n+1}} f(-2^n x)$$

$$+ \frac{2^{n+1} + 1}{2^{2n+3}} \left(3f(2^n x) + f(-2^n x) - f\left(2^{n+1}x\right)\right)$$

$$- \frac{2^{n+1} - 1}{2^{2n+3}} \left(3f(-2^n x) + f(2^n x) - f\left(-2^{n+1}x\right)\right),$$

we can easily verify the inequality (8.20) for $n + 1$ which ends the proof.   □

In the following theorem, S.-M. Jung [167] proved the Hyers–Ulam stability of the functional equation of Deeba under the approximately even condition.

Theorem 8.16 (Jung). *Assume that a function* $f : E_1 \to E_2$ *satisfies the system of inequalities*

$$\|f(x + y + z) + f(x) + f(y) + f(z)$$
$$- f(x + y) - f(y + z) - f(z + x)\| \leq \delta, \tag{8.21}$$
$$\|f(x) - f(-x)\| \leq \theta$$

for some $\delta, \theta \geq 0$ *and for all* $x, y, z \in E_1$. *Then there exists a unique quadratic function* $Q : E_1 \to E_2$ *which satisfies* (8.18) *and the inequality*

$$\|f(x) - Q(x)\| \leq 3\delta \tag{8.22}$$

for all $x \in E_1$.

Proof. It follows from (8.20) and the second condition in (8.21) that

$$\| f(x) - 2^{-2n} f(2^n x) \| \le 3\delta \sum_{k=1}^{n} 2^{-k} + \frac{2^n - 1}{2^{2n+1}} \theta. \qquad (a)$$

By (a) we have

$$\| 2^{-2n} f(2^n x) - 2^{-2m} f(2^m x) \|$$
$$= 2^{-2m} \| 2^{-2(n-m)} f(2^{n-m} \cdot 2^m x) - f(2^m x) \|$$
$$\le 2^{-2m} \left(3\delta \sum_{k=1}^{n-m} 2^{-k} + \frac{2^{n-m} - 1}{2^{2(n-m)+1}} \theta \right) \qquad (b)$$

for $n \ge m$. Since the right-hand side of the inequality (b) tends to 0 as m tends to ∞, the sequence $\{ 2^{-2n} f(2^n x) \}$ is a Cauchy sequence.

Therefore, we may apply a direct method to the definition of Q. Define

$$Q(x) = \lim_{n \to \infty} 2^{-2n} f(2^n x)$$

for all $x \in E_1$. From the first condition in (8.21), it follows that

$$\| Q(x + y + z) + Q(x) + Q(y) + Q(z)$$
$$- Q(x + y) - Q(y + z) - Q(z + x) \| \le 2^{-2n} \delta$$

for all $x, y, z \in E_1$ and for all $n \in \mathbb{N}$. Therefore, by letting $n \to \infty$ in the last inequality, it is clear that Q is a solution of (8.18). Analogously, by the second condition in (8.21), we can show that Q is even. By putting $z = -y$ in (8.18) and by taking account of $Q(0) = 0$, we see that Q is quadratic as an even solution of (8.18). According to (a), the inequality (8.22) holds true.

Now, let $Q' : E_1 \to E_2$ be another quadratic function which satisfies (8.18) and (8.22). Obviously, we have

$$Q(2^n x) = 4^n Q(x) \quad \text{and} \quad Q'(2^n x) = 4^n Q'(x)$$

for all $x \in E_1$ and $n \in \mathbb{N}$. Hence, it follows from (8.22) that

$$\| Q(x) - Q'(x) \| = 4^{-n} \| Q(2^n x) - Q'(2^n x) \|$$
$$\le 4^{-n} \left(\| Q(2^n x) - f(2^n x) \| + \| f(2^n x) - Q'(2^n x) \| \right)$$
$$\le 6\delta / 4^n$$

for all $x \in E_1$ and $n \in \mathbb{N}$. By letting $n \to \infty$ in the preceding inequality, we immediately see the uniqueness of Q.   □

From the direct combination of the inequalities in (8.21), it follows that the function $f : E_1 \to E_2$ in Theorem 8.16 satisfies the inequality

$$\|f(x + y) + f(x - y) - 2f(x) - 2f(y)\| \le \delta + \theta + \|f(0)\| \le 2\delta + \theta.$$

According to Theorem 8.1, there is a unique quadratic function $Q : E_1 \to E_2$ such that

$$\|f(x) - Q(x)\| \le \delta + (1/2)\theta.$$

We see that the last inequality contains a θ term which appeared as the upper bound for the second inequality in (8.21). The advantage of the inequality (8.22) compared to the last inequality is that the right-hand side of (8.22) contains no θ term.

Similarly as in the proof of Theorem 8.16, the Hyers–Ulam stability for equation (8.18) under the approximately odd condition is proved (see [167, Theorem 7]).

Theorem 8.17 (Jung). *Assume that a function* $f : E_1 \to E_2$ *satisfies the system of inequalities*

$$\begin{aligned} \|f(x + y + z) + f(x) + f(y) + f(z) \\ -f(x + y) - f(y + z) - f(z + x)\| \le \delta, \\ \|f(x) + f(-x)\| \le \theta \end{aligned} \tag{8.23}$$

for some $\delta, \theta \ge 0$ *and for all* $x, y, z \in E_1$. *Then there exists a unique additive function* $A : E_1 \to E_2$ *satisfying the inequality*

$$\|f(x) - A(x)\| \le 3\delta \tag{8.24}$$

for all $x \in E_1$.

Proof. From (8.20) and the second condition in (8.23), we get

$$\|f(x) - 2^{-n} f(2^n x)\| \le 3\delta \sum_{k=1}^{n} 2^{-k} + \frac{2^n - 1}{2^{2n+1}} \theta. \tag{a}$$

The sequence $\{2^{-n} f(2^n x)\}$ is a Cauchy sequence because, for $n \ge m$,

$$\begin{aligned} \|2^{-n} f(2^n x) &- 2^{-m} f(2^m x)\| \\ &= 2^{-m} \|2^{-(n-m)} f(2^{n-m} \cdot 2^m x) - f(2^m x)\| \\ &\le 2^{-m} \left(3\delta \sum_{k=1}^{n-m} 2^{-k} + \frac{2^{n-m} - 1}{2^{2(n-m)+1}} \theta \right) \\ &\to 0 \quad \text{as } m \to \infty. \end{aligned}$$

Now, define

$$A(x) = \lim_{n \to \infty} 2^{-n} f(2^n x)$$

for all $x \in E_1$. Similarly as in the proof of Theorem 8.16, due to (8.23), we see that the function A satisfies (8.18) and is odd. Putting $z = -y$ in (8.18), considering the oddness of A and letting $u = x + y, v = x - y$ yield

$$2A((u+v)/2) = A(u) + A(v).$$

According to [278], since $A(0) = 0$, the function A is additive. The validity of the inequality (8.24) follows directly from (a) and the definition of A.

Now, let $A' : E_1 \to E_2$ be another additive function which satisfies (8.24). It then follows from (8.24) that

$$
\begin{aligned}
\|A(x) - A'(x)\| &= 2^{-n}\|A(2^n x) - A'(2^n x)\| \\
&\le 2^{-n}(\|A(2^n x) - f(2^n x)\| + \|f(2^n x) - A'(2^n x)\|) \\
&\le 6\delta/2^n
\end{aligned}
$$

for all $x \in E_1$ and $n \in \mathbb{N}$. This implies the uniqueness of A.   □

The approximately even condition in (8.21) guarantees the "quadratic" property of Q, whereas the approximately odd condition in (8.23) guarantees the "additive" behavior of A.

G.-H. Kim [215] proved the Hyers–Ulam–Rassias stability of the quadratic functional equation (8.18), and I.-S. Chang and H.-M. Kim [61] generalized the preceding theorems of Jung.

8.5 Quadratic Functional Equation of Pexider Type

In this section, we will prove the Hyers–Ulam–Rassias stability of the *quadratic functional equation of Pexider type*:

$$f_1(x + y) + f_2(x - y) = f_3(x) + f_4(y) \tag{8.25}$$

This equation of Pexider type is useful to characterize the quasi-inner product spaces. B. R. Ebanks, Pl. Kannappan, and P. K. Sahoo [96] proved the following theorem concerning the general solution of equation (8.25) (see also [172]).

The general solution $f_1, f_2, f_3, f_4 : \mathbb{R} \to \mathbb{R}$ *of the functional equation* (8.25) *is given by*

$$
\begin{aligned}
f_1(x) &= B(x, x) + A_1(x) + A_2(x) + (a - b), \\
f_2(x) &= B(x, x) + A_1(x) - A_2(x) - (a + b), \\
f_3(x) &= 2B(x, x) + 2A_1(x) - (2b + c), \\
f_4(x) &= 2B(x, x) + 2A_2(x) + c,
\end{aligned}
$$

where $B : \mathbb{R}^2 \to \mathbb{R}$ *is a symmetric biadditive function,* $A_1, A_2 : \mathbb{R} \to \mathbb{R}$ *are additive functions, and* a, b, c *are real constants.*

Throughout this section, assume that E_1 and E_2 are a real normed space and a real Banach space, respectively. Let $\varphi : E_1^2 \to [0, \infty)$ be a given function with the properties:

(i) $\varphi(y, x) = \varphi(x, y)$;
(ii) $\varphi(x, -y) = \varphi(x, y)$;
(iii) it holds true that

$$\sum_{i=0}^{\infty} 2^{2i} \varphi(2^{-i}x, 2^{-i}y) < \infty \text{ for all } x, y \in E_1.$$

We now define

$$\varphi_e(x, y) = 2\varphi(x, y) + 4\varphi(x, 0) + 3\varphi(x + y, 0) + 3\varphi(x - y, 0)$$
$$+ 6\varphi((x + y)/2, (x + y)/2) + 2\varphi((x - y)/2, (x - y)/2)$$
$$+ 20\varphi(0, 0),$$
$$\varphi_o(x, y) = 2\varphi(x, y) + 2\varphi(x, x) + 2\varphi(y, y) + 2\varphi(2x, 0) + 2\varphi(2y, 0)$$
$$+ 2\varphi(x + y, 0) + 2\varphi(x - y, 0) + 14\varphi(0, 0),$$
$$\Phi(x, y) = \sum_{i=0}^{\infty} 2^{2(i+1)} \varphi_e(2^{-i-1}x, 2^{-i-1}y),$$
$$\Phi'(x, y) = \sum_{i=0}^{\infty} 2^{i+1} \varphi_o(2^{-i-2}(x + y), 2^{-i-2}(x - y)),$$

and

$$\Phi''(x, y) = \sum_{i=0}^{\infty} 2^{i+1} \varphi_o(2^{-i-2}(2x + y), 2^{-i-2}y)$$

for all $x, y \in E_1$. These definitions will be used in the following theorem. We see that φ_o satisfies the conditions *(i)*, *(ii)*, and *(iii)* instead of φ.

S.-M. Jung proved the Hyers–Ulam–Rassias stability of the quadratic functional equation of Pexider type which includes the following theorem as a special case (see [173, Theorem 5]).

Theorem 8.18 (Jung). *If given functions $f_1, f_2, f_3, f_4 : E_1 \to E_2$ satisfy the inequality*

$$\|f_1(x + y) + f_2(x - y) - f_3(x) - f_4(y)\| \le \varphi(x, y) \qquad (8.26)$$

for all $x, y \in E_1$, then there exist a quadratic function $Q : E_1 \to E_2$ and additive functions $A_1, A_2 : E_1 \to E_2$ such that

$$\|f_1(x) - Q(x) - A_1(x) - A_2(x) - f_1(0)\|$$
$$\le (1/8)\Phi(x, x) + (1/4)\Phi'(x, x) + (1/4)\Phi''(x, x)$$
$$+ 3\varphi(x/2, x/2) + (5/2)\varphi(x, 0) + (11/2)\varphi(0, 0),$$

$$\|f_2(x) - Q(x) - A_1(x) + A_2(x) - f_2(0)\|$$
$$\leq (1/8)\Phi(x,x) + (1/4)\Phi'(x,x) + (1/4)\Phi''(x,x)$$
$$+ \varphi(x/2, x/2) + (5/2)\varphi(x,0) + (7/2)\varphi(0,0),$$
$$\|f_3(x) - 2Q(x) - 2A_1(x) - f_3(0)\|$$
$$\leq (1/4)\Phi(x,x) + (1/2)\Phi'(x,x) + 2\varphi(x,0) + 2\varphi(0,0),$$
$$\|f_4(x) - 2Q(x) - 2A_2(x) - f_4(0)\| \leq (1/4)\Phi(x,x) + (1/2)\Phi''(x,x)$$

for all $x \in E_1$.

Proof. Let us define $F_i(x) = f_i(x) - f_i(0)$, and by F_i^e and F_i^o denote the even part and the odd part of F_i for $i \in \{1, 2, 3, 4\}$. Then, we get $F_i(0) = F_i^e(0) = F_i^o(0) = 0$ for $i \in \{1, 2, 3, 4\}$. Putting $x = y = 0$ in (8.26) and subtracting the resulting inequality from the original one yield

$$\|F_1(x+y) + F_2(x-y) - F_3(x) - F_4(y)\| \leq \varphi(x,y) + \varphi(0,0) \qquad (a)$$

for all $x, y \in E_1$. If we replace x and y in (a) with $-x$ and $-y$, respectively, and we add (subtract) the resulting inequality to (from) the original one, then we obtain

$$\|F_1^e(x+y) + F_2^e(x-y) - F_3^e(x) - F_4^e(y)\| \leq \varphi(x,y) + \varphi(0,0) \qquad (b)$$

and

$$\|F_1^o(x+y) + F_2^o(x-y) - F_3^o(x) - F_4^o(y)\| \leq \varphi(x,y) + \varphi(0,0) \qquad (c)$$

for all $x, y \in E_1$.

If we put $y = 0$, $x = 0$ (and replace y with x), $y = x$, or if we put $y = -x$ in (b), then we get

$$\|F_1^e(x) + F_2^e(x) - F_3^e(x)\| \leq \varphi(x,0) + \varphi(0,0), \qquad (d)$$

$$\|F_1^e(x) + F_2^e(x) - F_4^e(x)\| \leq \varphi(x,0) + \varphi(0,0), \qquad (e)$$

$$\|F_1^e(2x) - F_3^e(x) - F_4^e(x)\| \leq \varphi(x,x) + \varphi(0,0), \qquad (f)$$

or

$$\|F_2^e(2x) - F_3^e(x) - F_4^e(x)\| \leq \varphi(x,x) + \varphi(0,0) \qquad (g)$$

for all $x \in E_1$.

In view of (d) and (e), we see that

$$\|F_3^e(x) - F_4^e(x)\|$$
$$\leq \|F_1^e(x) + F_2^e(x) - F_4^e(x)\| + \|F_3^e(x) - F_1^e(x) - F_2^e(x)\|$$
$$\leq 2\varphi(x,0) + 2\varphi(0,0), \qquad (h)$$

and it follows from (f) and (g) that

$$\|F_1^e(x) - F_2^e(x)\| \le 2\varphi(x/2, x/2) + 2\varphi(0, 0) \qquad (i)$$

for any $x \in E_1$. By using (b), (h), and (i), we have

$$\begin{aligned}
&\|F_2^e(x+y) + F_2^e(x-y) - F_4^e(x) - F_4^e(y)\| \\
&\quad \le \|F_1^e(x+y) + F_2^e(x-y) - F_3^e(x) - F_4^e(y)\| \\
&\qquad + \|F_2^e(x+y) - F_1^e(x+y)\| + \|F_3^e(x) - F_4^e(x)\| \\
&\quad \le \varphi(x, y) + 2\varphi(x, 0) + 2\varphi\big((x+y)/2, (x+y)/2\big) + 5\varphi(0, 0). \qquad (j)
\end{aligned}$$

Putting $y = 0$ in (j) yields

$$\|2F_2^e(x) - F_4^e(x)\| \le 3\varphi(x, 0) + 2\varphi(x/2, x/2) + 5\varphi(0, 0). \qquad (k)$$

Hence, (j) and (k) imply

$$\begin{aligned}
&\|F_4^e(x+y) + F_4^e(x-y) - 2F_4^e(x) - 2F_4^e(y)\| \\
&\quad \le 2\|F_2^e(x+y) + F_2^e(x-y) - F_4^e(x) - F_4^e(y)\| \\
&\qquad + \|F_4^e(x+y) - 2F_2^e(x+y)\| + \|F_4^e(x-y) - 2F_2^e(x-y)\| \\
&\quad \le \varphi_e(x, y)
\end{aligned}$$

for all $x, y \in E_1$.

In view of (iii), we can apply [173, Theorem 3] to this case. Hence, there exists a unique quadratic function $Q : E_1 \to E_2$ such that

$$\|F_4^e(x) - 2Q(x)\| \le (1/4)\Phi(x, x) \qquad (l)$$

for all $x \in E_1$. (In order to obtain the last inequality, we need to make a routine computation.)

On account of (h), (i), (k), and (l), we get

$$\begin{aligned}
&\|F_1^e(x) - Q(x)\| \\
&\quad \le \|F_1^e(x) - F_2^e(x)\| + \|F_2^e(x) - (1/2)F_4^e(x)\| \\
&\qquad + \|(1/2)F_4^e(x) - Q(x)\| \\
&\quad \le (1/8)\Phi(x, x) + 3\varphi(x/2, x/2) + (3/2)\varphi(x, 0) + (9/2)\varphi(0, 0), \qquad (m) \\
&\|F_2^e(x) - Q(x)\| \\
&\quad \le \|F_2^e(x) - (1/2)F_4^e(x)\| + \|(1/2)F_4^e(x) - Q(x)\| \\
&\quad \le (1/8)\Phi(x, x) + \varphi(x/2, x/2) + (3/2)\varphi(x, 0) + (5/2)\varphi(0, 0), \qquad (n)
\end{aligned}$$

and

$$\|F_3^e(x) - 2Q(x)\|$$
$$\leq \|F_3^e(x) - F_4^e(x)\| + \|F_4^e(x) - 2Q(x)\|$$
$$\leq (1/4)\Phi(x,x) + 2\varphi(x,0) + 2\varphi(0,0) \tag{o}$$

for any $x \in E_1$.

If we put $y = 0$, $x = 0$ (and replace y with x), $y = x$, or if we put $y = -x$ in (c) separately, then we obtain

$$\|F_1^o(x) + F_2^o(x) - F_3^o(x)\| \leq \varphi(x,0) + \varphi(0,0), \tag{p}$$

$$\|F_1^o(x) - F_2^o(x) - F_4^o(x)\| \leq \varphi(x,0) + \varphi(0,0), \tag{q}$$

$$\|F_1^o(2x) - F_3^o(x) - F_4^o(x)\| \leq \varphi(x,x) + \varphi(0,0), \tag{r}$$

or

$$\|F_2^o(2x) - F_3^o(x) + F_4^o(x)\| \leq \varphi(x,x) + \varphi(0,0) \tag{s}$$

for all $x \in E_1$.

Due to (p) and (q), we have

$$\|2F_1^o(x) - F_3^o(x) - F_4^o(x)\|$$
$$\leq \|F_1^o(x) + F_2^o(x) - F_3^o(x)\| + \|F_1^o(x) - F_2^o(x) - F_4^o(x)\|$$
$$\leq 2\varphi(x,0) + 2\varphi(0,0) \tag{t}$$

and

$$\|2F_2^o(x) - F_3^o(x) + F_4^o(x)\|$$
$$\leq \|F_1^o(x) + F_2^o(x) - F_3^o(x)\| + \|F_2^o(x) + F_4^o(x) - F_1^o(x)\|$$
$$\leq 2\varphi(x,0) + 2\varphi(0,0) \tag{u}$$

for each $x \in E_1$.

Combining (r) with (t) yields

$$\|F_3^o(2x) + F_4^o(2x) - 2F_3^o(x) - 2F_4^o(x)\|$$
$$\leq \|F_3^o(2x) + F_4^o(2x) - 2F_1^o(2x)\|$$
$$\quad + \|2F_1^o(2x) - 2F_3^o(x) - 2F_4^o(x)\|$$
$$\leq 2\varphi(x,x) + 2\varphi(2x,0) + 4\varphi(0,0). \tag{v}$$

Analogously, by (s) and (u), we get

$$\|F_3^o(2x) - F_4^o(2x) - 2F_3^o(x) + 2F_4^o(x)\|$$
$$\leq \|F_3^o(2x) - F_4^o(2x) - 2F_2^o(2x)\|$$
$$\quad + \|2F_2^o(2x) - 2F_3^o(x) + 2F_4^o(x)\|$$
$$\leq 2\varphi(x,x) + 2\varphi(2x,0) + 4\varphi(0,0) \tag{w}$$

for any $x \in E_1$. It now follows from (v) and (w) that

$$\|F_3^o(2x) - 2F_3^o(x)\|$$
$$\leq \|(1/2)F_3^o(2x) + (1/2)F_4^o(2x) - F_3^o(x) - F_4^o(x)\|$$
$$+ \|(1/2)F_3^o(2x) - (1/2)F_4^o(2x) - F_3^o(x) + F_4^o(x)\|$$
$$\leq 2\varphi(x,x) + 2\varphi(2x,0) + 4\varphi(0,0) \qquad (x)$$

and

$$\|F_4^o(2x) - 2F_4^o(x)\| \leq 2\varphi(x,x) + 2\varphi(2x,0) + 4\varphi(0,0) \qquad (y)$$

for all $x \in E_1$.

In view of (c), (t), (u), (x), and (y), we have

$$\|F_3^o(x+y) + F_4^o(x+y) + F_3^o(x-y)$$
$$- F_4^o(x-y) - F_3^o(2x) - F_4^o(2y)\|$$
$$\leq \|2F_1^o(x+y) + 2F_2^o(x-y) - 2F_3^o(x) - 2F_4^o(y)\|$$
$$+ \|F_3^o(x+y) + F_4^o(x+y) - 2F_1^o(x+y)\|$$
$$+ \|F_3^o(x-y) - F_4^o(x-y) - 2F_2^o(x-y)\|$$
$$+ \|2F_3^o(x) - F_3^o(2x)\| + \|2F_4^o(y) - F_4^o(2y)\|$$
$$\leq \varphi_o(x,y) \qquad (z)$$

for all $x, y \in E_1$. If we replace y in (z) with $-y$ and then add (subtract) the resulting inequality to (from) (z), then we get

$$\|F_3^o(x+y) + F_3^o(x-y) - F_3^o(2x)\| \leq \varphi_o(x,y) \qquad (\alpha)$$

and

$$\|F_4^o(x+y) - F_4^o(x-y) - F_4^o(2y)\| \leq \varphi_o(x,y) \qquad (\beta)$$

for any $x, y \in E_1$.

Letting $u = x + y$ and $v = x - y$ in (α) yields

$$\|F_3^o(u) + F_3^o(v) - F_3^o(u+v)\| \leq \varphi_o((u+v)/2, (u-v)/2)$$

for all $u, v \in E_1$. According to Corollary 2.19, there exists a unique additive function $A_1 : E_1 \to E_2$ such that

$$\|F_3^o(x) - 2A_1(x)\| \leq (1/2)\Phi'(x,x) \qquad (\gamma)$$

for all $x \in E_1$.

Putting $u = x - y$ and $v = 2y$ in (β) yields

$$\|F_4^o(u + v) - F_4^o(u) - F_4^o(v)\| \leq \varphi_0(u + v/2, v/2)$$

for all $u, v \in E_1$. According to Corollary 2.19 again, there exists a unique additive function $A_2 : E_1 \to E_2$ such that

$$\|F_4^o(x) - 2A_2(x)\| \leq (1/2)\Phi''(x, x) \qquad (\delta)$$

for any $x \in E_1$.

From (t), (u), (γ), and (δ), it follows that

$$\|F_1^o(x) - A_1(x) - A_2(x)\|$$
$$\leq \|F_1^o(x) - (1/2)F_3^o(x) - (1/2)F_4^o(x)\|$$
$$\quad + \|(1/2)F_3^o(x) - A_1(x)\| + \|(1/2)F_4^o(x) - A_2(x)\|$$
$$\leq (1/4)\Phi'(x, x) + (1/4)\Phi''(x, x) + \varphi(x, 0) + \varphi(0, 0) \qquad (\epsilon)$$

and

$$\|F_2^o(x) - A_1(x) + A_2(x)\|$$
$$\leq \|F_2^o(x) - (1/2)F_3^o(x) + (1/2)F_4^o(x)\|$$
$$\quad + \|(1/2)F_3^o(x) - A_1(x)\| + \|A_2(x) - (1/2)F_4^o(x)\|$$
$$\leq (1/4)\Phi'(x, x) + (1/4)\Phi''(x, x) + \varphi(x, 0) + \varphi(0, 0) \qquad (\zeta)$$

for each $x \in E_1$.

Finally, the assertion of this theorem is true in view of the inequalities (l), (m), (n), (o), (γ), (δ), (ϵ), and (ζ). $\square$

Later, K.-W. Jun and Y.-H. Lee [153] proved Theorem 8.18 again in a setting

$$\|f_1(x + y) + f_2(x - y) - 2f_3(x) - 2f_4(y)\| \leq \varphi(x, y)$$

instead of the inequality (8.26).

In the next corollary, we introduce the Hyers–Ulam stability of the quadratic functional equation of Pexider type. We refer the reader to [173, Corollary 6] or [200, Theorem 3] for the proof.

Corollary 8.19. *If functions $f_1, f_2, f_3, f_4 : E_1 \to E_2$ satisfy the inequality*

$$\|f_1(x + y) + f_2(x - y) - f_3(x) - f_4(y)\| \leq \delta$$

for some $\delta \geq 0$ and for all $x, y \in E_1$, then there exist a unique quadratic function $Q : E_1 \to E_2$ and exactly two additive functions $A_1, A_2 : E_1 \to E_2$ such that

$$\|f_1(x) - Q(x) - A_1(x) - A_2(x) - f_1(0)\| \le (137/3)\delta,$$
$$\|f_2(x) - Q(x) - A_1(x) + A_2(x) - f_2(0)\| \le (125/3)\delta,$$
$$\|f_3(x) - 2Q(x) - 2A_1(x) - f_3(0)\| \le (136/3)\delta,$$
$$\|f_4(x) - 2Q(x) - 2A_2(x) - f_4(0)\| \le (124/3)\delta$$

for all $x \in E_1$.

D. Yang [359] improved Corollary 8.19 by replacing the domain of the functions f_i with a 2-divisible abelian group and by obtaining sharper estimates.

Chapter 9
Exponential Functional Equations

The exponential function $f(x) = e^x$ is a powerful tool in each field of natural sciences and engineering since many natural phenomena well-known to us can be described best of all by means of it. The famous *exponential functional equation* $f(x + y) = f(x)f(y)$ simplifies the elegant property of the exponential function, for example, $e^{x+y} = e^x e^y$. In Section 9.1, the superstability of the exponential functional equation will be proved. Section 9.2 deals with the stability of the exponential equation in the sense of R. Ger. Stability problems of the exponential functional equation on a restricted domain and asymptotic behaviors of exponential functions are discussed in Section 9.3. Another exponential functional equation $f(xy) = f(x)^y$ will be introduced in Section 9.4.

9.1 Superstability

The function $f(x) = a^x$ is said to be an exponential function, where $a > 0$ is a fixed real number. The exponent law of exponential functions is well represented by the *exponential functional equation*

$$f(x + y) = f(x)f(y).$$

Hence, we call every solution function of the exponential functional equation an *exponential function*. We now introduce the general solution of the exponential functional equation (see [90, Theorem 6.4]).

A function $f : \mathbb{R} \to \mathbb{C}$ is an exponential function if and only if either

$$f(x) = e^{A(x)+ia(x)} \text{ for all } x \in \mathbb{R} \quad or \quad f(x) = 0 \text{ for all } x \in \mathbb{R},$$

where $A : \mathbb{R} \to \mathbb{R}$ is an additive function and $a : \mathbb{R} \to \mathbb{R}$ satisfies

$$a(x + y) \equiv a(x) + a(y) \mod 2\pi$$

for all $x, y \in \mathbb{R}$.

S.-M. Jung, *Hyers–Ulam–Rassias Stability of Functional Equations in Nonlinear Analysis*, Springer Optimization and Its Applications 48, DOI 10.1007/978-1-4419-9637-4_9, © Springer Science+Business Media, LLC 2011

Indeed, a function $f : \mathbb{R} \to \mathbb{R}$ continuous at a point is an exponential function if and only if $f(x) = a^x$ for all $x \in \mathbb{R}$ or $f(x) = 0$ for all $x \in \mathbb{R}$, where $a > 0$ is a constant.

J. Baker, J. Lawrence, and F. Zorzitto [17] have proved the superstability of the exponential functional equation:

If a real-valued function f defined on a real vector space satisfies the functional inequality

$$|f(x + y) - f(x)f(y)| \leq \delta$$

for some $\delta > 0$ and for all x and y, then f is either bounded or exponential.

This theorem was the first result concerning the superstability phenomenon of functional equations.

J. Baker [16] generalized this famous result as follows:

Theorem 9.1 (Baker). *Let $(G, \cdot)$ be a semigroup and let $\delta > 0$ be given. If a function $f : G \to \mathbb{C}$ satisfies the inequality*

$$|f(x \cdot y) - f(x)f(y)| \leq \delta \tag{9.1}$$

for all $x, y \in G$, then either $|f(x)| \leq \left(1 + \sqrt{1 + 4\delta}\right)/2$ for all $x \in G$ or $f(x \cdot y) = f(x)f(y)$ for all $x, y \in G$.

Proof. If we put $\varepsilon = \left(1 + \sqrt{1 + 4\delta}\right)/2$, then $\varepsilon^2 - \varepsilon = \delta$ and $\varepsilon > 1$. Suppose there exists an $a \in G$ such that $|f(a)| > \varepsilon$, say $|f(a)| = \varepsilon + p$ for some $p > 0$. It then follows from (9.1) that

$$\begin{aligned}
\left|f\left(a^2\right)\right| &\geq \left|f(a)^2 - \left(f(a)^2 - f\left(a^2\right)\right)\right| \\
&\geq \left|f(a)^2\right| - \left|f(a)^2 - f\left(a^2\right)\right| \\
&\geq |f(a)|^2 - \delta \\
&= \varepsilon + p + (2\varepsilon - 1)p + p^2 \\
&> \varepsilon + 2p,
\end{aligned}$$

where we use the notations $a^2 = a \cdot a$, $a^3 = a \cdot a^2$, etc. Now, make the induction hypothesis

$$\left|f\left(a^{2^n}\right)\right| > \varepsilon + (n + 1)p. \tag{a}$$

Then, by (9.1) and (a), we get

$$\begin{aligned}
\left|f\left(a^{2^{n+1}}\right)\right| &= \left|f\left(a^{2^n}\right)f\left(a^{2^n}\right) - \left(f\left(a^{2^n}\right)f\left(a^{2^n}\right) - f\left(a^{2^n} \cdot a^{2^n}\right)\right)\right| \\
&\geq \left|f\left(a^{2^n}\right)\right|^2 - \delta \\
&> \left(\varepsilon + (n + 1)p\right)^2 - \left(\varepsilon^2 - \varepsilon\right) \\
&> \varepsilon + (n + 2)p,
\end{aligned}$$

and (a) is established for all $n \in \mathbb{N}$.

For every $x, y, z \in G$ we have

$$|f(x \cdot y \cdot z) - f(x \cdot y) f(z)| \leq \delta$$

and

$$|f(x \cdot y \cdot z) - f(x) f(y \cdot z)| \leq \delta.$$

Thus,

$$|f(x \cdot y) f(z) - f(x) f(y \cdot z)| \leq 2\delta.$$

Hence,

$$\begin{aligned} |f(x \cdot y) f(z) - f(x) f(y) f(z)| &\leq |f(x \cdot y) f(z) - f(x) f(y \cdot z)| \\ &+ |f(x) f(y \cdot z) - f(x) f(y) f(z)| \\ &\leq 2\delta + |f(x)|\delta \end{aligned}$$

or

$$|f(x \cdot y) - f(x) f(y)| |f(z)| \leq 2\delta + |f(x)|\delta.$$

In particular,

$$|f(x \cdot y) - f(x) f(y)| \leq \left(2\delta + |f(x)|\delta \right) / |f(a^{2^n})|$$

for all $x, y \in G$ and any $n \in \mathbb{N}$. Letting $n \to \infty$ and considering (a), we conclude that $f(x \cdot y) = f(x) f(y)$ for all $x, y \in G$. □

In the proof of the preceding theorem, the multiplicative property of the norm was crucial. Indeed, the proof above works also for functions $f : G \to E$, where E is a normed algebra in which the norm is multiplicative, i.e., $\|xy\| = \|x\| \|y\|$ for any $x, y \in E$. Examples of such real normed algebras are the quaternions and the Cayley numbers.

J. Baker [16] gave the following example to present that the theorem is false if the algebra does not have the multiplicative norm.

Example. Given $\delta > 0$, choose an ε with $|\varepsilon - \varepsilon^2| = \delta$. Let $M_2(\mathbb{C})$ denote the space of 2×2 complex matrices with the usual norm. Let us define $f : \mathbb{R} \to M_2(\mathbb{C})$ by

$$f(x) = \begin{pmatrix} e^x & 0 \\ 0 & \varepsilon \end{pmatrix}$$

for all $x \in \mathbb{R}$. Then f is unbounded and it satisfies $\|f(x + y) - f(x) f(y)\| = \delta$ for all $x, y \in \mathbb{R}$. However, f is not exponential.

L. Székelyhidi [341] has generalized the result of Baker, Lawrence, and Zorzitto [17] in another way. Let $(G, \cdot)$ be a semigroup and let V be a vector space of complex-valued functions on G. V is called *right invariant* if f belongs to V implies that the function $f(x \cdot y)$ belongs to V for each fixed $y \in G$. Similarly, we may define *left invariant vector spaces*, and we call V *invariant* if it is right and left invariant. Following Székelyhidi, a function $m : G \to \mathbb{C}$ is called an *exponential* if $m(x \cdot y) = m(x)m(y)$ for any $x, y \in G$.

The main theorem of Székelyhidi [341] is the following:

Theorem 9.2 (Székelyhidi). *Let $(G, \cdot)$ be a semigroup and V be a right invariant vector space of complex-valued functions on G. Let $f, m : G \to \mathbb{C}$ be functions such that the function $\varphi_y(x) = f(x \cdot y) - f(x)m(y)$ belongs to V for each fixed $y \in G$. Then either f belongs to V or m is an exponential.*

Proof. Assume that m is not an exponential. Then there exist $y, z \in G$ such that $m(y \cdot z) - m(y)m(z) \neq 0$. However, we have

$$
\begin{aligned}
f(x \cdot y \cdot z) - f(x \cdot y)m(z) = \big(& f(x \cdot y \cdot z) - f(x)m(y \cdot z) \big) \\
& - m(z)\big(f(x \cdot y) - f(x)m(y) \big) \\
& + f(x)\big(m(y \cdot z) - m(y)m(z) \big)
\end{aligned}
$$

and hence

$$
\begin{aligned}
f(x) = \big[& \big(f(x \cdot y \cdot z) - f(x \cdot y)m(z) \big) - \big(f(x \cdot y \cdot z) - f(x)m(y \cdot z) \big) \\
& + m(z)\big(f(x \cdot y) - f(x)m(y) \big) \big] \cdot \big(m(y \cdot z) - m(y)m(z) \big)^{-1}
\end{aligned}
$$

for any $x \in G$. Now, the right-hand side, as a function of x, belongs to V, and hence so does f. $\qquad\square$

Székelyhidi [341] presented the following corollary.

Corollary 9.3. *Let $(G, \cdot)$ be a semigroup with identity 1 and V be an invariant vector space of complex-valued functions on G. Let $f, m : G \to \mathbb{C}$ be functions such that the functions*

$$
\varphi_y(x) = f(x \cdot y) - f(x)m(y) \quad \text{and} \quad \psi_x(y) = f(x \cdot y) - f(x)m(y)
$$

belong to V for each fixed $y \in G$ and $x \in G$, respectively. Then either f belongs to V or m is an exponential and $f(x) = f(1)m(x)$ for any $x \in G$.

Proof. Suppose that f does not belong to V. Then, by the preceding theorem, m is an exponential. On the other hand, the function $\psi_1(y) = f(y) - f(1)m(y)$ belongs to V and we have

$$
\begin{aligned}
& f(x \cdot y) - f(1)m(x \cdot y) \\
& = f(x \cdot y) - f(x)m(y) + \big(f(x) - f(1)m(x) \big)m(y)
\end{aligned}
$$

for all $x, y \in G$. If there exists an $x_0 \in G$ such that $f(x_0) \neq f(1)m(x_0)$, then m belongs to V and so does f, which is a contradiction. Hence, $f(x) = f(1)m(x)$ holds true for all $x \in G$. $\qquad\square$

Applying Corollary 9.3, Székelyhidi [341] generalized the theorem of Baker, Lawrence, and Zorzitto as one can see in the following corollary.

Corollary 9.4. *Let $(G, \cdot)$ be an abelian group with identity and let $f, m : G \to \mathbb{C}$ be functions such that there exist functions $M_1, M_2 : G \to [0, \infty)$ with*

$$|f(x \cdot y) - f(x)m(y)| \leq \min \{M_1(x), M_2(y)\}$$

for all $x, y \in G$. Then either f is bounded or m is an exponential and $f(x) = f(1)m(x)$ for all $x \in G$.

Proof. Let V be the space of bounded complex-valued functions defined on G. Obviously, V is an invariant vector space and we can apply Corollary 9.3. $\square$

During the thirty-first International Symposium on Functional Equations, Th. M. Rassias [291] introduced the term *mixed stability* of the function $f : E \to \mathbb{R}$ (or $\mathbb{C}$), where E is a Banach space, with respect to two operations "addition" and "multiplication" among any two elements of the set $\{x, y, f(x), f(y)\}$. Especially, he raised an open problem concerning the behavior of solutions of the inequality

$$|f(x + y) - f(x)f(y)| \leq \theta (\|x\|^p + \|y\|^p)$$

(see also [312]).

In connection with this open problem, S.-M. Jung [164] generalized the theorem of Baker, Lawrence, and Zorzitto; more precisely, he proved the superstability of the exponential equation when the Cauchy difference $f(x + y) - f(x)f(y)$ is not bounded. Let $H : [0, \infty)^2 \to [0, \infty)$ be a monotonically increasing function (in both variables) for which there exist, for given $u, v \geq 0$, an $\alpha = \alpha(u, v) > 0$ and a $w_0 = w_0(u, v) > 0$ such that

$$H(u, v + w) \leq \alpha(u, v)H(w, w) \tag{9.2}$$

for all $w \geq w_0$.

Theorem 9.5 (Jung). *Let E be a complex normed space and let $f : E \to \mathbb{C}$ be a function for which there exist a $z \in E$ $(z \neq 0)$ and a real number β $(0 < \beta < 1)$ such that*

$$\sum_{k=1}^{\infty} H(k\|z\|, \|z\|) |f(z)|^{-k-1} < \beta \tag{9.3}$$

and

$$H(n\|z\|, n\|z\|) = o(|f(z)|^n) \quad as \quad n \to \infty. \tag{9.4}$$

Moreover, assume that f satisfies

$$|f(x + y) - f(x)f(y)| \leq H(\|x\|, \|y\|) \tag{9.5}$$

for all $x, y \in E$. Then f is an exponential function.

Proof. We use induction on n to prove

$$\left| f(nz) - f(z)^n \right| \leq \sum_{k=1}^{n-1} H\left(k \|z\|, \|z\| \right) |f(z)|^{n-k-1} \tag{a}$$

for all integers $n \geq 2$. In view of (9.5), (a) is true for $n = 2$. If we assume that (a) is valid for some integer $n \geq 2$, then we get for $n + 1$

$$\left| f\left((n+1)z\right) - f(z)^{n+1} \right|$$
$$\leq \left| f\left((n+1)z\right) - f(nz)f(z) \right| + |f(z)| \left| f(nz) - f(z)^n \right|$$
$$\leq H\left(n\|z\|, \|z\| \right) + \sum_{k=1}^{n-1} H\left(k\|z\|, \|z\| \right) |f(z)|^{n-k}$$
$$\leq \sum_{k=1}^{n} H\left(k\|z\|, \|z\| \right) |f(z)|^{n+1-k-1}$$

by using (9.5) and (a).

Multiplying both sides of (a) by $|f(z)|^{-n}$ and using (9.3) yield

$$\left| f(nz)f(z)^{-n} - 1 \right| \leq \sum_{k=1}^{n-1} H\left(k\|z\|, \|z\| \right) |f(z)|^{-k-1} < \beta,$$

which implies

$$\lim_{n \to \infty} \left| f(nz)f(z)^{-n} \right| \geq 1 - \beta > 0. \tag{b}$$

Hence, it follows from (9.4) and (b) that

$$H\left(n\|z\|, n\|z\| \right) = o\left(|f(nz)| \right) \quad \text{as } n \to \infty. \tag{c}$$

Now, let $x, y \in E$ be fixed arbitrarily, and choose an $n \in \mathbb{N}$ such that

$$H\left(\|x\|, \|y + nz\| \right) \leq H\left(\|x\|, \|y\| + n\|z\| \right) \leq \alpha\left(\|x\|, \|y\| \right) H\left(n\|z\|, n\|z\| \right)$$

holds true (see (9.2)). We then have

$$|f(nz)| \, |f(x+y) - f(x)f(y)| \leq |f(x+y)f(nz) - f(x+y+nz)|$$
$$+ |f(x+y+nz) - f(x)f(y+nz)|$$
$$+ |f(x)| \, |f(y+nz) - f(y)f(nz)|$$
$$\leq H\left(\|x+y\|, n\|z\| \right) + H\left(\|x\|, \|y+nz\| \right)$$
$$+ |f(x)| H\left(\|y\|, n\|z\| \right)$$

and hence

$$|f(nz)| |f(x + y) - f(x)f(y)| \le CH\big(n\|z\|, n\|z\|\big) \qquad (d)$$

for some $C > \max\{\alpha(\|x + y\|, 0), \ \alpha(\|x\|, \|y\|), \ \alpha(\|y\|, 0)\}$. Letting $n \to \infty$ in (d) and comparing this with (c), we conclude that f is an exponential function. $\square$

P. Găvruta [116] also gave an answer to the problem suggested by Rassias concerning the mixed stability.

Theorem 9.6 (Găvruta). *Let E_1 and E_2 be a real normed space and a normed algebra with multiplicative norm, respectively. If a function $f : E_1 \to E_2$ satisfies the inequality*

$$\|f(x + y) - f(x)f(y)\| \le \theta\big(\|x\|^p + \|y\|^p\big) \qquad (9.6)$$

for all $x, y \in E_1$ and for some $p > 0$ and $\theta > 0$, then either

$$\|f(x)\| \le \delta\|x\|^p \quad \text{for all } x \in E_1 \text{ with } \|x\| \ge 1$$

or f is an exponential function, where $\delta = (1/2)\big(2^p + (4^p + 8\theta)^{1/2}\big)$.

Proof. Assume that there exists an $x_0 \in E_1$ with $\|x_0\| \ge 1$ and $\|f(x_0)\| > \delta\|x_0\|^p$. Then, there exists an $\varepsilon > 0$ such that

$$\|f(x_0)\| > (\delta + \varepsilon)\|x_0\|^p.$$

It follows from (9.6) that

$$\|f(2x_0) - f(x_0)^2\| \le 2\theta\|x_0\|^p.$$

Since E_2 is a normed algebra with multiplicative norm, we have

$$\begin{aligned}
\|f(2x_0)\| &\ge \|f(x_0)\|^2 - \|f(2x_0) - f(x_0)^2\| \\
&> (\delta + \varepsilon)^2\|x_0\|^{2p} - 2\theta\|x_0\|^p \\
&\ge \big((\delta + \varepsilon)^2 - 2\theta\big)\|x_0\|^p.
\end{aligned}$$

By the definition of δ, we get

$$\delta^2 = 2^p\delta + 2\theta \quad \text{and} \quad \delta > 2^p$$

and hence

$$\|f(2x_0)\| > (\delta + 2\varepsilon)2^p\|x_0\|^p.$$

We will now prove that

$$\|f(2^n x_0)\| > (\delta + 2^n\varepsilon)\|2^n x_0\|^p \qquad (a)$$

for all $n \in \mathbb{N}$. It follows from (9.6) that

$$\left\| f\left(2^{n+1} x_0\right) - f(2^n x_0)^2 \right\| \leq 2\theta \left\| 2^n x_0 \right\|^p.$$

By applying (a), we obtain

$$
\begin{aligned}
\left\| f\left(2^{n+1} x_0\right) \right\| &\geq \left\| f(2^n x_0) \right\|^2 - \left\| f\left(2^{n+1} x_0\right) - f(2^n x_0)^2 \right\| \\
&> (\delta + 2^n \varepsilon)^2 \left\| 2^n x_0 \right\|^{2p} - 2\theta \left\| 2^n x_0 \right\|^p \\
&\geq \left((\delta + 2^n \varepsilon)^2 - 2\theta\right) \left\| 2^n x_0 \right\|^p \\
&> (\delta + 2^{n+1} \varepsilon) 2^p \left\| 2^n x_0 \right\|^p,
\end{aligned}
$$

which proves the validity of the inequality (a).

If we set $x_n = 2^n x_0$, then $\| x_n \| \geq 1$ for any $n \in \mathbb{N}$. By (a), we have

$$\lim_{n \to \infty} \| x_n \|^p / \| f(x_n) \| = 0. \tag{b}$$

Choose $x, y, z \in E_1$ with $f(z) \neq 0$. It then follows from (9.6) that

$$
\begin{aligned}
\| f(z) f(x + y) - f(x + y + z) \| &\leq \theta \left(\| z \|^p + \| x + y \|^p \right), \\
\| f(x + y + z) - f(x) f(y + z) \| &\leq \theta \left(\| x \|^p + \| y + z \|^p \right).
\end{aligned}
$$

Hence, we have

$$
\begin{aligned}
\| f(z) f(x + y) &- f(x) f(y + z) \| \\
&\leq \theta \left(\| z \|^p + \| x \|^p + \| x + y \|^p + \| y + z \|^p \right).
\end{aligned}
\tag{c}
$$

In view of (9.6), we get

$$\| f(x) f(y + z) - f(x) f(y) f(z) \| \leq \theta \| f(x) \| \left(\| y \|^p + \| z \|^p \right),$$

which together with (c) yields

$$\| f(z) f(x + y) - f(x) f(y) f(z) \| \leq \theta \varphi(x, y, z), \tag{d}$$

where

$$\varphi(x, y, z) = \| x \|^p + \| z \|^p + \| x + y \|^p + \| y + z \|^p + \| f(x) \| \left(\| y \|^p + \| z \|^p \right).$$

Since E_2 is a normed algebra with multiplicative norm, it follows from (d) that

$$\| f(x + y) - f(x) f(y) \| \leq \theta \varphi(x, y, z) / \| f(z) \|.$$

If we put $z = x_n$, then it follows from (b) that

$$\lim_{n\to\infty} \varphi(x, y, x_n)/\|f(x_n)\| = 0$$

and consequently we obtain $f(x + y) = f(x)f(y)$. □

9.2 Stability in the Sense of Ger

The group structure in the range space of the exponential functional equation is the "multiplication." R. Ger [122] pointed out that the superstability phenomenon of the functional inequality (9.1) is caused by the fact that the natural group structure in the range space is disregarded. Thus, it seems more natural to suggest the stability problem in the following form:

$$\left| \frac{f(x + y)}{f(x)f(y)} - 1 \right| \le \delta. \tag{9.7}$$

If, for each function $f : (G, +) \to E\backslash\{0\}$ satisfying the inequality (9.7) for some $\delta > 0$ and for all $x, y \in G$, there exists an exponential function $M : G \to E\backslash\{0\}$ such that

$$\|f(x)/M(x) - 1\| \le \Phi(\delta) \quad \text{and} \quad \|M(x)/f(x) - 1\| \le \Psi(\delta)$$

for all $x \in G$, where $\Phi(\delta)$ and $\Psi(\delta)$ depend on δ only, then the exponential functional equation is said to be *stable in the sense of Ger*.

Let $(G, +)$ be an amenable semigroup (for the definition of amenability see Section 2.5), and let $\delta \in [0, 1)$ be a fixed number. Ger [122] proved that if a function $f : G \to \mathbb{C}\backslash\{0\}$ satisfies the inequality (9.7) for all $x, y \in G$, then there exists an exponential function $M : G \to \mathbb{C}\backslash\{0\}$ such that

$$\max\{|f(x)/M(x) - 1|, |M(x)/f(x) - 1|\} \le (2 - \delta)/(1 - \delta)$$

for all $x \in G$.

We notice that the bound $(2 - \delta)/(1 - \delta)$ in the above inequality does not tend to zero even though δ does tend to zero. R. Ger and P. Šemrl [123] resolved this shortcoming and proved the following theorem.

Theorem 9.7 (Ger and Šemrl). *Let $(G, +)$ be a cancellative abelian semigroup, and let δ be a given number with $0 \le \delta < 1$. Assume that a function $f : G \to \mathbb{C}\backslash\{0\}$ satisfies the inequality (9.7) for all $x, y \in G$. Then there exists a unique exponential function $M : G \to \mathbb{C}\backslash\{0\}$ such that*

$$\max\{|f(x)/M(x) - 1|, |M(x)/f(x) - 1|\}$$
$$\le \left(1 + (1 - \delta)^{-2} - 2\big((1 + \delta)/(1 - \delta)\big)^{1/2} \right)^{1/2}$$

for each $x \in G$.

Proof. Every nonzero complex number λ can be uniquely expressed as

$$\lambda = |\lambda| \exp\left(i \arg(\lambda)\right),$$

where $-\pi < \arg(\lambda) \leq \pi$. Then, (9.7) yields

$$\left| \frac{|f(x+y)|}{|f(x)||f(y)|} \exp\left(i\left(\arg f(x+y) - \arg f(x) - \arg f(y)\right)\right) - 1 \right| \leq \delta$$

for every $x, y \in G$. It follows that

$$1 - \delta \leq \frac{|f(x+y)|}{|f(x)||f(y)|} \leq 1 + \delta \qquad\qquad (a)$$

and

$$\arg f(x+y) - \arg f(x) - \arg f(y) \in 2\pi \mathbb{Z} + \left[-\sin^{-1}\delta, \sin^{-1}\delta\right]$$

for all $x, y \in G$. As $\delta < 1$, it necessarily holds true that $\sin^{-1}\delta < \pi/2$.

This congruence, together with Corollary 2.48, implies that there exists a function $p : G \to \mathbb{R}$ such that

$$p(x+y) - p(x) - p(y) \in 2\pi\mathbb{Z} \qquad\qquad (b)$$

for all $x, y \in G$, and

$$|\arg f(x) - p(x)| \leq \sin^{-1}\delta \qquad\qquad (c)$$

for any $x \in G$.

Put $h(x) = |f(x)|$. It then follows from (a) that

$$1 - \delta \leq \frac{h(x+y)}{h(x)h(y)} \leq 1 + \delta$$

for any $x, y \in G$. Consequently, we have

$$|\ln h(x+y) - \ln h(x) - \ln h(y)| \leq -\ln(1-\delta)$$

for $x, y \in G$. According to an extended version of Theorem 2.3, there exists an additive function $a : G \to \mathbb{R}$ such that

$$|\ln h(x) - a(x)| \leq -\ln(1-\delta) \qquad\qquad (d)$$

for every $x \in G$.

Let us define a function $M : G \to \mathbb{C} \setminus \{0\}$ by

$$M(x) = \exp\left(a(x) + ip(x)\right)$$

for each $x \in G$. It follows from the additivity of a and (b) that M is an exponential function. Moreover, we have

$$|f(x)/M(x) - 1| = \left| \exp\left(\ln h(x) - a(x) \right) \exp\left(i \left(\arg f(x) - p(x) \right) \right) - 1 \right|$$

for every $x \in G$. Applying (c) and (d) we see that the number $f(x)/M(x)$ belongs to the set

$$\Omega = \left\{ \lambda \in \mathbb{C} \mid 1 - \delta \leq |\lambda| \leq (1-\delta)^{-1}, \; -\sin^{-1}\delta \leq \arg(\lambda) \leq \sin^{-1}\delta \right\}.$$

Obviously,

$$\sup \left\{ |\lambda - 1| \mid \lambda \in \Omega \right\} = \left| (1-\delta)^{-1} \exp\left(i \sin^{-1}\delta \right) - 1 \right|$$
$$= \left(1 + (1-\delta)^{-2} - 2\left((1+\delta)/(1-\delta) \right)^{1/2} \right)^{1/2}.$$

The proof of the inequality for $|M(x)/f(x) - 1|$ continues through in exactly the same way. $\qquad \square$

It was shown that the assumption $\delta < 1$ is indispensable in the above stability result.

9.3 Stability on a Restricted Domain

This section presents stability problems of the exponential functional equation on a restricted domain, and these results will be applied to the study of asymptotic properties of exponential functions. More precisely, it will be proved that a function $f : E \to \mathbb{C}$ is an exponential function if and only if $f(x + y) - f(x)f(y) \to 0$ as $\|x\| + \|y\| \to \infty$ under some suitable conditions, where E is a real (or complex) normed space. Moreover, we also present that a function $f : E \to \mathbb{C} \setminus \{0\}$ is exponential if and only if $f(x + y)/f(x)f(y) \to 1$ as $\|x\| + \|y\| \to \infty$.

Let $B = \left\{ (x, y) \in E^2 \mid \|x\| < d \text{ and } \|y\| < d \right\}$ for a given $d > 0$. S.-M. Jung [180] proved the superstability of the exponential functional equation on a restricted domain:

Theorem 9.8. *Let E be a real (or complex) normed space. Assume that a function $f : E \to \mathbb{C}$ satisfies the inequality*

$$|f(x + y) - f(x)f(y)| \leq \delta$$

for some $\delta > 0$ and for all $(x, y) \in E^2 \setminus B$. If there exists a number $C > 0$ such that

$$\sup_{\|x\| \leq d} |f(x)| \leq C, \tag{9.8}$$

then either f is an exponential function or f is bounded. In the latter case,

$$|f(x)| \leq (1/2)\left(1 + (1 + 4(1 + 2C)\delta)^{1/2}\right)$$

for all $x \in E$.

Proof. Suppose that (x, y) is an arbitrary point of B. In the case of $x = 0$, choose a $z \in E$ with $\|z\| = d$. Otherwise, let us choose $z = (1 + d/\|x\|)x$. Then, we have

$$(x - z, y + z), \ (z, x - z), \ (y, z) \in E^2 \setminus B.$$

With such a z, by using the relation

$$
\begin{aligned}
f(x + y) - f(x)f(y) = &\left(f(x + y) - f(x - z)f(y + z)\right) \\
&- f(y)\left(f(x) - f(z)f(x - z)\right) \\
&+ f(x - z)\left(f(y + z) - f(y)f(z)\right),
\end{aligned}
$$

we obtain

$$|f(x + y) - f(x)f(y)| \leq \left(1 + |f(y)| + |f(x - z)|\right)\delta. \qquad (a)$$

Hence, it follows from (a) and (9.8) that

$$|f(x + y) - f(x)f(y)| \leq (1 + 2C)\delta$$

for all $(x, y) \in E^2$. Now, the assertion follows from Theorem 9.1. $\qquad \square$

Jung [180] was able to deduce an asymptotic result analogous to an asymptotic behavior of additive functions (see Theorem 2.34 or Corollary 2.35).

Corollary 9.9. *Let E be a real (or complex) normed space. Suppose a function $f : E \to \mathbb{C}$ is unbounded. Moreover, assume that there exists, for infinitely many $n \in \mathbb{N}$, a constant $C_n > 0$ such that*

$$\sup_{\|x\| \leq n} |f(x)| \leq C_n. \qquad (9.9)$$

Then f is an exponential function if and only if

$$f(x + y) - f(x)f(y) \to 0 \quad as \ \|x\| + \|y\| \to \infty. \qquad (9.10)$$

Proof. Assume that the asymptotic condition (9.10) holds true. Let $\delta > 0$ be given. On account of (9.9) and (9.10), there exists a sufficiently large $n \in \mathbb{N}$ such that the inequality (9.9) holds true and

$$|f(x + y) - f(x)f(y)| \leq \delta$$

for $\|x\| \geq n$ or $\|y\| \geq n$. It then follows from Theorem 9.8 that either

$$|f(x)| \leq (1/2)\big(1 + \sqrt{1 + 4(1 + 2C_n)\delta}\big) \quad \text{for all } x \in E,$$

or f is exponential. Since f is unbounded, f should be exponential. The reverse assertion is trivial. □

Jung [180] has also proved the stability (in the sense of Ger) of the exponential functional equation on restricted domains.

Theorem 9.10 (Jung). *Let d and B be given as in Theorem 9.8. Let E be a real (or complex) normed space. If a function $f : E \to \mathbb{C}\backslash\{0\}$ satisfies the inequality (9.7) for some $\delta \in [0, 1/2)$ and for all $(x, y) \in E^2\backslash B$, then there exists an exponential function $M : E \to \mathbb{C}\backslash\{0\}$ such that*

$$\max\big\{|M(x)/f(x) - 1|, |f(x)/M(x) - 1|\big\}$$

$$\leq \left(1 + \frac{(1+\delta)^2}{(1-\delta)^4} - 2(1 - 4\delta^2)\Big(\frac{1+\delta}{1-\delta}\Big)^{3/2}\right)^{1/2}$$

for all $x \in E$.

Proof. Suppose (x, y) belongs to B. Applying the same argument given in the proof of Theorem 9.8, we can choose a $z \in E$ such that

$$(x - z, y + z), \ (z, x - z), \ (y, z) \in E^2\backslash B.$$

By using (9.7) and the relation

$$\frac{f(x + y)}{f(x)f(y)} = \frac{f(x + y)}{f(x - z)f(y + z)} \cdot \frac{f(z)f(x - z)}{f(x)} \cdot \frac{f(y + z)}{f(y)f(z)},$$

we get

$$\frac{(1-\delta)^2}{1+\delta} \leq \left|\frac{f(x + y)}{f(x)f(y)}\right| \leq \frac{(1+\delta)^2}{1-\delta}. \tag{a}$$

According to (9.7), the complex number $f(x+y)f(x-z)^{-1}f(y+z)^{-1}$ is separated from 1 by a distance δ at most. Hence, it holds true that

$$\arg \frac{f(x + y)}{f(x - z)f(y + z)} \in 2\pi\mathbb{Z} + \big[-\sin^{-1}\delta, \ \sin^{-1}\delta\big],$$

where $0 \leq \sin^{-1}\delta < \pi/6$ because of $0 \leq \delta < 1/2$. We obtain analogous relations for $f(x)f(z)^{-1}f(x-z)^{-1}$ and $f(y+z)f(y)^{-1}f(z)^{-1}$ in the same way. Therefore, we have

$$\arg f(x + y) - \arg f(x) - \arg f(y) \in 2\pi\mathbb{Z} + \big[-3\sin^{-1}\delta, \ 3\sin^{-1}\delta\big] \tag{b}$$

for any $(x, y) \in B$.

Indeed, the relations (a) and (b) hold true for all $(x, y) \in E^2$. Taking logarithms on both sides of (a) yields

$$\left| \ln |f(x + y)| - \ln |f(x)| - \ln |f(y)| \right| \leq \ln \frac{1 + \delta}{(1 - \delta)^2}$$

for all $x, y \in E$. In view of Theorem 2.3, there exists a unique additive function $a : E \to \mathbb{R}$ such that

$$\left| a(x) - \ln |f(x)| \right| \leq \ln \frac{1 + \delta}{(1 - \delta)^2} \tag{c}$$

for all $x \in E$.

On account of (b) and Corollary 2.48, there exists a function $b : E \to \mathbb{R}$ such that

$$b(x + y) - b(x) - b(y) \in \mathbb{Z}, \tag{d}$$

for any $x, y \in E$, and

$$| \arg f(x) - 2\pi b(x) | \leq 3 \sin^{-1} \delta \tag{e}$$

for every $x \in E$.

Now, let us define a function $M : E \to \mathbb{C} \setminus \{0\}$ by

$$M(x) = e^{a(x) + 2\pi i b(x)}$$

for each $x \in E$. From the additivity of a and (d), it follows that M is exponential. Let $x \in E$ be given. Since $f(x)$ can be expressed as $e^{\ln |f(x)| + i \arg f(x)}$, we have

$$|M(x)/f(x) - 1| = \left| e^{a(x) - \ln |f(x)|} e^{i(2\pi b(x) - \arg f(x))} - 1 \right|.$$

Now, let us define

$$D = \left\{ re^{i\theta} \in \mathbb{C} \;\middle|\; \frac{(1 - \delta)^2}{1 + \delta} \leq r \leq \frac{1 + \delta}{(1 - \delta)^2}, \; -3 \sin^{-1} \delta \leq \theta \leq 3 \sin^{-1} \delta \right\}.$$

If we set

$$z_1 = \frac{(1 - \delta)^2}{1 + \delta} e^{i 3 \sin^{-1} \delta} \quad \text{and} \quad z_2 = \frac{1 + \delta}{(1 - \delta)^2} e^{i 3 \sin^{-1} \delta},$$

then

$$|z_1 - 1|^2 = \frac{(1 - \delta)^4}{(1 + \delta)^2} - 2 \frac{(1 - \delta)^{5/2}}{(1 + \delta)^{1/2}} (1 - 4\delta^2) + 1,$$

$$|z_2 - 1|^2 = \frac{(1 + \delta)^2}{(1 - \delta)^4} - 2 \left(\frac{1 + \delta}{1 - \delta} \right)^{3/2} (1 - 4\delta^2) + 1.$$

We now have

$$|z_2 - 1|^2 - |z_1 - 1|^2$$

$$= \frac{(1+\delta)^2}{(1-\delta)^4}\left(1 - \frac{(1-\delta)^8}{(1+\delta)^4}\right) + 2(1 - 4\delta^2)\frac{(1-\delta)^{5/2}}{(1+\delta)^{1/2}}\left(1 - \frac{(1+\delta)^2}{(1-\delta)^4}\right)$$

$$= \frac{(1+\delta)^2}{(1-\delta)^4}\left(1 - \frac{(1-\delta)^4}{(1+\delta)^2}\right)\left(\frac{(1-\delta)^4}{(1+\delta)^2} - 2(1 - 4\delta^2)\frac{(1-\delta)^{5/2}}{(1+\delta)^{1/2}} + 1\right).$$

On the other hand, since $(1 - 4\delta^2)\sqrt{1 - \delta^2} \le 1$, we have $(1 - 4\delta^2)(1 - \delta)^{1/2} \le (1 + \delta)^{-1/2}$ and hence,

$$(1 - 4\delta^2)\frac{(1-\delta)^{5/2}}{(1+\delta)^{1/2}} \le \frac{(1-\delta)^2}{1+\delta} \quad \text{or} \quad -2(1 - 4\delta^2)\frac{(1-\delta)^{5/2}}{(1+\delta)^{1/2}} \ge -2\frac{(1-\delta)^2}{1+\delta}.$$

Thus, we get

$$|z_2 - 1|^2 - |z_1 - 1|^2 \ge \frac{(1+\delta)^2}{(1-\delta)^4}\left(1 - \frac{(1-\delta)^4}{(1+\delta)^2}\right)\left(\frac{(1-\delta)^2}{1+\delta} - 1\right)^2 \ge 0,$$

and hence we conclude that $|z_2 - 1| = \max\{|z - 1| \mid z \in D\}$.

Therefore, it follows from (c) and (e) that

$$|M(x)/f(x) - 1| \le |z_2 - 1|$$

$$= \left(1 + \frac{(1+\delta)^2}{(1-\delta)^4} - 2(1 - 4\delta^2)\left(\frac{1+\delta}{1-\delta}\right)^{3/2}\right)^{1/2}.$$

The proof of the inequality for $|f(x)/M(x) - 1|$ goes through in the same way. Hence, we omit the proof.   □

9.4 Exponential Functional Equation of Other Type

Every complex-valued function of the form $f(x) = a^x$ ($x \in \mathbb{C}$), where $a > 0$ is a given number, is a solution of the functional equation

$$f(xy) = f(x)^y. \tag{9.11}$$

Hence, the above functional equation may be regarded as a variation of the exponential functional equation.

If a function $f : \mathbb{R} \to (0, \infty)$ satisfies the equation (9.11) for all $x, y \in \mathbb{R}$, it then follows from (9.11) that

$$f(x)^y = f(y)^x$$

for all $x, y \in \mathbb{R}$. Thus, there exists a real number $a = f(1) > 0$ such that $f(x) = a^x$ for all $x \in \mathbb{R}$. On the other hand, each function $f : \mathbb{R} \to (0, \infty)$ of the form $f(x) = a^x$ satisfies the equation (9.11) for any $x, y \in \mathbb{R}$. Therefore, the general solution $f : \mathbb{R} \to (0, \infty)$ of the functional equation (9.11) is $f(x) = a^x$, where $a > 0$ is a given number.

Let us introduce some convenient notations:

$$e_i(x) = x^{2^i}, \quad p_n(x) = \prod_{i=0}^{n-1} e_i(x), \quad \text{and} \quad \alpha(x) = \sum_{n=1}^{\infty} p_n(x)^{-1}$$

for each $x \in \mathbb{R}$ and $n \in \mathbb{N}_0$.

S.-M. Jung [165] proved the stability of the equation (9.11) in the sense of Ger.

Theorem 9.11 (Jung). *Let $\delta \in (0, 1)$ be a given number. If a function $f : (0, \infty) \to (0, \infty)$ satisfies the inequality*

$$|f(xy)/f(x)^y - 1| \le \delta \tag{9.12}$$

for all $x, y > 0$, then there exists a unique constant $a > 0$ such that

$$(1 - \delta)^{\alpha(x)} \le a^x/f(x) \le (1 + \delta)^{\alpha(x)} \tag{9.13}$$

for all $x > 1$.

Proof. Substituting $e_{n-1}(x)$ for x and y in (9.12) yields

$$1 - \delta \le \frac{f(e_n(x))}{f(e_{n-1}(x))^{e_{n-1}(x)}} \le 1 + \delta$$

for every $x > 0$ and $n \in \mathbb{N}$. For $n > m \ge 0$, we obtain

$$\frac{f(e_n(x))}{f(e_m(x))^{e_m(x)e_{m+1}(x)\cdots e_{n-1}(x)}}$$

$$= \frac{f(e_n(x))}{f(e_{n-1}(x))^{e_{n-1}(x)}} \left(\frac{f(e_{n-1}(x))}{f(e_{n-2}(x))^{e_{n-2}(x)}} \right)^{e_{n-1}(x)}$$

$$\cdots \left(\frac{f(e_{m+1}(x))}{f(e_m(x))^{e_m(x)}} \right)^{e_{m+1}(x)e_{m+2}(x)\cdots e_{n-1}(x)}$$

and

$$\left(\frac{f(e_n(x))}{f(e_m(x))^{e_m(x)e_{m+1}(x)\cdots e_{n-1}(x)}} \right)^{1/p_n(x)} = \frac{f(e_n(x))^{1/p_n(x)}}{f(e_m(x))^{1/p_m(x)}}.$$

Hence, we have

$$(1 - \delta)^{\alpha_n(x) - \alpha_m(x)} \leq \frac{f(e_n(x))^{1/p_n(x)}}{f(e_m(x))^{1/p_m(x)}} \leq (1 + \delta)^{\alpha_n(x) - \alpha_m(x)}, \qquad (a)$$

where

$$\alpha_n(x) = \sum_{i=1}^{n} p_i(x)^{-1}.$$

Since the sequence $\{\alpha_n(x)\}$ converges as $n \to \infty$ for $x > 1$, it follows from (a) that the sequence $\{\ln f(e_n(x))^{1/p_n(x)}\}$ is a Cauchy sequence for each $x > 1$. Therefore, we can define

$$L(x) = \lim_{n \to \infty} \ln f(e_n(x))^{1/p_n(x)}$$

and

$$F(x) = e^{L(x)}$$

for all $x > 1$. In fact, we get

$$F(x) = \lim_{n \to \infty} f(e_n(x))^{1/p_n(x)} \qquad (b)$$

for any $x > 1$.

Substituting $e_n(x)$ and $e_n(y)$ for x and y in (9.12), respectively, yields

$$(1 - \delta) f(e_n(x))^{e_n(y)} \leq f(e_n(xy)) \leq (1 + \delta) f(e_n(x))^{e_n(y)} \qquad (c)$$

for all $x, y > 0$ and for any $n \in \mathbb{N}$. Hence, by (b) and (c), we have

$$\lim_{n \to \infty} (1 - \delta)^{1/p_n(xy)} f(e_n(x))^{e_n(y)/p_n(xy)} \leq F(xy)$$

and

$$F(xy) \leq \lim_{n \to \infty} (1 + \delta)^{1/p_n(xy)} f(e_n(x))^{e_n(y)/p_n(xy)}$$

for all $x, y > 0$ with $xy > 1$. Since $p_n(xy) \to \infty$ as $n \to \infty$ ($xy > 1$) and $e_n(y)/p_n(xy) = y/p_n(x)$ for any $n \in \mathbb{N}$, it follows from (b) and the preceding inequalities that $F(xy) = F(x)^y$ for $x > 1$ and $y > 0$ with $xy > 1$. Therefore, there exists a constant $a > 0$ such that $F(x) = a^x$ for all $x > 1$ (see above). By putting $m = 0$ in (a) and using (b), we easily see the validity of (9.13).

Assume now that b is another positive constant such that

$$(1 - \delta)^{\alpha(x)} \leq b^x/f(x) \leq (1 + \delta)^{\alpha(x)}$$

for all $x > 1$. Without loss of generality, let $b > a$. Then, the following inequalities

$$\left(\frac{1-\delta}{1+\delta}\right)^{\alpha(x)} \leq \left(\frac{b}{a}\right)^x = \frac{b^x}{f(x)} \frac{f(x)}{a^x} \leq \left(\frac{1+\delta}{1-\delta}\right)^{\alpha(x)}$$

lead to a contradiction, since $\alpha(x) < 1$ for $x > 2$ and $(b/a)^x \to \infty$ as $x \to \infty$. Therefore, we conclude that there exists a unique constant $a > 0$ satisfying the relations in (9.13). $\qquad\square$

Let $d > 0$ be given, and let $\delta \in (0, 1)$ satisfy the condition

$$1 + \delta < 2(1 - \delta)^d. \tag{9.14}$$

It is not difficult to demonstrate that there exists at least one $\delta \in (0, 1)$ satisfying the condition (9.14) for any $d > 0$.

The following theorem is also due to Jung [165].

Theorem 9.12. *Let a function* $f : (0, \infty) \to (0, \infty)$ *satisfy the inequality* (9.12) *for all* $x, y > 0$ *with* $x + y \geq d$. *Then there exists a unique constant* $a > 0$ *such that*

$$\left(2 - \frac{1+\delta}{(1-\delta)^d}\right)^{\alpha(x)} \leq \frac{a^x}{f(x)} \leq \left(\frac{1+\delta}{(1-\delta)^d}\right)^{\alpha(x)} \tag{9.15}$$

for any $x > 1$.

Proof. Let $x, y > 0$ be given with $x + y < d$. Choose a $t > 0$ satisfying $tx \geq d$. From (9.12) and the relation

$$\frac{f(xy)}{f(x)^y} = \frac{f(xy)}{f(tx)^{y/t}} \left(\frac{f(tx)^{1/t}}{f(x)}\right)^y,$$

it follows that

$$\frac{1-\delta}{(1+\delta)^d} \leq \frac{f(xy)}{f(x)^y} \leq \frac{1+\delta}{(1-\delta)^d}.$$

Hence, the inequality

$$|f(xy)/f(x)^y - 1| \leq \frac{1+\delta}{(1-\delta)^d} - 1 \tag{a}$$

holds true for $x, y > 0$ with $x + y < d$. In view of the hypothesis, the inequality (a) holds true for all $x, y > 0$. The condition (9.14) guarantees that the value of the right-hand side of (a) belongs to $(0, 1)$. Hence, on account of (a) and Theorem 9.11, there exists a unique constant $a > 0$ satisfying (9.15) for each $x > 1$. $\qquad\square$

The following corollary is an easily applicable version of Theorem 9.12 (see [165]).

Corollary 9.13. *Let $\varepsilon > 1$ be given. If a function $f : (0, \infty) \to (0, \infty)$ satisfies the inequality (9.12) for all $x, y > 0$ with $x + y \geq d$, then there exists a unique constant $a > 0$ such that*

$$\left(2 - \frac{1 + \delta}{(1 - \delta)^d}\right)^{\alpha(\varepsilon)} \leq \frac{a^x}{f(x)} \leq \left(\frac{1 + \delta}{(1 - \delta)^d}\right)^{\alpha(\varepsilon)}$$

for all $x \geq \varepsilon$.

Now, we provide a sufficient condition for a function $f : (0, \infty) \to (0, \infty)$ to be asymptotically exponential (ref. [165]).

Corollary 9.14. *If a function $f : (0, \infty) \to (0, \infty)$ satisfies the condition*

$$\left| f(xy)/f(x)^y - 1 \right| = o\left((x + y)^{-1}\right) \quad as \ x + y \to \infty, \tag{9.16}$$

then there exists a unique constant $a > 0$ such that $f(x) = a^x$ for any $x > 1$.

Proof. Let $\varepsilon > 1$ be arbitrary. According to (9.16), there exists a sequence $\{\delta_n\}$, monotonically decreasing to zero, such that

$$\left| f(xy)/f(x)^y - 1 \right| \leq \delta_n/n$$

for all $x, y > 0$ with $x + y \geq n$. By Corollary 9.13, there exists a unique constant $a_n > 0$ such that

$$\left(2 - \frac{1 + \delta_n/n}{(1 - \delta_n/n)^n}\right)^{\alpha(\varepsilon)} \leq \frac{a_n^x}{f(x)} \leq \left(\frac{1 + \delta_n/n}{(1 - \delta_n/n)^n}\right)^{\alpha(\varepsilon)} \tag{a}$$

for every $x \geq \varepsilon$. Obviously, there exists an infinite subset I of $\mathbb{N}$ such that the sequence $\{(1 + \delta_n/n)/(1 - \delta_n/n)^n\}$ decreases monotonically to 1 as n tends to infinity through I. Let $m, n \in I$ satisfy $n > m$. In view of (a) and the above consideration, we get

$$\left(2 - \frac{1 + \delta_m/m}{(1 - \delta_m/m)^m}\right)^{\alpha(\varepsilon)} \leq \frac{a_n^x}{f(x)} \leq \left(\frac{1 + \delta_m/m}{(1 - \delta_m/m)^m}\right)^{\alpha(\varepsilon)}$$

for any $x \geq \varepsilon$, which implies $a_m = a_n = a$ for all $m, n \in I$. Letting $n \to \infty$ through I in (a) and using the above consideration again, we can conclude that $f(x) = a^x$ for any $x \geq \varepsilon$. Since $\varepsilon > 1$ was given arbitrarily, $f(x) = a^x$ holds true for all $x > 1$. $\qquad \square$

Chapter 10
Multiplicative Functional Equations

The *multiplicative functional equation* $f(xy) = f(x)f(y)$ may be identified with the exponential functional equation if the domain of functions involved is a semigroup. However, if the domain space is a field or an algebra, then the former is obviously different from the latter. It is well-known that the general solution $f : \mathbb{R} \to \mathbb{R}$ of the multiplicative functional equation $f(xy) = f(x)f(y)$ is $f(x) = 0$, $f(x) = 1$, $f(x) = e^{A(\ln |x|)}|\text{sign}(x)|$, and $f(x) = e^{A(\ln |x|)}\text{sign}(x)$ for all $x \in \mathbb{R}$, where $A : \mathbb{R} \to \mathbb{R}$ is an additive function and $\text{sign} : \mathbb{R} \to \{-1, 0, 1\}$ is the *sign function*. If we impose the continuity on solution functions $f : \mathbb{R} \to \mathbb{R}$ of the multiplicative equation, then $f(x) = 0$, $f(x) = 1$, $f(x) = |x|^\alpha$, and $f(x) = |x|^\alpha \text{sign}(x)$ for all $x \in \mathbb{R}$, where α is a positive real constant. The first section deals with the superstability of the multiplicative Cauchy equation and a functional equation connected with the Reynolds operator. In Section 10.2, the results on δ-multiplicative functionals on complex Banach algebras will be discussed in connection with the AMNM algebras which will be described in Section 10.3. Another multiplicative functional equation $f(x^y) = f(x)^y$ for real-valued functions defined on $\mathbb{R}$ will be discussed in Section 10.4. This functional equation is superstable in the sense of Ger. In the last section, we will prove that a new multiplicative functional equation $f(x + y) = f(x)f(y)f(1/x + 1/y)$ is stable in the sense of Ger.

10.1 Superstability

If the domain of functions involved is a semigroup, the multiplicative Cauchy functional equation

$$f(xy) = f(x)f(y)$$

may be identified with the exponential Cauchy functional equation $f(x + y) = f(x)f(y)$. Therefore, in view of Theorem 9.1, it is obvious that the multiplicative Cauchy equation is superstable.

S.-M. Jung, *Hyers–Ulam–Rassias Stability of Functional Equations in Nonlinear Analysis*, Springer Optimization and Its Applications 48, DOI 10.1007/978-1-4419-9637-4_10, © Springer Science+Business Media, LLC 2011

Let $(G, \cdot)$ be a semigroup and let $\delta > 0$ be given. If a function $f : G \to \mathbb{C}$ satisfies the inequality

$$|f(x \cdot y) - f(x)f(y)| \le \delta$$

for all $x, y \in G$, then either $|f(x)| \le \left(1 + \sqrt{1 + 4\delta}\right)/2$ for all $x \in G$ or $f(x \cdot y) = f(x)f(y)$ for all $x, y \in G$.

Let $(G, \circ)$ be an abelian semigroup and let $g : G \to G$ be a given function. We now consider a functional equation connected with the Reynolds operator,

$$f(x \circ g(y)) = f(x)f(y) \tag{10.1}$$

for all $x, y \in G$. If g is the identity function, then the functional equation (10.1) reduces to the multiplicative Cauchy functional equation. If $(G, \circ) = (\mathbb{R}, \cdot)$ and $g = f$, then the equation (10.1) reduces to the functional equation

$$f(xf(y)) = f(x)f(y),$$

whose origin is in the averaging theory applied to turbulent fluid motion. This functional equation is connected with some linear operators, i.e., the Reynolds operator, the averaging operator, and the multiplicatively symmetric operator.

In 2007, A. Najdecki [257] proved that the functional equation (10.1) is superstable.

Theorem 10.1 (Najdecki). *Let $(G, \circ)$ be an abelian semigroup, let $g : G \to G$ be a given function, and let $\mathbb{K}$ denote either $\mathbb{R}$ or $\mathbb{C}$. If a function $f : G \to \mathbb{K}$ satisfies the inequality*

$$\left|f(x \circ g(y)) - f(x)f(y)\right| \le \delta \tag{10.2}$$

for all $x, y \in G$ and for some $\delta > 0$, then either f is bounded or f is a solution of the functional equation (10.1).

Proof. Suppose that f is unbounded. Then we can choose a sequence $\{x_n\}$ of elements of G such that $0 \ne |f(x_n)| \to \infty$ as $n \to \infty$. Put $y = x_n$ in (10.2) to obtain

$$\left|f(x \circ g(x_n))/f(x_n) - f(x)\right| \le \delta/|f(x_n)|. \tag{a}$$

Since $|f(x_n)| \to \infty$ as $n \to \infty$, it follows from (a) that

$$f(x) = \lim_{n \to \infty} f(x \circ g(x_n))/f(x_n) \tag{b}$$

for all $x \in G$.

Replacing x in (10.2) with $x \circ g(x_n)$ yields

$$\left|f(x \circ g(x_n) \circ g(y)) - f(x \circ g(x_n))f(y)\right| \le \delta \tag{c}$$

for any $x, y \in G$. It follows from (c) that

$$\lim_{n\to\infty} \frac{f\big(x \circ g(y) \circ g(x_n)\big) - f\big(x \circ g(x_n)\big) f(y)}{f(x_n)} = 0. \qquad (d)$$

Thus, by (b) and (d), we obtain

$$
\begin{aligned}
f\big(x \circ g(y)\big) &= \lim_{n\to\infty} \frac{f\big(x \circ g(y) \circ g(x_n)\big)}{f(x_n)} \\
&= \lim_{n\to\infty} \frac{f\big(x \circ g(y) \circ g(x_n)\big) - f\big(x \circ g(x_n)\big) f(y)}{f(x_n)} \\
&\quad + \lim_{n\to\infty} \frac{f\big(x \circ g(x_n)\big)}{f(x_n)} f(y) \\
&= f(x) f(y)
\end{aligned}
$$

for all $x, y \in G$. $\qquad\qquad\square$

10.2 δ-Multiplicative Functionals

In this section, let E denote a commutative complex Banach algebra and let E^* denote the dual space of E.

For any linear functional ϕ on E we define

$$\check{\phi}(x, y) = \phi(xy) - \phi(x)\phi(y)$$

for all $x, y \in E$. If the norm of the bilinear functional $\check{\phi}$ is less than or equal to $\delta > 0$, then ϕ is called δ-multiplicative.

We introduce basic lemmas presented by B. E. Johnson [146].

Lemma 10.2. Let $\delta > 0$ be given. If $\phi \in E^*$ is δ-multiplicative, then $\|\phi\| \leq 1 + \delta$.

Proof. By the hypothesis, we have

$$|\phi(xy) - \phi(x)\phi(y)| \leq \delta \|x\| \|y\| \qquad (a)$$

for all $x, y \in E$. Given ε with $0 < \varepsilon < \|\phi\|$, choose an x with $\|x\| = 1$ such that $\|\phi\| - \varepsilon < |\phi(x)|$ and put $y = x$ in (a). Then, we get

$$\big|\phi(x)^2 - \phi\big(x^2\big)\big| \leq \delta$$

and hence

$$\big(\|\phi\| - \varepsilon\big)^2 < |\phi(x)|^2 \leq \big|\phi\big(x^2\big)\big| + \delta \leq \|\phi\| + \delta,$$

since $\|x^2\| \le \|x\|^2 = 1$. Letting $\varepsilon \to 0$ in the above inequality and solving the resulting inequality, we obtain

$$\|\phi\| \le (1/2)\left(1 + \sqrt{1 + 4\delta}\,\right) \le 1 + \delta,$$

which ends the proof. $\square$

K. Jarosz [145] proved a proposition in which the same bound for ϕ is obtained as in Lemma 10.2, but without the assumption of continuity for the linear functional ϕ. The following lemma follows from Lemma 10.2 by a straightforward calculation. Hence, we omit the proof (ref. [146]).

Lemma 10.3. *Let $\delta > 0$ be given, let a linear functional ϕ on E be δ-multiplicative, and let ψ be a continuous linear functional on E ($\psi \in E^*$). Then $\phi + \psi$ is $(\delta + (3 + 2\delta)\|\psi\| + \|\psi\|^2)$-multiplicative. If $\lambda \in \mathbb{C}$, then $(1 + \lambda)\phi$ is $|1 + \lambda|(\delta + |\lambda|(1 + \delta)^2)$-multiplicative.*

If the algebra E lacks an identity, E can be extended by adjoining an identity to E, and the extended one is denoted by E^1. Each linear functional ϕ on E can also be extended to a linear functional $\tilde{\phi}$ on E^1 by putting $\tilde{\phi}(1) = 1$. Then, ϕ is δ-multiplicative if and only if $\tilde{\phi}$ is. In some cases, an algebra which does not have an identity may have an approximate identity, which may be defined as follows: A net $\{e_\alpha\}$ in E is an *approximate identity* for E if for each $x \in E$ there exists a subnet $\{e_\beta\}$ of $\{e_\alpha\}$ such that

$$\lim_{\beta} e_\beta x = x.$$

Johnson estimated in [146] lower bounds for the norm of a δ-multiplicative functional (cf. [137]).

Lemma 10.4. *(i) Let E possess an identity 1 and let ϕ be a δ-multiplicative linear functional on E with $0 < \delta < 1/4$. Then either $|1 - \phi(1)| < 2\delta$, in which case $\|\phi\| > 1 - 2\delta$, or $\|\phi\| < 2\delta$.*

(ii) Let $K > 0$ and let E have an approximate identity $\{e_\alpha\}$ with $\|e_\alpha\| \le K$ for all α. Let $0 < \delta < (4K^2)^{-1}$ and let ϕ be a δ-multiplicative linear functional on E. Then either $\limsup_{\alpha} |1 - \phi(e_\alpha)| < 2\delta K^2$, in which case $\|\phi\| > 1/K - 2\delta K$, or $\limsup_{\alpha} |\phi(e_\alpha)| < 2\delta K^2$, in which case $\|\phi\| < 2\delta K$.

Proof. Note that the first statement is a special case of the second. Let L be the set of points of accumulation of the bounded net $\{\phi(e_\alpha)\}$ and let $\ell, \ell' \in L$. We have

$$|\phi(e_\alpha e_\beta) - \phi(e_\alpha)\phi(e_\beta)| \le \delta K^2.$$

If we take limits on suitable subnets, first for α and then for β, we get

$$|\phi(e_\beta) - \ell\phi(e_\beta)| \le \delta K^2 \quad \text{and} \quad |\ell'(1 - \ell)| \le \delta K^2 < 1/4.$$

For the special case where $\ell = \ell'$, we have either $|1 - \ell| < 2\delta K^2$ or $|\ell| \leq (1/2)(\sqrt{1 + 4\delta K^2} - 1) < 2\delta K^2$. If $|1 - \ell| < 2\delta K^2$ and $|\ell'| < 2\delta K^2$, then $|\ell(1 - \ell')| > 1/4$, so either we have the first alternative for all $\ell \in L$ or the second for all $\ell \in L$.

The first alternative gives

$$\limsup_{\alpha} |1 - \phi(e_\alpha)| < 2\delta K^2,$$

so

$$K\|\phi\| \geq \limsup_{\alpha} |\phi(e_\alpha)| > 1 - 2\delta K^2.$$

The second gives

$$\limsup_{\alpha} |\phi(e_\alpha)| < 2\delta K^2.$$

If $\ell \in L$, taking the limit on a suitable subnet in the inequality

$$|\phi(xe_\alpha) - \phi(x)\phi(e_\alpha)| \leq \delta\|x\|K$$

yields $|\phi(x)| \leq \delta\|x\|K(1 - \ell)^{-1}$ and hence $\|\phi\| \leq \delta K(1 - \ell)^{-1} < 2\delta K$, which ends the proof. $\square$

Using Lemmas 10.3 and 10.4, we can prove the following lemma (see [137]).

Lemma 10.5. *Suppose that E has an identity 1 and let $\phi \in E^*$ be δ-multiplicative with $0 < \delta < 1/4$. If $\|\phi\| > 2\delta$, then the functionals $\psi_1 = \phi/\|\phi\|$ and $\psi_2 = \phi/\phi(1)$ are η-multiplicative, lie within a distance of $2\delta(1 + \delta)(1 - 2\delta)^{-1}$ of ϕ, and have $\|\psi_1\| = 1$ and $\psi_2(1) = 1$, where $\eta = (8/3)(1 + 2\delta)(3\delta + 2\delta^2 + 2\delta^3)$.*

Proof. The arguments for ψ_1 and ψ_2 are the same. We consider ψ_1. By the hypothesis, ϕ satisfies the first alternative of Lemma 10.4 (i), so that $\|\phi\| > 1 - 2\delta$. Hence, we get $(1 - \|\phi\|)/\|\phi\| < 2\delta/(1 - 2\delta)$. Now, use Lemma 10.2 to obtain $\|\psi_1 - \phi\| < 2\delta(1 + \delta)/(1 - 2\delta)$. Next, apply Lemma 10.3 with $\lambda + 1 = \|\phi\|^{-1}$, so that $\lambda = (1 - \|\phi\|)/\|\phi\| < 2\delta/(1 - 2\delta)$. According to Lemma 10.3, ψ_1 is $|1 + \lambda|(\delta + |\lambda|(1 + \delta)^2)$-multiplicative. Thus,

$$|1 + \lambda|(\delta + |\lambda|(1 + \delta)^2) < (1 - 2\delta)^{-2}(3\delta + 2\delta^2 + 2\delta^3).$$

Since $0 < \delta < 1/4$, we have $(1 - 2\delta)^{-2} < (8/3)(1 + 2\delta)$ which completes the proof. $\square$

B. E. Johnson [146] proved Lemma 10.5 with $\eta = (1 + 2\delta)(2\delta + 4\delta^2 + 4\delta^3)$. Let us define

$$D = \{\phi \in E^* \mid \|\phi\| = |\phi(1)| = \phi(1)\}$$

and

$$W(x) = \{\phi(x) \mid \phi \in D\}$$

for each $x \in E$. The set $W(x)$ is called the *numerical range* of x.

Johnson [146] also presented the following lemma.

Lemma 10.6. *Assume that E has an identity 1. Let $b \in E$ and $\mu \in \mathbb{R}$ be given. If $\Re(z) \leq \mu$ for all $z \in W(b)$, then $\left\| (\lambda 1 - b)^{-1} \right\| \leq \left(\Re(\lambda) - \mu \right)^{-1}$ for any $\lambda \in \mathbb{C}$ with $\Re(\lambda) > \mu$.*

Proof. Let $a \in E$ with $\|a\| = 1$ and let $\phi \in E^*$ with $\|\phi\| = 1 = \phi(a)$. Then, $g(c) = \phi(ac)$ is an element of D. We have

$$\|a(\lambda 1 - b)\| \geq \left| \phi\big(a(\lambda 1 - b)\big) \right| = |\lambda - g(b)| \geq \Re\big(\lambda - g(b)\big) \geq \Re(\lambda) - \mu.$$

By [22, Theorem 6], $(\lambda 1 - b)^{-1}$ exists. By putting

$$a = (\lambda 1 - b)^{-1} \left\| (\lambda 1 - b)^{-1} \right\|^{-1},$$

we achieve the required inequality.   □

Using the previous lemmas, Johnson [146] proved the following theorem.

Theorem 10.7 (Johnson). *Assume that E has an identity 1. Let $\delta \in (0, 1)$ be fixed. Suppose that $\phi \in E^*$ is δ-multiplicative with $\phi(1) = 1$. Then there exists an element ψ of D such that $\|\phi - \psi\| \leq \delta(2 + \delta)$.*

Proof. The inequality

$$\left| \phi(\lambda 1 - b)\phi\big((\lambda 1 - b)^{-1}\big) - 1 \right| \leq \delta \|\lambda 1 - b\| \left\| (\lambda 1 - b)^{-1} \right\|$$

together with Lemma 10.2 implies that if λ, μ, and b are as in Lemma 10.6, then

$$
\begin{aligned}
\left(\Re(\lambda) - \mu \right)^{-1} (1 + \delta) |\phi(\lambda 1 - b)| &\geq \left\| (\lambda 1 - b)^{-1} \right\| (1 + \delta) |\phi(\lambda 1 - b)| \\
&\geq \|\phi\| \left\| (\lambda 1 - b)^{-1} \right\| |\phi(\lambda 1 - b)| \\
&\geq |\phi(\lambda 1 - b)| \left| \phi\big((\lambda 1 - b)^{-1}\big) \right| \\
&\geq 1 - \delta \|\lambda 1 - b\| \left\| (\lambda 1 - b)^{-1} \right\| \\
&\geq 1 - \delta \|\lambda 1 - b\| \left(\Re(\lambda) - \mu \right)^{-1},
\end{aligned}
$$

so

$$|\lambda - \phi(b)| = |\phi(\lambda 1 - b)| \geq (1 + \delta)^{-1} \left(\Re(\lambda) - \mu - \delta \|\lambda 1 - b\| \right).$$

Suppose that $\Re(\phi(b)) > \mu$, then we can put $\lambda = \phi(b)$ in this inequality and consider Lemma 10.2 to see that

$$\Re(\phi(b)) - \mu \le \delta \, \|\phi(b)1 - b\| \le \delta(2 + \delta)\|b\|.$$

Clearly, this inequality also holds true for $\Re(\phi(b)) \le \mu$.

Put $\delta_1 = \delta(2+\delta)$ and let B^* denote the closed unit ball in E^*. We want to present that $D \cap (\phi + \delta_1 B^*) \ne \emptyset$. Suppose on the contrary that this set is empty. Then, we can apply the Hahn–Banach theorem to E^* in the w^*-topology to find a hyperplane strongly separating the compact convex sets D and $\phi + \delta_1 B^*$. Thus, there exist an $a \in E$, which we can assume has norm 1, and a $\mu \in \mathbb{R}$ with $\Re(f(a)) < \mu$ for all $f \in D$ and $\Re(f(a)) > \mu$ for all $f \in \phi + \delta_1 B^*$. Taking a $g \in B^*$ with $g(a) = 1$ and putting $f = \phi - \delta_1 g$ yield $\Re(\phi(a)) - \delta_1 = \Re(f(a)) > \mu$ and hence, $\Re(\phi(a)) - \mu > \delta(2+\delta)\|a\|$ which is contrary to the inequality at the end of the previous paragraph. Hence, we have verified that the set $D \cap (\phi + \delta_1 B^*)$ is not empty. This completes the proof.   □

10.3 Theory of AMNM Algebras

As in the previous section, let E be a commutative complex Banach algebra and let E^* denote the space of continuous linear functionals on E.

We will denote by $\hat{E}$ the set of characters of E. For each $\phi \in E^*$ we put

$$d(\phi) = \inf\{\|\phi - \psi\| \mid \psi \in \hat{E} \cup \{0\}\}.$$

We state that E is an *algebra in which approximately multiplicative functionals are near multiplicative functionals*, or E is *AMNM* for short, if for each $\varepsilon > 0$ there exists a $\delta > 0$ such that $d(\phi) < \varepsilon$ whenever ϕ is a δ-multiplicative linear functional.

Johnson [146] proved the following:

Lemma 10.8. *Let E have an identity. Then the following statements are equivalent:*

(i) *E is AMNM;*

(ii) *For every sequence $\{\phi_n\}$ in E^* with $\|\check{\phi}_n\| \to 0$ there exists a sequence $\{\psi_n\}$ in $\hat{E} \cup \{0\}$ with $\|\phi_n - \psi_n\| \to 0$;*

(iii) *For every sequence $\{\phi_n\}$ in E^* with $\|\check{\phi}_n\| \to 0$ there exists a subsequence $\{\phi_{n_i}\}$ and a sequence $\{\psi_i\}$ in $\hat{E} \cup \{0\}$ with $\|\phi_{n_i} - \psi_i\| \to 0$;*

(iv) *For every sequence $\{\phi_n\}$ in E^* with $\|\check{\phi}_n\| \to 0$ and $\inf_n \|\phi_n\| > 0$ there exists a sequence $\{\psi_n\}$ in $\hat{E}$ with $\|\phi_n - \psi_n\| \to 0$;*

(v) *For every sequence $\{\phi_n\}$ in E^* with $\|\check{\phi}_n\| \to 0$ and $\phi_n(1) = 1 = \|\phi_n\|$ there exists a sequence $\{\psi_n\}$ in $\hat{E}$ with $\|\phi_n - \psi_n\| \to 0$;*

(vi) *For any $\varepsilon > 0$ there exists a $\delta > 0$ such that if $\phi \in E^*$ with $\phi(1) = 1 = \|\phi\|$ and $\|\check{\phi}\| < \delta$, then $d(\phi) < \varepsilon$.*

Conditions (i) to (iv) are equivalent even though E lacks an identity. If E has an approximate identity of norm 1, then (i) to (iv) are equivalent to the following:

(vii) For any $\varepsilon > 0$ there exists a $\delta > 0$ such that if $\phi \in E^*$ with $\|\phi\| = 1$ and $\|\check{\phi}\| < \delta$, then $d(\phi) < \varepsilon$.

Proof. The implications $(i) \Rightarrow (ii) \Rightarrow (iii) \Rightarrow (i)$, $(i) \Rightarrow (iv) \Rightarrow (iii)$, and $(i) \Rightarrow (v)$, (vi), and (vii) are elementary. The implications $(v) \Rightarrow (iv)$, $(vi) \Rightarrow (i)$, and $(vii) \Rightarrow (i)$ follow from Lemmas 10.3 and 10.4, and Theorem 10.7.   □

Johnson [146] proved that every finite-dimensional commutative complex Banach algebra is AMNM as we see in the following theorem.

Theorem 10.9. *If the dimension of E is finite, then E is AMNM.*

Proof. Let $\{\phi_n\}$ be a sequence in E^* with $\|\check{\phi}_n\| \to 0$. Then, by Lemma 10.2, $\{\phi_n\}$ is bounded and, thus, has a convergent subsequence $\{\phi_{n_i}\}$ with limit ψ. By continuity of the function $\phi \mapsto \check{\phi}$, we see $\psi \in \hat{E} \cup \{0\}$. Hence, the condition (iii) of Lemma 10.8 is satisfied with $\psi_i = \psi$.   □

A subset J of E is said to be an *ideal* if

(I1) J is a subspace of E (in the vector space sense),
(I2) $xy \in J$ whenever $x \in E$ and $y \in J$.

Johnson [146] provided the following theorem.

Theorem 10.10. *Let J be a closed ideal in E.*

(i) If J and E/J are AMNM, then so is E.
(ii) If E is AMNM, then so is J.
(iii) If E is AMNM and J has a bounded approximate identity, then E/J is AMNM.

Proof. (i) Let $\{\phi_n\}$ be a sequence in E^* with $\|\check{\phi}_n\| \to 0$. Using a subsequence if necessary, we may assume that either $\|\phi_n|_J\| \to 0$ or that there exists an $\eta > 0$ with $\|\phi_n|_J\| > \eta$ for $n \in \mathbb{N}$. Consider the second of these cases first. Since $\|\check{\phi}_n|_J\| \leq \|\check{\phi}_n\|$, we note that there are $\psi'_n \in \hat{J}$ with $\|\phi_n|_J - \psi'_n\| \to 0$. The elements ψ'_n of $\hat{J}$ can be extended to elements ψ_n of $\hat{E}$ by

$$\psi_n(x) = \psi'_n(xj)/\psi'_n(j)$$

for all $x \in E$ and $j \in J \setminus \operatorname{Ker} \psi'_n$. Let $j, k \in J \setminus \operatorname{Ker} \psi'_n$. Then, $\psi'_n(xjk) = \psi'_n(xj)\psi'_n(k) = \psi'_n(xk)\psi'_n(j)$, so that

$$\psi'_n(xjk)/\psi'_n(jk) = \psi'_n(xj)/\psi'_n(j) = \psi'_n(xk)/\psi'_n(k).$$

Hence, the definition of ψ_n is independent of the choice of j.

By the Hahn–Banach theorem, the functional $(\phi_n - \psi_n)|_J$ can be extended to an element θ_n of E^* with $\|\theta_n\| = \|(\phi_n - \psi_n)|_J\|$, so that $\|\theta_n\| \to 0$. Thus, by Lemma 10.3, $\|\check{\vartheta}_n\| \to 0$, where $\vartheta_n = \phi_n - \theta_n$ for $n \in \mathbb{N}$. Let $j_n \in J$ with $\|j_n\| = 1$ and $|\phi_n(j_n)| > \eta$. Then, it holds true that

$$\vartheta_n(x) = \vartheta_n(xj_n)/\vartheta_n(j_n) - \check{\vartheta}_n(x, j_n)/\vartheta_n(j_n) = \psi_n(x) - \check{\vartheta}_n(x, j_n)/\vartheta_n(j_n)$$

for each $x \in E$. Thus, $(\vartheta_n - \psi_n)(x) \to 0$ uniformly for $\|x\| \leq 1$, i.e., $\|\vartheta_n - \psi_n\| \to 0$ and hence $\|\phi_n - \psi_n\| \to 0$. Now, consider the case where $\|\phi_n|_J\| \to 0$ and let θ_n be an extension of $\phi_n|_J$ to E with $\|\theta_n\| = \|\phi_n|_J\|$. Put $\vartheta_n = \phi_n - \theta_n$. Then, $\|\check{\vartheta}_n\| \to 0$ by Lemma 10.3 and $\vartheta_n = 0$ on J. Hence, we may consider $\{\vartheta_n\}$ as a sequence in E/J. Since E/J is AMNM, there exists a sequence $\{\psi_n\}$ in $\widehat{E/J} \cup \{0\} \subset \hat{E} \cup \{0\}$ with $\|\vartheta_n - \psi_n\| \to 0$, and so $\|\phi_n - \psi_n\| \to 0$.

(ii) Suppose that E is AMNM and $\{\phi_n\}$ is a sequence in J^* with $\|\check{\phi}_n\| \to 0$ and $k = \inf_n \|\phi_n\| > 0$. Let $\{j_n\}$ be a sequence in J with $\|j_n\| < 2/k$ and $\phi_n(j_n) = 1$ for $n \in \mathbb{N}$. Put

$$\Phi_n(x) = \phi_n(xj_n)$$

for all $x \in E$ and $n \in \mathbb{N}$. Then, $\Phi_n \in E^*$ and for $x, y \in E$

$$\begin{aligned}\Phi_n(xy) - \Phi_n(x)\Phi_n(y) &= \phi_n(xyj_n) - \phi_n(xj_n)\phi_n(yj_n) \\ &= \check{\phi}_n(xj_n, yj_n) - \check{\phi}_n(xyj_n, j_n),\end{aligned}$$

so $\check{\Phi}_n(x, y) \to 0$ uniformly for $\|x\|, \|y\| \leq 1$. Thus, there exists a sequence $\{\Psi_n\}$ in $\hat{E} \cup \{0\}$ with $\|\Phi_n - \Psi_n\| \to 0$, and we have $\|\phi_n - \psi_n\| \to 0$ with $\psi_n = \Psi_n|_J \in \hat{J} \cup \{0\}$.

(iii) Finally, suppose that E is AMNM, J has a bounded approximate identity, and $\{\phi_n\}$ is a sequence in $(E/J)^* \subset E^*$ with $\|\check{\phi}_n\| \to 0$. Then, there is a sequence $\{\psi_n\}$ in $\hat{E} \cup \{0\}$ with $\|\phi_n - \psi_n\| \to 0$. By Lemma 10.4 (ii) with E replaced by J and taking the limit as $\delta \to 0$, we have either $\|\psi_n|_J\| \geq 1/K$ or $\|\psi_n|_J\| = 0$. As $\|\psi_n|_J\| \to 0$, we get $\psi_n|_J = 0$ eventually. Hence, ψ_n is a multiplicative element of $(E/J)^*$. □

B. E. Johnson [146] proved the following:

Corollary 10.11. *Let J be a closed ideal in E for which E/J is finite-dimensional. Then E is AMNM if and only if J is.*

Proof. By Theorem 10.9, E/J is AMNM and the corollary follows from Theorem 10.10 (i) and (ii). □

A topological vector space X is said to be *locally compact* if 0 has a neighborhood of which closure is compact. Johnson also presented in the same paper that the following commutative Banach algebras are AMNM:

(i) $C_0(X)$, where X is a locally compact Hausdorff space;
(ii) ℓ^p for $p \in [1, \infty]$;

(iii) $L^1(G)$, *where G is a locally compact abelian group;*

(iv) *the algebra $\ell^1(\mathbb{Z}^+)$ of power series $\sum_{n=0}^{\infty} a_n z^n$ with $\sum |a_n| < \infty$;*

(v) *the convolution algebra $L^1(0, \infty)$;*

(vi) *the disc algebra, i.e., the algebra of the continuous functions on the closed unit disc in $\mathbb{C}$ which are analytic in the interior of the disc.*

Now, let us discuss the general case of functions between Banach algebras E_1 and E_2. We denote the space of bounded linear functions from E_1 into E_2 by $L(E_1, E_2)$. For $T \in L(E_1, E_2)$, let us define

$$\check{T}(x, y) = T(xy) - T(x)T(y)$$

for all $x, y \in E_1$.

(E_1, E_2) is said to be an *AMNM pair* (*almost multiplicative maps are near multiplicative maps*) if for each $\varepsilon > 0$ and $K > 0$ there is a $\delta > 0$ such that if $T \in L(E_1, E_2)$ with $\|T\| \leq K$ and $\|\check{T}\| \leq \delta$, then there exists a multiplicative function $T' \in L(E_1, E_2)$ with $\|T - T'\| \leq \varepsilon$.

The main theorem of Johnson [147] states that if E_1 is amenable and E_2 is a dual space, then the pair (E_1, E_2) is AMNM. He also proved that the following pairs are AMNM:

(i) E_1 *is a finite-dimensional semi-simple algebra and E_2 is a Banach algebra;*

(ii) $E_1 = \ell^1(\mathbb{Z}^+)$ *or $E_1 =$ disc algebra and $E_2 = C(X)$, where X is a compact Hausdorff space;*

(iii) $E_1 = E_2 =$ *algebra of all bounded linear functions on a separable Hilbert space.*

10.4   Functional Equation $f(x^y) = f(x)^y$

In Section 9.4, the functional equation $f(xy) = f(x)^y$ was introduced as a variation of the exponential functional equation $f(x + y) = f(x)f(y)$. Similarly, the functional equation

$$f(x^y) = f(x)^y \tag{10.3}$$

may be regarded as a multiplicative functional equation.

In the following theorem, the functional equation (10.3) will be solved in the class of functions $f : (0, \infty) \to \mathbb{R}$ (see [171]).

Theorem 10.12. *A function $f : (0, \infty) \to \mathbb{R}$ satisfies the functional equation (10.3) for all $x, y > 0$ if and only if there exist real numbers c and d such that f has one of the following eight forms:*

(i) $f(x) = x^c$ *for $0 < x < 1$, and $f(x) = 0$ for $x \geq 1$;*

(ii) $f(x) = x^c$ *for $0 < x < 1$, $f(1) = 0$, and $f(x) = x^d$ for $x > 1$;*

(iii) $f(x) = x^c$ for $0 < x \le 1$, and $f(x) = 0$ for $x > 1$;
(iv) $f(x) = x^c$ for $0 < x \le 1$, and $f(x) = x^d$ for $x > 1$;
(v) $f(x) = 0$ for $x > 0$;
(vi) $f(x) = 0$ for $0 < x \le 1$, and $f(x) = x^d$ for $x > 1$;
(vii) $f(x) = 0$ for $x \ne 1$, and $f(1) = 1$;
(viii) $f(x) = 0$ for $0 < x < 1$, and $f(x) = x^d$ for $x \ge 1$.

Proof. First, let f satisfy the functional equation (10.3) for all $x, y > 0$. By putting $x = 1$ in (10.3) we immediately obtain $f(1) = 0$ or $f(1) = 1$. Assume that there exist some $p, q > 0$ with $p = q^2$ and $f(p) < 0$. Then, it follows from (10.3) that $0 \le f(q)^2 = f(q^2) = f(p) < 0$ which leads to a contradiction. Hence, $f(x) \ge 0$ holds true for all $x > 0$.

Assume that there exists some $0 < a < 1$ such that $f(a) > 0$. Put $s = a^y$ for $y > 0$. It then follows from (10.3) that $f(s) = f(a)^y = s^{\log_a f(a)}$ for $0 < s < 1$, since $\{a^y \mid y > 0\} = (0, 1)$. If we put $c = \log_a f(a)$, then $f(x) = x^c$ for all $0 < x < 1$.

Assume now that there exists some $b > 1$ such that $f(b) > 0$. Similarly, we obtain $f(s) = s^{\log_b f(b)}$ for $s > 1$. Using the notation $d = \log_b f(b)$ yields $f(x) = x^d$ for $x > 1$.

Cumulatively, f has one of the eight forms (*i*) to (*viii*).

Finally, it is easy to observe that if f has one of the given eight forms, then it satisfies the functional equation (10.3) for all $x, y > 0$. □

The functional equation (10.3) is different from the multiplicative functional equation $f(xy) = f(x) f(y)$. If we define a function $f : (0, \infty) \to \mathbb{R}$ by

$$f(x) = \begin{cases} x & \text{(for } 0 < x \le 1), \\ x^2 & \text{(for } x > 1), \end{cases}$$

then f does not satisfy the equation $f(xy) = f(x) f(y)$, whereas it is a solution of equation (10.3) according to Theorem 10.12 (*iv*).

Similarly, we can prove the following corollary. Hence, we omit the proof (or see [171]).

Corollary 10.13. *A function $f : (0, \infty) \to \mathbb{R}$ satisfies the functional equation (10.3) for all $x > 0$ and $y \in \mathbb{R}$ if and only if there exists a real number c such that $f(x) = x^c$ for all $x > 0$.*

The group structure of the range space of the exponential functional equation $f(x + y) = f(x) f(y)$ was taken into account, and the stability in the sense of R. Ger was introduced in Section 9.2.

It is interesting to note that we still have a superstability phenomenon for the functional equation (10.3) even though the relevant inequality is established in the spirit of R. Ger, whereas the exponential equation $f(x + y) = f(x) f(y)$ is stable in the similar setting.

Jung [171] provided the following theorem.

Theorem 10.14. *Suppose δ_1 and δ_2 are given with $0 \le \delta_1 < 1$ and $\delta_2 \ge 0$. Let a function $f : (0, \infty) \to (0, \infty)$ satisfy the inequalities*

$$1 - \delta_1 \le \frac{f(x^y)}{f(x)^y} \le 1 + \delta_2 \tag{10.4}$$

for all $x, y > 0$. Then there exist real numbers c and d such that

$$f(x) = \begin{cases} x^c & (for\ 0 < x \le 1), \\ x^d & (for\ x > 1). \end{cases}$$

Proof. Let $x > 0$ be given. From (10.4) it follows that

$$(1 - \delta_1)^{m/n} \le \frac{f\big((x^m)^{n/m}\big)^{m/n}}{f(x^m)} \le (1 + \delta_2)^{m/n} \tag{a}$$

for all $m, n \in \mathbb{N}$. Let $m, n \in \mathbb{N}$ satisfy $n \ge m$. By (a), we have

$$\big|(1/n) \ln f(x^n) - (1/m) \ln f(x^m)\big|$$
$$= (1/m)\big|(m/n) \ln f\big((x^m)^{n/m}\big) - \ln f(x^m)\big|$$
$$= (1/m)\left|\ln \frac{f\big((x^m)^{n/m}\big)^{m/n}}{f(x^m)}\right|$$
$$\to 0 \quad \text{as}\ m \to \infty.$$

Thus, $\{(1/n) \ln f(x^n)\}$ is a Cauchy sequence, and we may define

$$L(x) = \lim_{n \to \infty} (1/n) \ln f(x^n)$$

and

$$F(x) = \exp\big(L(x)\big)$$

for all $x > 0$. Indeed, it holds true that

$$F(x) = \lim_{n \to \infty} f(x^n)^{1/n}. \tag{b}$$

Substituting x^n for x in (10.4) yields

$$1 - \delta_1 \le \frac{f(x^{ny})}{f(x^n)^y} \le 1 + \delta_2$$

and hence

$$(1 - \delta_1)^{1/n} \leq \frac{f(x^{ny})^{1/n}}{f(x^n)^{y/n}} \leq (1 + \delta_2)^{1/n}.$$

Therefore,

$$\lim_{n \to \infty} f(x^{ny})^{1/n} f(x^n)^{-y/n} = 1. \qquad (c)$$

From (b) and (c), it follows that

$$F(x^y) = \lim_{n \to \infty} f(x^{yn})^{1/n} = \lim_{n \to \infty} f(x^{yn})^{1/n} f(x^n)^{-y/n} f(x^n)^{y/n}$$

and

$$F(x^y) = F(x)^y \qquad (d)$$

for all $x, y > 0$. Obviously, (10.4) and (b) imply $F(x) = f(x)$ for all $x > 0$. Since $f(x) > 0$ holds true for all $x > 0$, by Theorem 10.12 (iv), there are real numbers c and d such that $f(x) = x^c$ for $0 < x \leq 1$ and $f(x) = x^d$ for $x > 1$. $\qquad \square$

Analogously, Jung [171] proved the following corollary.

Corollary 10.15. *Assume that a function $f : (0, \infty) \to (0, \infty)$ satisfies the inequality (10.4) for some $0 \leq \delta_1 < 1$, $\delta_2 \geq 0$ and for all $x > 0$ and all $y \in \mathbb{R}$. Then there exists a real number c such that $f(x) = x^c$ for all $x > 0$.*

Proof. Obviously, we can apply the definition (b) in the proof of Theorem 10.14 to the proof of the corollary. As mentioned in the proof of Theorem 10.14, we have $F(x) = f(x) > 0$ for all $x > 0$. Hence, for any $x > 0$ there exist an $\varepsilon > 0$ and an $m \in \mathbb{N}$ such that $f(x^n)^{1/n} > \varepsilon$ for any $n \geq m$.

Now, let $y < 0$ be given. Since the function $x \mapsto x^y$ is continuous on the interval $[\varepsilon/2, \infty)$, the equality (d) in the proof of Theorem 10.14 is true for $y < 0$.

Therefore, we can conclude that F satisfies the functional equation (10.3) for $x > 0$ and $y \in \mathbb{R}$. By Corollary 10.13, the assertion is obvious. $\qquad \square$

Now, we study the superstability of the functional equation (10.3) on a restricted domain (ref. [171]).

Theorem 10.16. *Let δ_1 and δ_2 be given with $0 \leq \delta_1 < 1$ and $\delta_2 \geq 0$. Given $m > 0$, suppose a function $f : (0, \infty) \to (0, \infty)$ satisfies the inequality (10.4) for all pairs (x, y) with $x > m$ or $|y| > m$. Then there exists a real number c such that $f(x) = x^c$ for all $x > 0$.*

Proof. Suppose $0 < x \leq m$ and $-m \leq y \leq m$. We can choose an $n \in \mathbb{N}$ satisfying $nx > m$. Using the equality

$$\frac{f(x^y)}{f(x)^y} = \frac{f\left((nx)^{y \log_{nx} x}\right)}{f(nx)^{y \log_{nx} x}} \cdot \frac{f(nx)^{y \log_{nx} x}}{f\left((nx)^{\log_{nx} x}\right)^y}$$

and the inequality (10.4) yields

$$\frac{1 - \delta_1}{(1 + \delta_2)^m} \leq \frac{f(x^y)}{f(x)^y} \leq \frac{1 + \delta_2}{(1 - \delta_1)^m} \quad \text{for } y \geq 0$$

or

$$(1 - \delta_1)^{m+1} \leq \frac{f(x^y)}{f(x)^y} \leq (1 + \delta_2)^{m+1} \quad \text{for } y < 0.$$

In both cases we can find some δ_3 and δ_4 ($\delta_1 \leq \delta_3 < 1$, $\delta_4 \geq \delta_2$) such that f satisfies the inequality

$$1 - \delta_3 \leq \frac{f(x^y)}{f(x)^y} \leq 1 + \delta_4 \tag{a}$$

for any pair (x, y) with $0 < x \leq m$ and $|y| \leq m$. Hence, f satisfies the inequality (a) for all $x > 0$ and $y \in \mathbb{R}$. Therefore, the assertion of our theorem is an immediate consequence of Corollary 10.15. $\qquad\square$

Using the result of Theorem 10.16, we can easily obtain an interesting relationship between the power functions and the asymptotic behavior of the equation (10.3) as we observe in the following corollary:

Corollary 10.17. *Let $f : (0, \infty) \to (0, \infty)$ be a function. There exists a real number c such that $f(x) = x^c$ for all $x > 0$ if and only if $f(x^y)/f(x)^y \to 1$ as $|x| + |y| \to \infty$.*

10.5 Functional Equation $f(x + y) = f(x)f(y)f(1/x + 1/y)$

K. J. Heuvers introduced in his paper [132] a new type of logarithmic functional equation

$$f(x + y) = f(x) + f(y) + f(1/x + 1/y)$$

and proved that this equation is equivalent to the "original" logarithmic equation $f(xy) = f(x) + f(y)$ in the class of functions $f : (0, \infty) \to \mathbb{R}$.

If we slightly modify the functional equation of Heuvers, we may obtain a new functional equation

$$f(x + y) = f(x)f(y)f(1/x + 1/y), \tag{10.5}$$

which we may call a multiplicative functional equation because the function $f(x) = x^a$ is a solution of this equation.

By using the theorem of Heuvers (see Theorem 11.9 or [132]), we can easily prove that if both the domain and range of relevant functions are positive real numbers, then the functional equation (10.5) is equivalent to the "original" multiplicative equation $f(xy) = f(x)f(y)$.

By modifying an idea of Heuvers, S.-M. Jung [175] proved that the functional equation (10.5) and the equation $f(xy) = f(x)f(y)$ are equivalent to each other in the class of functions $f : \mathbb{R}\setminus\{0\} \to \mathbb{R}$. Moreover, he also investigated a stability problem of the functional equation (10.5) in the sense of Ger. We will now introduce his results.

Every solution of the "original" multiplicative (logarithmic) functional equation is called a multiplicative (logarithmic) function. For more information on multiplicative functions or logarithmic functions, one can refer to [3].

First, we will introduce a lemma essential to prove Theorems 10.21 and 10.23 which are the main theorems of this section. The proof of the following lemma is elementary.

Lemma 10.18. *It holds true that*

$$\{(u, v) \mid u = x^{-1} + y^{-1}, \ v = 1 - x(x + y)^{-1}y^{-1},$$
$$x, y \in \mathbb{R}\setminus\{0\} \ \text{with} \ x + y \neq 0\}$$
$$\supset \{(u, v) \mid u \in \mathbb{R}\setminus\{0\}, \ v \in \mathbb{R} \ \text{with} \ u + v \neq 1 \ \text{and} \ u(1 - v) > 0\}.$$

Proof. Let us consider the system of equations

$$\begin{cases} x^{-1} + y^{-1} = u, \\ x(x + y)^{-1}y^{-1} = 1 - v \end{cases} \tag{a}$$

with variables x and y, where $u \in \mathbb{R}\setminus\{0\}$ and $v \in \mathbb{R}$ with $u + v \neq 1$ and $u(1-v) > 0$. It suffices to prove that the system has at least one solution (x, y) with $x, y, x + y \in \mathbb{R}\setminus\{0\}$.

Combining both equations in the system yields a quadratic equation

$$(u^2 - u + uv)x^2 - 2ux + 1 = 0.$$

Applying the quadratic formula, we find the solutions of the above equation:

$$x = \frac{u \pm \sqrt{u(1 - v)}}{u(u + v - 1)} \neq 0 \quad \text{and} \quad y = \pm\frac{1}{\sqrt{u(1 - v)}} \neq 0.$$

Moreover, we see by the first equation of the system (a) that $x + y \neq 0$ because of $u \in \mathbb{R}\setminus\{0\}$. $\square$

We will verify in the following lemma that if a function $f : \mathbb{R} \to \mathbb{R}$ is a solution of the functional equation (10.5) and $f(x) = 0$ for some $x \in \mathbb{R}\setminus\{0\}$, then f is a null function.

Lemma 10.19. *If a function $f : \mathbb{R} \to \mathbb{R}$ satisfies the equation (10.5) for all $x, y \in \mathbb{R}\setminus\{0\}$ and if there is an $x_0 \neq 0$ with $f(x_0) = 0$, then $f(x) = 0$ for any $x \in \mathbb{R}$.*

Proof. Put $x = x_0$ in (10.5) to obtain

$$f(x_0 + y) = f(x_0) f(y) f(1/x_0 + 1/y) = 0$$

for each $y \in \mathbb{R} \setminus \{0\}$.   □

Two sets of solutions of the functional equation (10.5) with $f(1) = 1$ resp. $f(1) = -1$ are equivalent to each other. In particular, we introduce the following lemma whose proof is trivial.

Lemma 10.20. *If a function $f : \mathbb{R} \to \mathbb{R}$ is a solution of the functional equation (10.5) for all $x, y \in \mathbb{R} \setminus \{0\}$ and if $f(1) = -1$, then the function $g : \mathbb{R} \to \mathbb{R}$ defined by $g(x) = -f(x)$ is also a solution of the functional equation (10.5) for any $x, y \in \mathbb{R} \setminus \{0\}$ with $g(1) = 1$.*

In the following theorem, we will prove that the multiplicative equation (10.5) is equivalent to the "original" one, $f(xy) = f(x)f(y)$, in the class of functions $f : \mathbb{R} \to \mathbb{R}$.

Theorem 10.21 (Jung). *A function $f : \mathbb{R} \to \mathbb{R}$ satisfies the functional equation (10.5) for every $x, y \in \mathbb{R} \setminus \{0\}$ if and only if there exist a constant $\sigma \in \{-1, 1\}$ and a multiplicative function $m : \mathbb{R} \to \mathbb{R}$ (i.e., $m(xy) = m(x)m(y)$ for all $x, y \in \mathbb{R} \setminus \{0\}$) such that $f(x) = \sigma m(x)$ for all $x \in \mathbb{R} \setminus \{0\}$.*

Proof. If there exists an $x_0 \neq 0$ with $f(x_0) = 0$, then Lemma 10.19 implies that $f(x) = 0$ for every $x \in \mathbb{R}$. In this case, we may choose $\sigma = 1$ and a multiplicative function $m \equiv 0$ such that $f(x) = \sigma m(x)$ for all $x \in \mathbb{R} \setminus \{0\}$.

Assume now that $f(x) \neq 0$ for all $x \in \mathbb{R} \setminus \{0\}$. Put $y = 1/x$ in (10.5) to obtain

$$f(1/x) = 1/f(x) \tag{a}$$

for each $x \in \mathbb{R} \setminus \{0\}$. With $x = 1$, (a) implies $f(1) = 1$ or $f(1) = -1$.

In view of Lemma 10.20, we may without loss of generality assume that

$$f(1) = 1. \tag{b}$$

It follows from (a) that

$$f\big(1/(x + y)\big) f(x + y) = f\big(1/(y + z)\big) f(y + z) = 1$$

for $x + y, y + z \in \mathbb{R} \setminus \{0\}$. Hence, by using (10.5), we have

$$
\begin{aligned}
f&\big(1/(x + y) + 1/z\big) f(1/x + 1/y) \\
&= f\big(1/(x + y)\big) f(1/z) f(x + y + z) f(1/x) f(1/y) f(x + y) \\
&= f\big(1/(y + z)\big) f(1/x) f(x + y + z) f(1/y) f(1/z) f(y + z) \\
&= f\big(1/(y + z) + 1/x\big) f(1/y + 1/z) \tag{c}
\end{aligned}
$$

for all $x, y, z \in \mathbb{R} \setminus \{0\}$ with $x + y \neq 0$ and $y + z \neq 0$.

If we set

$$u = 1/x + 1/y \quad \text{and} \quad v = 1/(x + y) + 1/z, \tag{d}$$

then

$$\begin{aligned}
uv &= (1/x + 1/y)\big(1/(x + y) + 1/z\big) \\
&= 1/(yz) + 1/(xy) + 1/(xz) \\
&= (1/(y + z))(1/y + 1/z) + (1/x)(1/y + 1/z) \\
&= \big(1/(y + z) + 1/x\big)(1/y + 1/z).
\end{aligned}$$

If we additionally set

$$1/y + 1/z = 1 \tag{e}$$

in (c), then (b), (c), (d), and (e) imply that the function f satisfies

$$f(uv) = f(u)f(v) \tag{f}$$

for all $u \in \mathbb{R} \setminus \{0\}$, $v \in \mathbb{R}$ with $u + v \neq 1$ and $u(1 - v) > 0$ (see Lemma 10.18 and the fact that $u = 1/x + 1/y$ and $v = 1/(x + y) + 1/z = 1 - x(x + y)^{-1}y^{-1}$ for some $x, y \in \mathbb{R} \setminus \{0\}$ with $x + y \neq 0$).

Let $\alpha \approx -1.324717956\ldots$ be a real solution of the cubic equation $x^3 - x + 1 = 0$; more precisely, let

$$\alpha = \big(-1/2 + (23/108)^{1/2}\big)^{1/3} + \big(-1/2 - (23/108)^{1/2}\big)^{1/3}.$$

Using (f) and (a) yields

$$f(u) = f(u^2/u) = f(u^2)f(1/u) = f(u^2)/f(u)$$

or

$$f(u^2) = f(u)^2 \tag{g}$$

for any $u < 0$ ($u \neq \alpha$) or $u > 1$ ($u \neq \alpha$ means $u^2 + 1/u \neq 1$). Using (a) and (g), we have

$$f(1/u^2) = 1/f(u^2) = 1/f(u)^2 = f(1/u)^2$$

for $u < 0$ ($u \neq \alpha$) or $u > 1$. Hence, (b) and (g) yield that f satisfies (g) for all $u \in \mathbb{R} \setminus \{0\}$. (The validity of (g) for $u = 1/\alpha \neq \alpha$ or $u \geq 1$ implies that $f(1/u^2) = f(1/u)^2$ for $1/u = \alpha$ or $0 < 1/u \leq 1$.)

Let $s > t \geq 1$ be given. It then follows from (f), (g), and (a) that

$$f(st) = f(s^2 t/s) = f(s^2)f(t/s) = f(s)^2 f(t)f(1/s) = f(s)f(t) \tag{h}$$

for all $s \geq 1$ and $t \geq 1$. (We may replace each of s and t by the other when $t > s \geq 1$ and use (g) and (b) to prove (h) for $s = t$.) From (a) and (h), we get

$$f\big((1/s)(1/t)\big) = 1/f(st) = \big(1/f(s)\big)\big(1/f(t)\big) = f(1/s)f(1/t),$$

for all $s \geq 1$ and $t \geq 1$, or

$$f(st) = f(s)f(t)$$

for all $0 < s, t \leq 1$. For the case when $s \geq 1$ and $0 < t < 1$ ($t \geq 1$ and $0 < s < 1$), we may use (f) to obtain $f(st) = f(s)f(t)$. Altogether, we may conclude by considering (b) that f satisfies (f) for all pairs (u, v) of

$$\{(u, v) \mid u > 0, \ v > 0\}$$
$$\cup \{(u, v) \mid u \in \mathbb{R} \setminus \{0\}, \ v \in \mathbb{R}, \ u + v \neq 1, \ u(1 - v) > 0\}. \tag{i}$$

By (g), we have $f(u)^2 = f(u^2) = f((-u)^2) = f(-u)^2$ for all $u \in \mathbb{R} \setminus \{0\}$. Hence, we get

$$f(u) = -f(-u) \quad \text{or} \quad f(u) = f(-u) \tag{j}$$

for each $u \in \mathbb{R} \setminus \{0\}$.

If we assume that $f(u) = f(-u)$ for all $-1 \leq u < 0$ and that there exists a $u_0 < -1$ with $f(u_0) = -f(-u_0)$, then it follows from (a) that

$$f(1/u_0) = 1/f(u_0) = -1/f(-u_0) = -f(-1/u_0)$$

and $-1 < 1/u_0 < 0$ which are contrary to our assumption. (Due to Lemma 10.19, we can assume that $f(-1/u_0) \neq 0$.)

Now, suppose there exists a u_0 ($-1 \leq u_0 < 0$) with $f(u_0) = -f(-u_0)$. It then follows from (i) that

$$f(u_0 v) = f\big((-u_0)(-v)\big) = f(-u_0)f(-v) = -f(u_0)f(-v) = -f(-u_0 v)$$

for all $v < -1$ with $v \neq u_0 - 1$, i.e.,

$$f(u) = -f(-u) \tag{k}$$

for all $u := u_0 v > 1$ ($u \neq u_0^2 - u_0$). Using (a) and (k) yields

$$f(u) = 1/f(1/u) = -1/f(-1/u) = -f(-u)$$

for any $0 < u < 1$ ($u \neq (u_0^2 - u_0)^{-1}$). From (10.5), (a), (f), and (i) we get

$$f(x - 1) = f(x)f(-1)f(1/x - 1)$$
$$= f(x)f(-1)f(1 - x)f(1/x)$$

$$= f(-1)f(1-x)$$
$$= -f(-1)f(x-1)$$

for $0 < x < 1$, which implies that $f(-1) = -1$. Therefore,

$$f(1) = -f(-1).$$

Altogether, we see that if there exists a u_0 $(-1 \le u_0 < 0)$ with $f(u_0) = -f(-u_0)$, then f satisfies (k) for all $u > 0$ with possible exceptions at $u_0^2 - u_0$ and $(u_0^2 - u_0)^{-1}$.

Taking (j) into consideration, assume that

$$f(u_0^2 - u_0) = f(u_0 - u_0^2).$$

By (f) and (i), we have

$$f((u_0 - u_0^2)v) = f((u_0^2 - u_0)(-v))$$
$$= f(u_0^2 - u_0)f(-v)$$
$$= f(u_0 - u_0^2)f(-v)$$
$$= f(-(u_0 - u_0^2)v)$$

for all $v < -1$ with $v \ne u_0 - u_0^2 - 1$, i.e.,

$$f(u) = f(-u)$$

for each $u := (u_0 - u_0^2)v > u_0^2 - u_0$ with $u \ne (u_0 - u_0^2)(u_0 - u_0^2 - 1)$, which is contrary to the fact that (k) holds true for all $u > 0$ with possible exceptions at two points. (In view of Lemma 10.19, we may exclude the case when $f(u) = 0$ for some $u \ne 0$.)

Similarly, if we assume

$$f(1/(u_0^2 - u_0)) = f(1/(u_0 - u_0^2)),$$

then this assumption also leads to a contradiction. Hence, we can conclude that if there exists a u_0 $(-1 \le u_0 < 0)$ with $f(u_0) = -f(-u_0)$, then

$$f(u) = -f(-u)$$

for all $u > 0$. If we replace $-u$ with u, we will see that this equation is true also for all $u < 0$.

Therefore, f satisfies either

$$f(u) = -f(-u) \quad \text{for all } u < 0$$

or

$$f(u) = f(-u) \quad \text{for all } u < 0.$$

This fact together with (f) and (i) yields

$$f(uv) = f(u)f(v)$$

for all $u, v \in \mathbb{R} \setminus \{0\}$.

The proof of the reverse assertion is clear.   □

If a function $f : \mathbb{R} \to \mathbb{R}$ is a solution of the functional equation (10.5) for all $x, y \in \mathbb{R} \setminus \{0\}$ and additionally satisfies $f(0) \neq 0$, then we see by putting $y = -x$ in (10.5) and considering (j) in the proof of Theorem 10.21 that $f(x) \in \{-1, 1\}$ for all $x \in \mathbb{R} \setminus \{0\}$. Therefore, we have the following:

Corollary 10.22. *An unbounded function $f : \mathbb{R} \to \mathbb{R}$ is a solution of the functional equation (10.5) for all $x, y \in \mathbb{R} \setminus \{0\}$ if and only if there exist a constant $\sigma \in \{-1, 1\}$ and an unbounded multiplicative function $m : \mathbb{R} \to \mathbb{R}$ such that $f(x) = \sigma m(x)$ for all $x \in \mathbb{R}$.*

In the following theorem, we will prove the stability of the functional equation (10.5) in the sense of Ger.

Theorem 10.23 (Jung). *If a function $f : \mathbb{R} \to (0, \infty)$ satisfies the inequality*

$$\left| \frac{f(x + y)}{f(x)f(y)f(1/x + 1/y)} - 1 \right| \leq \delta \tag{10.6}$$

for some $0 \leq \delta < 1$ and for all $x, y \in \mathbb{R} \setminus \{0\}$, then there exists a unique multiplicative function $m : \mathbb{R} \setminus \{0\} \to (0, \infty)$ such that

$$\left(\frac{1 - \delta}{1 + \delta} \right)^{13/2} \leq \frac{m(x)}{f(x)} \leq \left(\frac{1 + \delta}{1 - \delta} \right)^{13/2}$$

for any $x \in \mathbb{R} \setminus \{0\}$. If f is additionally assumed to be unbounded, then the domain of m can be extended to the whole real space $\mathbb{R}$ with $m(0) = 0$.

Proof. It follows from (10.6) that

$$(1 + \delta)^{-1} \leq \frac{f(x)f(y)f(1/x + 1/y)}{f(x + y)} \leq (1 - \delta)^{-1} \tag{a}$$

for any $x, y \in \mathbb{R} \setminus \{0\}$. Putting $y = 1/x$ in (a) yields

$$(1 + \delta)^{-1} \leq f(x)f(1/x) \leq (1 - \delta)^{-1} \tag{b}$$

for each $x \in \mathbb{R} \setminus \{0\}$. With $x = 1$, (b) yields

$$(1 + \delta)^{-1/2} \le f(1) \le (1 - \delta)^{-1/2}. \tag{c}$$

From (10.6) we get

$$(1 - \delta)^2$$
$$\le \frac{f\left(1/(x + y) + 1/z\right) f(1/x + 1/y)}{f\left(1/(x + y)\right) f(1/z) f(x + y + z) f(1/x) f(1/y) f(x + y)}$$
$$\le (1 + \delta)^2 \tag{d}$$

and

$$(1 - \delta)^2$$
$$\le \frac{f\left(1/(y + z) + 1/x\right) f(1/y + 1/z)}{f\left(1/(y + z)\right) f(1/x) f(x + y + z) f(1/y) f(1/z) f(y + z)}$$
$$\le (1 + \delta)^2 \tag{e}$$

for all $x, y, z \in \mathbb{R} \setminus \{0\}$ with $x + y \ne 0$ resp. $y + z \ne 0$. If we divide the inequalities in (d) by those in (e) and consider (b), then

$$\left(\frac{1 - \delta}{1 + \delta}\right)^3 \le \frac{f\left(1/(x + y) + 1/z\right) f(1/x + 1/y)}{f\left(1/(y + z) + 1/x\right) f(1/y + 1/z)} \le \left(\frac{1 + \delta}{1 - \delta}\right)^3 \tag{f}$$

for all $x, y, z \in \mathbb{R} \setminus \{0\}$ with $x + y \ne 0$ and $y + z \ne 0$.

If we define u and v by (d) in the proof of Theorem 10.21, then we have

$$uv = \left(1/(y + z) + 1/x\right)(1/y + 1/z)$$

as we see in the proof of Theorem 10.21. If an additional condition (e) in the proof of Theorem 10.21 is also assumed, then (d) and (e) in the proof of Theorem 10.21, together with (c) and (f), imply that

$$\left(\frac{1 - \delta}{1 + \delta}\right)^{7/2} \le \frac{f(u)f(v)}{f(uv)} \le \left(\frac{1 + \delta}{1 - \delta}\right)^{7/2} \tag{g}$$

for all $u \in \mathbb{R} \setminus \{0\}$, $v \in \mathbb{R}$ with $u + v \ne 1$ and $u(1 - v) > 0$ (see Lemma 10.18 and the fact that $u = 1/x + 1/y$ and $v = 1/(x + y) + 1/z = 1 - x(x + y)^{-1} y^{-1}$ for some $x, y \in \mathbb{R} \setminus \{0\}$ with $x + y \ne 0$).

Let us define $\alpha \approx -1.324717956\ldots$ as in the proof of Theorem 10.21. From the relation

$$\frac{f(u^2)}{f(u)^2} = \frac{f(u^2)f(1/u)}{f(u)} \frac{1}{f(u)f(1/u)}$$

and from (g) and (b), it follows that

$$\left(\frac{1-\delta}{1+\delta}\right)^{9/2} \le \frac{f\left(u^2\right)}{f(u)^2} \le \left(\frac{1+\delta}{1-\delta}\right)^{9/2} \tag{h}$$

for any $u < 0$ ($u \ne \alpha$) or $u > 1$ ($u \ne \alpha$ implies $u^2 + 1/u \ne 1$). On account of (b) and (h), we obtain

$$\left(\frac{1-\delta}{1+\delta}\right)^{13/2} \le \frac{f\left(1/u^2\right)}{f(1/u)^2} = \frac{f\left(u^2\right)f\left(1/u^2\right)}{f(u)^2 f(1/u)^2} \frac{f(u)^2}{f\left(u^2\right)} \le \left(\frac{1+\delta}{1-\delta}\right)^{13/2}$$

for $u < 0$ ($u \ne \alpha$) or $u > 1$. This fact together with (h) and (c) yields that

$$\left(\frac{1-\delta}{1+\delta}\right)^{13/2} \le \frac{f\left(u^2\right)}{f(u)^2} \le \left(\frac{1+\delta}{1-\delta}\right)^{13/2} \tag{i}$$

for all $u \in \mathbb{R} \setminus \{0\}$, since $1/\alpha \ne \alpha$.

Let $s > t \ge 1$. With

$$\frac{f(s)f(t)}{f(st)} = \frac{f\left(s^2\right)f(t/s)}{f\left(s^2 \cdot t/s\right)} \frac{f(t)f(1/s)}{f(t/s)} \frac{f(s)^2}{f\left(s^2\right)} \frac{1}{f(s)f(1/s)},$$

(g), (h), (b), and (c) yield

$$\left(\frac{1-\delta}{1+\delta}\right)^{25/2} \le \frac{f(s)f(t)}{f(st)} \le \left(\frac{1+\delta}{1-\delta}\right)^{25/2} \tag{j}$$

for all $s \ge 1$ and $t \ge 1$. (We can replace each of s and t with the other when $t > s \ge 1$ and we apply (i) to the proof of (j) for the case $s = t$.) By

$$\frac{f(1/s)f(1/t)}{f\left(1/(st)\right)} = f(s)f(1/s)f(t)f(1/t)\frac{f(st)}{f(s)f(t)} \frac{1}{f(st)f\left(1/(st)\right)},$$

and using (b) and (j) we obtain

$$\left(\frac{1-\delta}{1+\delta}\right)^{29/2} \le \frac{f(1/s)f(1/t)}{f\left(1/(st)\right)} \le \left(\frac{1+\delta}{1-\delta}\right)^{29/2} \tag{k}$$

for all $s \ge 1$ and $t \ge 1$. Hence, by (g), (j), and (k) we conclude that

$$\left(\frac{1-\delta}{1+\delta}\right)^{29/2} \le \frac{f(u)f(v)}{f(uv)} \le \left(\frac{1+\delta}{1-\delta}\right)^{29/2} \tag{l}$$

for each pair (u, v) of

$$\{(u, v) \mid u > 0, \ v > 0\}$$
$$\cup \ \{(u, v) \mid u \in \mathbb{R} \setminus \{0\}, \ v \in \mathbb{R}, \ u + v \neq 1, \ u(1 - v) > 0\}. \tag{m}$$

(We can use (g) to verify inequalities in (l) either for $s \geq 1$ and $0 < t < 1$ or for $t \geq 1$ and $0 < s < 1$.)

The fact

$$\frac{f(u)^2}{f(-u)^2} = \frac{f(u)^2}{f(u^2)} \frac{f((-u)^2)}{f(-u)^2}$$

together with (i) implies that

$$\left(\frac{1 - \delta}{1 + \delta}\right)^{13/2} \leq \frac{f(u)}{f(-u)} \leq \left(\frac{1 + \delta}{1 - \delta}\right)^{13/2}$$

for every $u \in \mathbb{R} \setminus \{0\}$. This fact, (l), and (m) together with the relations

$$\frac{f(u)f(v)}{f(uv)} = \frac{f(-u)f(-v)}{f(uv)} \frac{f(u)}{f(-u)} \frac{f(v)}{f(-v)} \quad \text{(for } u, v < 0)$$

and

$$\frac{f(u)f(v)}{f(uv)} = \frac{f(u)f(-v)}{f(-uv)} \frac{f(-uv)}{f(uv)} \frac{f(v)}{f(-v)} \quad \text{(for } u > 0, \ v < 0)$$

imply

$$\left(\frac{1 - \delta}{1 + \delta}\right)^{55/2} \leq \frac{f(uv)}{f(u)f(v)} \leq \left(\frac{1 + \delta}{1 - \delta}\right)^{55/2} \tag{n}$$

for any $u, v \in \mathbb{R} \setminus \{0\}$.

We now claim that

$$\delta_1^{1 + 2 + \cdots + 2^{n-1}} \leq \frac{f(u^{2^n})}{f(u)^{2^n}} \leq \delta_2^{1 + 2 + \cdots + 2^{n-1}} \tag{o}$$

for all $u \in \mathbb{R} \setminus \{0\}$ and $n \in \mathbb{N}$, where we put

$$\delta_1 = \left(\frac{1 - \delta}{1 + \delta}\right)^{13/2} \quad \text{and} \quad \delta_2 = \left(\frac{1 + \delta}{1 - \delta}\right)^{13/2}.$$

Due to (i) our assertion is obvious for $n = 1$. Assume that (o) is true for some $n \geq 1$. Then, the relation

$$\frac{f(u^{2^{n+1}})}{f(u)^{2^{n+1}}} = \frac{f((u^{2^n})^2)}{f(u^{2^n})^2} \left(\frac{f(u^{2^n})}{f(u)^{2^n}}\right)^2$$

together with (i) and (o) gives

$$\delta_1^{1+2+\cdots+2^n} \leq \frac{f\left(u^{2^{n+1}}\right)}{f(u)^{2^{n+1}}} \leq \delta_2^{1+2+\cdots+2^n}$$

which proves the validity of (o) for all $u \in \mathbb{R} \setminus \{0\}$ and $n \in \mathbb{N}$.

Let us define functions $g_n : \mathbb{R} \setminus \{0\} \to \mathbb{R}$ by

$$g_n(u) = 2^{-n} \ln f\left(u^{2^n}\right)$$

for each $u \in \mathbb{R} \setminus \{0\}$ and $n \in \mathbb{N}$. Let $m, n \in \mathbb{N}$ be arbitrarily given with $n > m$. It then follows from (o) that

$$|g_n(u) - g_m(u)| = 2^{-m} \left| 2^{-(n-m)} \ln \frac{f\left(\left(u^{2^m}\right)^{2^{n-m}}\right)}{f\left(u^{2^m}\right)^{2^{n-m}}} \right| \to 0 \quad \text{as} \quad m \to \infty.$$

Hence, $\{g_n(u)\}$ is a Cauchy sequence for every fixed $u \in \mathbb{R} \setminus \{0\}$. Therefore, we can define functions $\ell : \mathbb{R} \setminus \{0\} \to \mathbb{R}$ and $m : \mathbb{R} \setminus \{0\} \to (0, \infty)$ by

$$\ell(u) = \lim_{n \to \infty} g_n(u) \quad \text{and} \quad m(u) = e^{\ell(u)}.$$

Indeed, we know that

$$m(u) = \lim_{n \to \infty} f\left(u^{2^n}\right)^{2^{-n}}$$

for every $u \in \mathbb{R} \setminus \{0\}$.

Replace u and v in (n) with u^{2^n} and v^{2^n}, respectively, and extract the 2^nth root of the resulting inequalities and then take the limit as $n \to \infty$ to obtain

$$m(uv) = m(u)m(v)$$

for all $u, v \in \mathbb{R} \setminus \{0\}$. Hence, we conclude by considering (o) that there exists a multiplicative function $m : \mathbb{R} \setminus \{0\} \to (0, \infty)$ with

$$\delta_1 \leq \frac{m(u)}{f(u)} \leq \delta_2 \tag{p}$$

for any $u \in \mathbb{R} \setminus \{0\}$.

Suppose $m' : \mathbb{R} \setminus \{0\} \to (0, \infty)$ is another multiplicative function satisfying (p) instead of m. Since m and m' are multiplicative, we see that

$$m\left(u^{2^n}\right) = m(u)^{2^n} \quad \text{and} \quad m'\left(u^{2^n}\right) = m'(u)^{2^n}.$$

Thus, it follows from (p) that

$$\frac{m(u)}{m'(u)} = \left(\frac{m\left(u^{2^n}\right)}{f\left(u^{2^n}\right)}\right)^{2^{-n}} \left(\frac{f\left(u^{2^n}\right)}{m'\left(u^{2^n}\right)}\right)^{2^{-n}} \to 1 \quad \text{as} \quad n \to \infty,$$

which implies the uniqueness of m.

By (p) we see that m is unbounded if and only if f is so. Hence, it is not difficult to show that if f is unbounded, then the domain of m can be extended to the whole real space $\mathbb{R}$ by defining $m(0) = 0$. □

Corollary 10.24. *If a function $f : \mathbb{R} \to (-\infty, 0)$ satisfies the functional inequality (10.6) for some $0 \le \delta < 1$ and for all $x, y \in \mathbb{R} \setminus \{0\}$, then there exists a unique multiplicative function $m : \mathbb{R} \setminus \{0\} \to (0, \infty)$ with*

$$-\left(\frac{1+\delta}{1-\delta}\right)^{13/2} \le \frac{m(x)}{f(x)} \le -\left(\frac{1-\delta}{1+\delta}\right)^{13/2}$$

for each $x \in \mathbb{R} \setminus \{0\}$. Moreover, if f is unbounded, then the domain of m can be extended to the whole real space $\mathbb{R}$ with $m(0) = 0$.

Chapter 11
Logarithmic Functional Equations

It is not difficult to demonstrate the Hyers–Ulam stability of the *logarithmic functional equation* $f(xy) = f(x) + f(y)$ for functions $f : (0, \infty) \to E$, where E is a Banach space. More precisely, if a function $f : (0, \infty) \to E$ satisfies the functional inequality $\| f(xy) - f(x) - f(y) \| \leq \delta$ for some $\delta > 0$ and for all $x, y > 0$, then there exists a unique logarithmic function $L : (0, \infty) \to E$ (this means that $L(xy) = L(x) + L(y)$ for all $x, y > 0$) such that $\| f(x) - L(x) \| \leq \delta$ for any $x > 0$. In this chapter, we will introduce a new functional equation $f(x^y) = yf(x)$ which has the logarithmic property in the sense that the logarithmic function $f(x) = \ln x$ ($x > 0$) is a solution of the equation. Moreover, the functional equation of Heuvers $f(x + y) = f(x) + f(y) + f(1/x + 1/y)$ will be discussed.

11.1 Functional Equation $f(x^y) = yf(x)$

The general solution $f : \mathbb{R} \setminus \{0\} \to \mathbb{R}$ of the *logarithmic functional equation*

$$f(xy) = f(x) + f(y) \tag{11.1}$$

is given by

$$f(x) = A\big(\ln |x| \big) \text{ for } x \in \mathbb{R} \setminus \{0\},$$

where $A : \mathbb{R} \to \mathbb{R}$ is an additive function. If a continuous function $f : \mathbb{R} \setminus \{0\} \to \mathbb{R}$ satisfies the logarithmic equation (11.1), then there exists a real constant c such that $f(x) = c \ln |x|$ for all $x \in \mathbb{R} \setminus \{0\}$.

The *logarithmic function* $f(x) = \ln x$ ($x > 0$) clearly satisfies the functional equation

$$f(x^y) = yf(x). \tag{11.2}$$

Therefore, we may regard the functional equation (11.2) as a sort of logarithmic functional equation.

S.-M. Jung [158] solved the functional equation (11.2) in the class of differentiable functions $f : (0, \infty) \to \mathbb{R}$ as we will indicate in the following theorem.

S.-M. Jung, *Hyers–Ulam–Rassias Stability of Functional Equations in Nonlinear Analysis*, Springer Optimization and Its Applications 48, DOI 10.1007/978-1-4419-9637-4_11, © Springer Science+Business Media, LLC 2011

Theorem 11.1. *A differentiable function* $f : (0, \infty) \to \mathbb{R}$ *satisfies the functional equation* (11.2) *for all* $x > 0$ *and* $y \in \mathbb{R}$ *if and only if* $f(x) = c \ln x$ *for any* $x > 0$, *where* c *is an arbitrary real constant.*

Proof. Let $f : (0, \infty) \to \mathbb{R}$ be a differentiable function and satisfy the functional equation (11.2) for all $x > 0$ and $y \in \mathbb{R}$. Obviously, putting $x = 1$ in (11.2) yields $f(1) = 0$.

Assume that there exists some $x_0 > 0$ ($x_0 \neq 1$) such that $f(x_0) = 0$. Then, the fact $\{x_0^y \mid -\infty < y < \infty\} = (0, \infty)$, together with (11.2), yields $f(x) = 0$ for all $x > 0$.

Assume now that $f(x) \neq 0$ for all $x > 0$ ($x \neq 1$). Differentiations of (11.2) with respect to x and y yield $f'(x^y) = x^{1-y} f'(x)$ and $f'(x^y) = x^{-y} (\ln x)^{-1} f(x)$, respectively. Since $f(x) \neq 0$ for all $x \neq 1$, it follows from the last two equations that

$$\frac{f'(x)}{f(x)} = \frac{1}{x \ln x}$$

for any $x > 0$ ($x \neq 1$). By integrating both sides of the last equation, we obtain

$$|f(x)| = \begin{cases} a_1 |\ln x| & (\text{for } 0 < x < 1), \\ a_2 |\ln x| & (\text{for } x > 1), \end{cases}$$

where a_1 and a_2 are positive real constants. The continuity of f implies that

$$f(x) = \begin{cases} c_1 \ln x & (\text{for } 0 < x \leq 1), \\ c_2 \ln x & (\text{for } x > 1), \end{cases}$$

where $c_1, c_2 \neq 0$ are arbitrary real constants. However, the differentiability of f at $x = 1$ implies $c_1 = c_2$. Hence, it must hold true

$$f(x) = c_1 \ln x,$$

for all $x > 0$, where $c_1 \neq 0$ is an arbitrary real constant. Therefore, we have $f(x) = c \ln x$ ($x > 0$) where c is an arbitrary (including 0) real constant.

The reverse assertion of our theorem is trivial. $\square$

Analogously as in the proof of Theorem 11.1, we can also prove that a differentiable function $f : (0, \infty) \to \mathbb{R}$ satisfies the functional equation (11.2) for all $x, y > 0$ if and only if $f(x) = c \ln x$ ($x > 0$), where c is an arbitrary real constant.

11.2 Superstability of Equation $f(x^y) = yf(x)$

It is obvious that the logarithmic functional equation $f(xy) = f(x) + f(y)$ is stable in the sense of Hyers and Ulam. We now prove the superstability of the functional equation (11.2) in the conventional setting (11.3) (see [158]).

Theorem 11.2. *If a function* $f : (0, \infty) \to \mathbb{C}$ *satisfies the functional inequality*

$$|f(x^y) - yf(x)| \leq \delta \tag{11.3}$$

for some $\delta \geq 0$ *and for all* $x > 0$ *and* $y \in \mathbb{R}$, *then either* f *is identically zero or it satisfies the functional equation* (11.2) *for all* $x > 0$ *and* $y \in \mathbb{R}$.

Proof. By taking $x = 1$ and letting $y \to \infty$ in (11.3), we get $f(1) = 0$. First, assume there exists an $x_0 > 0$ ($x_0 \neq 1$) such that $f(x_0) = 0$. Then the fact $\{x_0^y \mid -\infty < y < \infty\} = (0, \infty)$ and (11.3) imply that $|f(x)| \leq \delta$ for any $x > 0$, and this, together with (11.3) again, implies that f is identically zero.

Assume now that $f(x) \neq 0$ for any $x > 0$ ($x \neq 1$). Let $z > 1$ be fixed and suppose $x > 0$. We use induction on n to prove the inequality

$$\left| z^{-n} f\left(x^{z^n}\right) - f(x) \right| \leq \delta \sum_{i=1}^{n} z^{-i} \tag{a}$$

for all $n \in \mathbb{N}$. The validity of (a) for $n = 1$ follows immediately from (11.3). Now, assume (a) holds true for some integer $n > 0$. It then follows from (11.3) and (a) that

$$\left| z^{-(n+1)} f\left(x^{z^{n+1}}\right) - f(x) \right| \leq z^{-(n+1)} \left| f\left(x^{z^{n+1}}\right) - zf\left(x^{z^n}\right) \right|$$
$$+ \left| z^{-n} f\left(x^{z^n}\right) - f(x) \right|$$
$$\leq \delta \sum_{i=1}^{n+1} z^{-i},$$

which ends the proof of (a).

Let $m, n \in \mathbb{N}$ satisfy $n \geq m$. Using (a) yields

$$\left| z^{-n} f\left(x^{z^n}\right) - z^{-m} f\left(x^{z^m}\right) \right| = z^{-m} \left| z^{-(n-m)} f\left((x^{z^m})^{z^{n-m}}\right) - f\left(x^{z^m}\right) \right|$$
$$\leq \frac{\delta}{z^m(z-1)}$$
$$\to 0 \quad \text{as } m \to \infty.$$

Hence, $\{z^{-n} f(x^{z^n})\}$ is a Cauchy sequence, and we can define

$$L_z(x) = \lim_{n \to \infty} z^{-n} f\left(x^{z^n}\right)$$

for all $x > 0$.

Now, let $x > 0$ be given with $x \neq 1$. If we put $y = z^n$, let $n \to \infty$ in (11.3), and then take into account $f(x) \neq 0$, we easily see

$$\lim_{n \to \infty} \left| f\left(x^{z^n}\right) \right| = \infty. \tag{b}$$

Let $y \neq 0$ be given. Substituting x^{z^n} for x in (11.3), dividing by $|y f(x^{z^n})|$ both sides, and letting $n \to \infty$, together with (b) yield

$$\lim_{n\to\infty} f\left(x^{yz^n}\right) y^{-1} f\left(x^{z^n}\right)^{-1} = 1.$$

Therefore, we have

$$L_z(x^y) = \lim_{n\to\infty} z^{-n} f\left(x^{yz^n}\right) = \lim_{n\to\infty} z^{-n} f\left(x^{yz^n}\right) y^{-1} f\left(x^{z^n}\right)^{-1} y f\left(x^{z^n}\right)$$

and hence

$$L_z(x^y) = y L_z(x) \qquad\qquad (c)$$

for any $x > 0$ $(x \neq 1)$ and for all $y \neq 0$. Moreover, it follows from (a) that

$$|L_z(x) - f(x)| \leq \frac{\delta}{z-1} \qquad\qquad (d)$$

for all $x > 0$.

Furthermore, let $L : (0, \infty) \to \mathbb{C}$ be another function satisfying (c) and (d). Substituting y^n for y in (c) yields

$$L_z(x^{y^n}) = y^n L_z(x) \quad \text{and} \quad L(x^{y^n}) = y^n L(x) \qquad\qquad (e)$$

for all $x > 0$ $(x \neq 1)$ and for any $y \neq 0$. Let $y > 1$ be fixed in (e). Then, considering (d) and (e) we get

$$\begin{aligned}
|L_z(x) - L(x)| &= y^{-n} |L_z(x^{y^n}) - L(x^{y^n})| \\
&\leq y^{-n} |L_z(x^{y^n}) - f(x^{y^n})| + y^{-n} |f(x^{y^n}) - L(x^{y^n})| \\
&\leq y^{-n} \frac{2\delta}{z-1}
\end{aligned}$$

for all $n \in \mathbb{N}$. That is, $L_z(x) = L(x)$ for all $x > 0$ $(x \neq 1)$. This implies that if $z' > z > 1$, then

$$L_z(x) = L_{z'}(x) \qquad\qquad (f)$$

for all $x > 0$ $(x \neq 1)$ because $L_{z'}$ satisfies (c) and (d).

By letting $z \to \infty$ in (d) and by (c) and (f), we conclude that f itself satisfies the functional equation (11.2) for all $x > 0$ $(x \neq 1)$ and for any $y \neq 0$. Now, if $x = 1$, then $f(1) = 0$. Hence,

$$0 = f(1) = f(x^y) = y f(x)$$

for all $y \in \mathbb{R}$. Analogously, if $y = 0$, then

$$0 = f(1) = f(x^y) = yf(x),$$

for all $x > 0$, which ends the proof. $\square$

It is assumed in Theorem 11.2 that the inequality (11.3) holds true for all $x > 0$ and $y \in \mathbb{R}$. Can we also expect the superstability of the functional equation (11.2) in somewhat restricted variables? Similarly as in the proof of Theorem 11.2, we can prove the following theorems. Hence, we omit the proofs.

Theorem 11.3. *Assume $f : (0, \infty) \to \mathbb{C}$ satisfies the inequality (11.3) for some $\delta \geq 0$ and for all $x, y > 0$. Then we have the following possibilities:*

(i) *f is identically zero;*
(ii) *f is identically zero on $(0, 1)$ and satisfies the equation (11.2) for $x \geq 1$ and $y > 0$;*
(iii) *f is identically zero on $(1, \infty)$ and satisfies the equation (11.2) for $x \in (0, 1]$ and $y > 0$;*
(iv) *f satisfies the equation (11.2) for $x, y > 0$.*

For the particular case of $D = \{x \in (0, \infty) \mid f(x) \neq 0\} = (0, \infty) \backslash \{1\}$ in Theorem 11.3, Jung [158] obtained the following result.

Corollary 11.4. *Let a function $f : (0, \infty) \to \mathbb{C}$ satisfy the inequality (11.3) for some $\delta \geq 0$ and for all $x, y > 0$. If $D = (0, \infty)\backslash\{1\}$, then f satisfies the functional equation (11.2) for all $x, y > 0$.*

Jung [158] also obtained the following:

Theorem 11.5. *Let a function $f : (0, \infty) \to \mathbb{C}$ satisfy the inequality (11.3) for some $\delta \geq 0$ and for all $x, y > 1$. If there exists an $\varepsilon > 0$ such that*

$$\inf_{x \geq 1+\varepsilon} |f(x)| > 0, \tag{11.4}$$

then f satisfies the functional equation (11.2) for any $x, y > 1$.

By using the result of Theorem 11.5, we can prove the following corollary (ref. [158]).

Corollary 11.6. *Let $d > 2$ be fixed and let a function $f : (0, \infty) \to \mathbb{C}$ satisfy the inequality (11.3) for some $\delta \geq 0$ and for all $x, y > 0$ with $x + y \geq d$. If (11.4) holds true for some $\varepsilon > 0$, then f itself satisfies the functional equation (11.2) for all $x, y > 1$.*

Proof. Let $x, y > 1$ satisfy the condition $x + y < d$. Choose a number $z > 1$ with $zx \geq d$. It then follows from (11.3) that

$$|f(x^y) - yf(x)| \le \left| f(x^y) - \frac{y \ln x}{\ln zx} f(zx) \right| + y \left| \frac{\ln x}{\ln zx} f(zx) - f(x) \right|$$

$$\le \delta(1 + y)$$

$$< \delta d,$$

since

$$\frac{y \ln x}{\ln zx} > 0, \quad \frac{\ln x}{\ln zx} > 0, \quad \frac{y \ln x}{\ln zx} + zx > d, \quad \frac{\ln x}{\ln zx} + zx > d.$$

Our hypothesis implies that f satisfies

$$|f(x^y) - yf(x)| < \delta d$$

for all $x, y > 1$. Hence, by Theorem 11.5, f satisfies the functional equation (11.2) for all $x, y > 1$. $\square$

The assertion in Corollary 11.6 is also true under a weakened condition (11.5) instead of (11.3) as mentioned in the following corollary (cf. [158]).

Corollary 11.7. *Let* $f : (0, \infty) \to \mathbb{C}$ *be a function. Assume that* (11.4) *holds true for some* $\varepsilon > 0$. *If*

$$\lim_{n \to \infty} \sup_{x+y \ge n} |f(x^y) - yf(x)| < \infty, \tag{11.5}$$

then f *satisfies the functional equation* (11.2) *for all* $x, y > 1$.

Proof. According to (11.5), there is an $M > 0$ such that

$$\lim_{n \to \infty} \sup_{x+y \ge n} |f(x^y) - yf(x)| < M,$$

and hence, we can choose an n_0 such that

$$\sup_{x+y \ge n_0} |f(x^y) - yf(x)| < 2M. \tag{a}$$

Corollary 11.6, together with (a), implies that f satisfies the functional equation (11.2) for all $x, y > 1$. $\square$

We will now prove the superstability of the functional equation (11.2) in the sense of Ger. First of all, let δ be given with $0 < \delta < 1$, and define

$$\alpha_n(x, y) = \prod_{i=1}^{n} \left(1 - \delta x^{-y^i}\right), \quad \beta_n(x, y) = \prod_{i=1}^{n} \left(1 + \delta x^{-y^i}\right)$$

and

$$\alpha(x, y) = \lim_{n \to \infty} \alpha_n(x, y), \quad \beta(x, y) = \lim_{n \to \infty} \beta_n(x, y)$$

for any $x > 1$ and $y \geq 2$. It is not difficult to demonstrate that $\alpha(x, y) > 0$ for any $x > 1$ and $y \geq 2$.

Jung [158] provided the following theorem.

Theorem 11.8. *If a function* $f : (0, \infty) \to (0, \infty)$ *satisfies the inequality*

$$\left| \frac{f(x^y)}{yf(x)} - 1 \right| \leq \delta x^{-y} \tag{11.6}$$

for all $x, y > 0$ *and for some* $0 < \delta < 1$, *then* f *satisfies the functional equation* (11.2) *for all* $x > 1$ *and* $y > 0$.

Proof. Let $x\ (> 1)$ and $y\ (\geq 2)$ be fixed. We use induction on n to prove

$$\alpha_n(x, y) \leq \frac{f(x^{y^n})}{y^n f(x)} \leq \beta_n(x, y). \tag{a}$$

The inequality (a) for $n = 1$ is an immediate consequence of (11.6). Assume now that (a) is true for some integer $n > 0$. Then, by the equality

$$\frac{f(x^{y^{n+1}})}{y^{n+1} f(x)} = \frac{f(x^{y^{n+1}})}{yf(x^{y^n})} \cdot \frac{f(x^{y^n})}{y^n f(x)}$$

and by substituting x^{y^n} for x in (11.6) we get

$$\alpha_{n+1}(x, y) \leq \frac{f(x^{y^{n+1}})}{y^{n+1} f(x)} \leq \beta_{n+1}(x, y)$$

which ends the proof of (a).

Let $m, n \in \mathbb{N}$ be given with $n \geq m$. Since

$$\ln y^{-n} f(x^{y^n}) - \ln y^{-m} f(x^{y^m}) = \ln \frac{f((x^{y^m})^{y^{n-m}})}{y^{n-m} f(x^{y^m})},$$

(a) implies

$$\sum_{i=m+1}^{n} \ln(1 - \delta x^{-y^i}) \leq \ln y^{-n} f(x^{y^n}) - \ln y^{-m} f(x^{y^m})$$

$$\leq \sum_{i=m+1}^{n} \ln(1 + \delta x^{-y^i}).$$

Since the left- and right-hand sides of the above inequality tend to 0 as $m \to \infty$, the sequence $\{\ln y^{-n} f(x^{y^n})\}$ is a Cauchy sequence. Therefore, we can define

$$P_y(x) = \lim_{n \to \infty} \ln y^{-n} f(x^{y^n})$$

and

$$L_y(x) = \exp(P_y(x))$$

for all $x > 1$ and $y \geq 2$. Indeed, it is not difficult to demonstrate

$$L_y(x) = \lim_{n \to \infty} y^{-n} f(x^{y^n})$$

for any $x > 1$.

Substituting x^{y^n} ($x > 1$, $y \geq 2$) and z (> 0) for x and y in (11.6), respectively, and letting $n \to \infty$, then we obtain

$$\lim_{n \to \infty} f(x^{zy^n}) z^{-1} f(x^{y^n})^{-1} = 1. \tag{b}$$

Hence, by using (b), we have

$$L_y(x^z) = \lim_{n \to \infty} y^{-n} f(x^{zy^n}) = \lim_{n \to \infty} y^{-n} f(x^{zy^n}) z^{-1} f(x^{y^n})^{-1} z f(x^{y^n})$$

and

$$L_y(x^z) = z L_y(x) \tag{c}$$

for all $x > 1$ and $z > 0$. Clearly, it follows from (a) that L_y satisfies the inequality

$$\alpha(x, y) \leq \frac{L_y(x)}{f(x)} \leq \beta(x, y) \tag{d}$$

for $x > 1$.

Now, let $L : (1, \infty) \to (0, \infty)$ be a function which satisfies (d) for $x > 1$ and (c) for $x > 1$ and $z > 0$. (It follows from (d) that if $y \geq 2$, then $L_y(x) \neq 0$ and $L(x) \neq 0$ for all $x > 1$.) Then, it follows from (d) and (c) that

$$\frac{\alpha(x^{y^n}, y)}{\beta(x^{y^n}, y)} \leq \frac{L_y(x)}{L(x)} = \frac{L_y(x^{y^n})}{L(x^{y^n})} \leq \frac{\beta(x^{y^n}, y)}{\alpha(x^{y^n}, y)}$$

holds true for $x > 1$ and $y \geq 2$, which implies the uniqueness of L_y because $\alpha(x, y) \to 1$, $\beta(x, y) \to 1$ as $x \to \infty$. This implies that if $y' > y \geq 2$, then $L_{y'}(x) = L_y(x)$ for all $x > 1$, because $L_{y'}$ satisfies (d) for all $x > 1$ and (c) for $x > 1$ and $z > 0$. Therefore, by letting $y \to \infty$ in (d) and by considering $\alpha(x, y) \to 1$, $\beta(x, y) \to 1$ as $y \to \infty$ (when $x > 1$), we conclude that f itself satisfies (c) for all $x > 1$ and $z > 0$. $\square$

11.3 Functional Equation of Heuvers

K. J. Heuvers [132] introduced a new type of logarithmic functional equation

$$f(x + y) = f(x) + f(y) + f(1/x + 1/y), \qquad (11.7)$$

which will be called the *functional equation of Heuvers*, and proved that this equation is equivalent to the "original" logarithmic equation, $f(xy) = f(x) + f(y)$, in the class of functions $f : (0, \infty) \to \mathbb{R}$.

Theorem 11.9. *A function* $f : (0, \infty) \to \mathbb{R}$ *is a solution of the functional equation* (11.7) *if and only if* f *is a solution of the "original" logarithmic functional equation.*

Proof. If we set $y = 1/x$ in (11.7), then we have

$$f(1/x) = -f(x) \qquad (a)$$

and

$$f(1) = 0. \qquad (b)$$

We observe that

$$(Hf)(x, y) = f(x + y) - f(x) - f(y)$$

is a first Cauchy difference and

$$\begin{aligned}
(Hf)(x, y, z) &= (Hf)(x + y, z) - (Hf)(x, z) - (Hf)(y, z) \\
&= f(x + y + z) - f(x + y) - f(x + z) - f(y + z) \\
&\quad + f(x) + f(y) + f(z)
\end{aligned}$$

is a second Cauchy difference which is symmetric in x, y, and z (see [131]).
Thus, we get

$$\begin{aligned}
f\big(1/(x + y) + 1/z\big) &- f(1/x + 1/z) - f(1/y + 1/z) \\
&= f(x + y + z) - f(x + y) - f(x + z) - f(y + z) \\
&\quad + f(x) + f(y) + f(z) \\
&= f\big(1/(y + z) + 1/x\big) - f(1/y + 1/x) - f(1/z + 1/x).
\end{aligned}$$

Consequently, we obtain

$$f\big(1/(x+y)+1/z\big)+f(1/x+1/y) = f\big(1/(y+z)+1/x\big)+f(1/y+1/z). \quad (c)$$

Set $1/y + 1/z = 1$ in (c). Since y and z are positive, they are obviously greater than 1 and it follows that

$$\frac{1}{z} = 1 - \frac{1}{y} = \frac{y-1}{y} \quad \text{or} \quad z = \frac{y}{y-1}.$$

Moreover, we have

$$\frac{1}{x+y} + \frac{1}{z} = \frac{1}{x+y} + \frac{y-1}{y} = \frac{y^2 + xy - x}{y(x+y)},$$

$$\frac{1}{x} + \frac{1}{y} = \frac{x+y}{xy},$$

$$\frac{1}{y+z} + \frac{1}{x} = \frac{1}{y + y/(y-1)} + \frac{1}{x} = \frac{y-1}{y^2} + \frac{1}{x} = \frac{y^2 + xy - x}{xy^2}.$$

It follows from (b) and (c) that

$$f\left(\frac{x+y}{xy}\right) + f\left(\frac{y^2 + xy - x}{y(x+y)}\right) = f\left(\frac{y^2 + xy - x}{xy^2}\right) \qquad (d)$$

for all $x > 0$ and $y > 1$, since $1/y + 1/z = 1$.

Now, let

$$u = \frac{x+y}{xy} = \frac{1}{x} + \frac{1}{y} \quad \text{and} \quad v = \frac{y^2 + xy - x}{y(x+y)}.$$

It then follows that

$$uv = \frac{y^2 + xy - x}{xy^2}$$

and it follows from (d) that

$$f(uv) = f(u) + f(v).$$

Since $x > 0$ and $y > 1$, u and v are positive but there may be further restriction on them. We have

$$u = \frac{1}{x} + \frac{1}{y} \quad \text{and} \quad v = 1 - \frac{x/y}{x+y}.$$

Thus, we get

$$u - \frac{1}{x} = \frac{1}{y} = \frac{xu - 1}{x} \quad \text{or} \quad y = \frac{x}{xu - 1}.$$

Moreover, we obtain

$$x + y = x + \frac{x}{xu - 1} = \frac{x^2 u}{xu - 1} \quad \text{or} \quad \frac{1}{x+y} = \frac{xu - 1}{x^2 u}.$$

Thus, we have

$$v = 1 - \frac{(xu - 1)^2}{x^2 u} \quad \text{or} \quad uv = u - \frac{(xu - 1)^2}{x^2}.$$

Hence, we get

$$u - uv = u(1 - v) = \frac{(xu - 1)^2}{x^2} \quad \text{or} \quad y^2 = \frac{x^2}{(xu - 1)^2} = \frac{1}{u(1 - v)}.$$

Therefore, it follows that $0 < v < 1$ and $y = u^{-1/2}(1 - v)^{-1/2}$. Thus, we have

$$u - 1/y = u - u^{1/2}(1 - v)^{1/2} = 1/x = u^{1/2}\left(u^{1/2} - (1 - v)^{1/2}\right)$$

or

$$x = \frac{u^{1/2} + (1 - v)^{1/2}}{u^{1/2}(u - 1 + v)}.$$

Hence, if $u > 1$, then $u + v - 1 > 0$. Therefore, we get

$$f(uv) = f(u) + f(v)$$

for all $u > 1$ and $0 < v < 1$.

Since

$$f(u) = f(u) + f(1) = f(u \cdot 1) \quad \text{and} \quad f(v) = f(1) + f(v) = f(1 \cdot v),$$

we obtain

$$f(uv) = f(u) + f(v) \tag{e}$$

for all $u \geq 1$ and $0 < v \leq 1$.

Assume that $s \geq 1$. Then, $0 < 1/s \leq 1$ and $1 \leq s \leq s^2$ and it follows from (a) and (e) that

$$f(s) = f(s^2/s) = f(s^2) + f(1/s) = f(s^2) - f(s)$$

or

$$f(s^2) = 2f(s). \tag{f}$$

If $1 \leq s \leq t$, then $0 < s/t \leq 1$ and $1 \leq t \leq t^2$. Then, it follows from (a), (e), and (f) that

$$\begin{aligned}
f(st) &= f(s/t) + f(t^2) \\
&= f(s) + f(1/t) + f(t^2) \\
&= f(s) - f(t) + 2f(t) \\
&= f(s) + f(t) \tag{g}
\end{aligned}$$

for any $s \geq 1$ and $t \geq 1$.

Assume now that $0 < s, t < 1$. It then follows from (a) and (g) that

$$-f(st) = f(1/(st)) = f((1/s)(1/t)) = -f(s) - f(t).$$

Consequently, we have

$$f(uv) = f(u) + f(v)$$

for all $u > 0$ and $v > 0$. Thus, every solution of (11.7) is a solution of the "original" logarithmic functional equation.

Assume that f is a solution of the "original" logarithmic functional equation. Then, it holds true that $f(xy) = f(x) + f(y)$ for all $x > 0$ and $y > 0$. Hence, we get

$$f(x + y) = f(x(1/x + 1/y)y) = f(x) + f(1/x + 1/y) + f(y).$$

Therefore, each solution of the "original" logarithmic functional equation is also a solution of (11.7).   □

We will now apply Theorem 10.23 to the proof of the Hyers–Ulam stability of the functional equation of Heuvers. S.-M. Jung [175] contributed to the following theorem.

Theorem 11.10. *If a function* $f : \mathbb{R} \to \mathbb{R}$ *satisfies the functional inequality*

$$\left| f(x + y) - f(x) - f(y) - f(1/x + 1/y) \right| \le \delta \tag{11.8}$$

for some $0 \le \delta < \ln 2$ *and for all* $x, y \in \mathbb{R} \setminus \{0\}$, *then there exists a unique logarithmic function* $\ell : \mathbb{R} \setminus \{0\} \to \mathbb{R}$ *such that*

$$|f(x) - \ell(x)| \le (13/2)(\delta - \ln(2 - e^\delta)) \tag{11.9}$$

for each $x \in \mathbb{R} \setminus \{0\}$.

Proof. If we define a function $g : \mathbb{R} \to (0, \infty)$ by

$$g(x) = e^{f(x)}, \tag{a}$$

then it follows from (11.8) that

$$\left| \frac{g(x + y)}{g(x)g(y)g(1/x + 1/y)} - 1 \right| \le e^\delta - 1$$

for all $x, y \in \mathbb{R} \setminus \{0\}$. According to Theorem 10.23, there exists a multiplicative function $m : \mathbb{R} \setminus \{0\} \to (0, \infty)$ with

$$\left(\frac{2 - e^\delta}{e^\delta}\right)^{13/2} \leq \frac{m(x)}{g(x)} \leq \left(\frac{e^\delta}{2 - e^\delta}\right)^{13/2} \qquad (b)$$

for $x \in \mathbb{R} \setminus \{0\}$. Define a function $\ell : \mathbb{R} \setminus \{0\} \to \mathbb{R}$ by

$$\ell(x) = \ln m(x). \qquad (c)$$

Then, ℓ is a logarithmic function. We can conclude by (a), (b), and (c) that the inequality (11.9) holds true for any $x \in \mathbb{R} \setminus \{0\}$.

Let $\ell' : \mathbb{R} \setminus \{0\} \to \mathbb{R}$ be another logarithmic function satisfying (11.9) for each $x \in \mathbb{R} \setminus \{0\}$. Since ℓ and ℓ' are logarithmic, we get

$$\ell(x^{2^n}) = 2^n \ell(x) \quad \text{and} \quad \ell'(x^{2^n}) = 2^n \ell'(x)$$

for all $x \in \mathbb{R} \setminus \{0\}$ and $n \in \mathbb{N}$. Hence, it follows from (11.9) that

$$\begin{aligned}
|\ell(x) - \ell'(x)| &= 2^{-n}|\ell(x^{2^n}) - \ell'(x^{2^n})| \\
&\leq 2^{-n}|\ell(x^{2^n}) - f(x^{2^n})| + 2^{-n}|f(x^{2^n}) - \ell'(x^{2^n})| \\
&\to 0 \quad \text{as } n \to \infty,
\end{aligned}$$

which implies the uniqueness of ℓ. $\qquad \square$

The functional equation

$$f(x + y) = g(x) + h(y) + k(1/x + 1/y) \qquad (11.10)$$

is a Pexider generalization of the functional equation (11.7). K. J. Heuvers and Pl. Kannappan [133] investigated the solutions of the functional equation (11.10).

Theorem 11.11. *The twice-differentiable solution $f, g, h, k : (0, \infty) \to \mathbb{R}$ of the functional equation (11.10) is given by*

$$\begin{aligned}
f(x) &= -a \ln x + bx + c_1, \\
g(x) &= -a \ln x + bx - d/x + c_1 + c_3, \\
h(x) &= -a \ln x + bx - d/x - c_2 - c_3, \\
k(x) &= -a \ln x + dx + c_2,
\end{aligned} \qquad (11.11)$$

where a, b, c_1, c_2, c_3, and d are real constants.

Proof. If we differentiate (11.10) with respect to x and y, then we get

$$f''(x + y) = (1/xy)^2 k''(1/x + 1/y).$$

If we set $s = x + y$ and $t = 1/x + 1/y$ in the last equation, then we obtain $f''(s) = (t^2/s^2)k''(t)$ or

$$s^2 f''(s) = t^2 k''(t) \quad \text{for } st \geq 4. \tag{a}$$

If we set $s = 4$ in (a), then we have $t^2 k''(t) = a$ for $t \geq 1$. On the other hand, if we set $t = 4$ in (a), then we get $s^2 f''(s) = a$ for $s \geq 1$, where a is a constant.

For $0 < t < 1$ choose $s > 4$ so that $st \geq 4$. Then it follows from (a) that f and k satisfy the differential equation $f''(s) = a/s^2$ and $k''(t) = a/t^2$ for $s, t > 0$. If we integrate each function twice, then we obtain

$$f(x) = -a \ln x + bx + c_1 \quad \text{and} \quad k(x) = -a \ln x + dx + c_2.$$

Putting these into (11.10) yields

$$g(x) + h(y) = -a \ln x - a \ln y + bx + by - d/x - d/y + c_1 - c_2$$

or

$$g(x) + a \ln x - bx + d/x - c_1 = -\big(h(y) + a \ln y - by + d/y + c_2\big) = c_3,$$

where c_3 is a constant. Hence, we conclude that the twice-differentiable solution of the functional equation (11.10) is given by (11.11). $\qquad \square$

Chapter 12
Trigonometric Functional Equations

The famous addition and subtraction rules for trigonometric functions can be represented by using functional equations. Some of these equations will be introduced and the stability problems for them will be surveyed. Section 12.1 deals with the superstability phenomenon of the cosine functional equation (12.1) which stands for an addition theorem of cosine function. Similarly, the superstability of the sine functional equation (12.3) is proved in Section 12.2. In Section 12.3, trigonometric functional equations (12.8) and (12.9) with two unknown functions will be discussed. It is very interesting that these functional equations for complex-valued functions defined on an amenable group are not superstable, but they are stable in the sense of Hyers and Ulam, whereas the equations (12.1) and (12.3) are superstable. In Section 12.4, we will deal with the Hyers–Ulam stability of the Butler–Rassias functional equation.

12.1 Cosine Functional Equation

The addition rule $\cos(x + y) + \cos(x - y) = 2 \cos x \cos y$ for the cosine function may be symbolized by the functional equation

$$f(x + y) + f(x - y) = 2f(x)f(y). \tag{12.1}$$

This equation is called the *cosine functional equation* or *d'Alembert equation*.

In 1968, Pl. Kannappan [209] determined the general solution of the cosine functional equation (12.1):

Every nontrivial solution $f : \mathbb{R} \to \mathbb{C}$ of the functional equation (12.1) is given by
$$f(x) = (1/2)\big(m(x) + m(-x)\big),$$
where $m : \mathbb{R} \to \mathbb{C} \setminus \{0\}$ is an exponential function.

J. Baker [16] was the first person to prove the superstability for this equation, and later P. Găvruta [114] presented a short proof for the theorem. Before this proof provided by Găvruta, we will introduce a lemma of Baker which is necessary to prove the theorem of Baker and Găvruta (see [16]).

S.-M. Jung, *Hyers–Ulam–Rassias Stability of Functional Equations in Nonlinear Analysis*, Springer Optimization and Its Applications 48, DOI 10.1007/978-1-4419-9637-4_12, © Springer Science+Business Media, LLC 2011

Lemma 12.1. *Let $(G, +)$ be an abelian group and let a function $f : G \to \mathbb{C}$ satisfy the functional inequality*

$$|f(x + y) + f(x - y) - 2f(x)f(y)| \leq \delta \tag{12.2}$$

for all $x, y \in G$ and for some $\delta > 0$. If $|f(x)| > (1 + \sqrt{1 + 2\delta})/2$ for some $x \in G$, then $|f(2^n x)| \to \infty$ as $n \to \infty$.

Proof. Let $x \in G$ and $y = |f(x)| = \varepsilon + p$, where $\varepsilon = (1 + \sqrt{1 + 2\delta})/2$ and $p > 0$. For simplicity, let $\mu = \delta + \varepsilon$. Then, we have

$$\begin{aligned}
2y^2 - y - \mu &= 2(\varepsilon + p)^2 - \varepsilon - p - \delta - \varepsilon \\
&= 2(\varepsilon^2 - \varepsilon) + (4\varepsilon - 1)p + 2p^2 - \delta \\
&= (4\varepsilon - 1)p + 2p^2 \\
&> 3p,
\end{aligned}$$

since $\varepsilon > 1$ and $2(\varepsilon^2 - \varepsilon) = \delta$. Hence,

$$2y^2 - \mu > y + 3p. \tag{a}$$

If we put $y = x$ in (12.2), we get

$$\left| f(2x) + f(0) - 2f(x)^2 \right| \leq \delta,$$

and hence

$$|f(2x)| \geq \left| 2f(x)^2 - f(0) \right| - \delta \geq 2|f(x)|^2 - |f(0)| - \delta.$$

Since $|f(0)| \leq \varepsilon$, we have

$$|f(2x)| \geq 2|f(x)|^2 - \mu. \tag{b}$$

It follows from (a) and (b) that

$$|f(2x)| \geq 2|f(x)|^2 - \mu > |f(x)| + 3p > \varepsilon + 2p.$$

Assume that

$$|f(2^n x)| \geq \varepsilon + 2^n p \tag{c}$$

for some integer $n \geq 2$. Then, if we replace x and y in (12.2) with $2^n x$ simultaneously, it then follows from (c) that

$$\begin{aligned}
\left| f\left(2^{n+1}x \right) \right| &\geq \left| 2f(2^n x)^2 - f(0) \right| - \delta \\
&\geq 2|f(2^n x)|^2 - \varepsilon - \delta \\
&\geq 2\varepsilon^2 - 2\varepsilon - \delta + \varepsilon + 2^{n+2}\varepsilon p + 2^{2n+1}p^2 \\
&\geq \varepsilon + 2^{n+1}p,
\end{aligned}$$

since $\varepsilon > 1$ and $2\varepsilon^2 - 2\varepsilon - \delta = 0$. Hence, (c) holds true for all $n \in \mathbb{N}$. $\qquad\qquad\square$

In the following theorem, we show that the cosine functional equation (12.1) is superstable (ref. [16, 114]).

Theorem 12.2 (Baker and Gǎvruta). *Let $(G, +)$ be an abelian group and let a function $f : G \to \mathbb{C}$ satisfy the functional inequality (12.2) for all $x, y \in G$ and for some $\delta > 0$. Then either $|f(x)| \leq \left(1 + \sqrt{1 + 2\delta}\right)/2$ for any $x \in G$ or f satisfies the cosine functional equation (12.1) for all $x, y \in G$.*

Proof. If there exists an $x_0 \in G$ such that

$$|f(x_0)| > \left(1 + \sqrt{1 + 2\delta}\right)/2,$$

then there exists a sequence $\{x_n\}$ with

$$\lim_{n \to \infty} |f(x_n)| = \infty \qquad\qquad (a)$$

(see Lemma 12.1). Let $x, y \in G$ be given. From (12.2) it follows that

$$|2f(x_n)f(x) - f(x + x_n) - f(x - x_n)| \leq \delta$$

for all $n \in \mathbb{N}$, and (a) implies

$$f(x) = \lim_{n \to \infty} \frac{f(x + x_n) + f(x - x_n)}{2f(x_n)}. \qquad\qquad (b)$$

From (b) we get

$$2f(x)f(y) = \lim_{n \to \infty} A_n \qquad\qquad (c)$$

and

$$f(x + y) + f(x - y) = \lim_{n \to \infty} B_n, \qquad\qquad (d)$$

where

$$A_n = \frac{\left(f(x + x_n) + f(x - x_n)\right)\left(f(y + x_n) + f(y - x_n)\right)}{2f(x_n)^2}$$

and

$$B_n = (1/2)f(x_n)^{-2}\big(f(x + y + x_n) + f(x + y - x_n) \\ + f(x - y + x_n) + f(x - y - x_n)\big)f(x_n).$$

By (12.2), we have

$$|2f(x + x_n)f(y + x_n) - f(x + y + 2x_n) - f(x - y)| \leq \delta,$$
$$|2f(x - x_n)f(y + x_n) - f(x + y) - f(x - y - 2x_n)| \leq \delta,$$
$$|2f(x + x_n)f(y - x_n) - f(x + y) - f(x - y + 2x_n)| \leq \delta,$$
$$|2f(x - x_n)f(y - x_n) - f(x + y - 2x_n) - f(x - y)| \leq \delta,$$
$$|-2f(x + y + x_n)f(x_n) + f(x + y + 2x_n) + f(x + y)| \leq \delta,$$
$$|-2f(x + y - x_n)f(x_n) + f(x + y) + f(x + y - 2x_n)| \leq \delta,$$
$$|-2f(x - y + x_n)f(x_n) + f(x - y) + f(x - y + 2x_n)| \leq \delta,$$
$$|-2f(x - y - x_n)f(x_n) + f(x - y) + f(x - y - 2x_n)| \leq \delta$$

for every $n \in \mathbb{N}$, and hence

$$|A_n - B_n| \leq \frac{2\delta}{|f(x_n)|^2} \qquad\qquad (e)$$

for all $n \in \mathbb{N}$. The relations (a), (c), (d), and (e) imply that f satisfies the functional equation (12.1) for any $x, y \in G$. $\qquad\square$

R. Badora [9] assumed the Kannappan condition $f(x + y + z) = f(x + z + y)$ instead of the commutativity of the group G and proved the following theorem:

Let $(G, +)$ be a group. If a function $f : G \to \mathbb{C}$ satisfies

$$|f(x + y) + f(x - y) - 2f(x)f(y)| \leq \delta,$$
$$|f(x + y + z) - f(x + z + y)| \leq \varepsilon$$

for all $x, y, z \in G$ and for some $\delta, \varepsilon > 0$, then either f is bounded or f is a solution of the cosine functional equation (12.1).

In 2002, R. Badora and R. Ger [10] proved the superstability of the cosine functional equation.

Theorem 12.3. *Let $(G, +)$ be an abelian group and let $\varphi : G \to [0, \infty)$ be a given function. If a function $f : G \to \mathbb{C}\backslash\{0\}$ satisfies the inequality*

$$|f(x + y) + f(x - y) - 2f(x)f(y)| \leq \varphi(x)$$

for all $x, y \in G$, then either f is bounded or f is a solution of the cosine functional equation (12.1).

Pl. Kannappan and G.-H. Kim [212] investigated the superstability of the generalized cosine functional equations

$$f(x + y) + f(x - y) = 2f(x)g(y)$$

and

$$f(x + y) + f(x - y) = 2g(x)f(y)$$

(see also [218]).

12.2 Sine Functional Equation

The *sine functional equation*

$$f(x + y)f(x - y) = f(x)^2 - f(y)^2 \tag{12.3}$$

reminds us of one of the trigonometric formulas:

$$\sin(x + y)\sin(x - y) = \sin^2 x - \sin^2 y.$$

We now introduce a theorem concerning the general solution of the sine functional equation (12.3):

A function $f : \mathbb{R} \to \mathbb{C}$ is a solution of the sine functional equation (12.3) if and only if f is of the form

$$f(x) = 0, \quad f(x) = A(x), \quad \text{or} \quad f(x) = c(m(x) - m(-x)),$$

where $A : \mathbb{R} \to \mathbb{C}$ is a nonzero additive function, $m : \mathbb{R} \to \mathbb{C}\backslash\{0\}$ is an exponential function, and c is a nonzero constant.

P. W. Cholewa [69] observed the superstability phenomenon of the sine functional equation (12.3). For the proof of the theorem of Cholewa, we need the following three lemmas originating from the paper [69].

Lemma 12.4. *Let $\delta > 0$, let $(G, +)$ be an abelian group in which division by 2 is uniquely performable, and let a function $f : G \to \mathbb{C}$ satisfy the inequality*

$$\left| f(x + y)f(x - y) - f(x)^2 + f(y)^2 \right| \le \delta \tag{12.4}$$

for all $x, y \in G$. If f is unbounded, then $f(0) = 0$.

Proof. Put $y = x$ and then $u = 2x$ in (12.4). Then, we have

$$\left| f(u)f(0) - f(u/2)^2 + f(u/2)^2 \right| = |f(u)||f(0)| \le \delta$$

and since $|f(u)|$ can be as large as possible, we must have $f(0) = 0$. $\qquad\square$

Lemma 12.5. *The hypotheses in Lemma 12.4 are assumed. If f is unbounded, then the inequality*

$$|f(x + y) + f(x - y) - 2f(x)g(y)| \le \delta \tag{12.5}$$

holds true for all $x, y \in G$, where

$$g(x) = \frac{f(x+a) - f(x-a)}{2f(a)} \tag{12.6}$$

for all $x \in G$ and for some $a \in G$ with $|f(a)| \geq 4$.

Proof. If we put $x = (u+v)/2$ and $y = (u-v)/2$ in (12.4), then we have

$$\left| f(u)f(v) - f\big((u+v)/2\big)^2 + f\big((u-v)/2\big)^2 \right| \leq \delta \tag{a}$$

for every $u, v \in G$. Using (a) and (12.6) yields

$$|f(x+y) + f(x-y) - 2f(x)g(y)|$$
$$\leq \frac{1}{|f(a)|} \left| f(x+y)f(a) - f\left(\frac{x+y+a}{2}\right)^2 + f\left(\frac{x+y-a}{2}\right)^2 \right|$$
$$+ \frac{1}{|f(a)|} \left| f(x-y)f(a) - f\left(\frac{x-y+a}{2}\right)^2 + f\left(\frac{x-y-a}{2}\right)^2 \right|$$
$$+ \frac{1}{|f(a)|} \left| f\left(\frac{x+y+a}{2}\right)^2 - f\left(\frac{x-y-a}{2}\right)^2 - f(x)f(y+a) \right|$$
$$+ \frac{1}{|f(a)|} \left| f(x)f(y-a) - f\left(\frac{x+y-a}{2}\right)^2 + f\left(\frac{x-y+a}{2}\right)^2 \right|$$
$$+ \left| 2f(x)\frac{f(y+a) - f(y-a)}{2f(a)} - 2f(x)g(y) \right|$$
$$\leq \frac{4\delta}{|f(a)|}$$
$$\leq \delta$$

for all $x, y \in G$.   □

Lemma 12.6. *The hypotheses in Lemma 12.4 are assumed. If f is unbounded, then f satisfies the equation*

$$f(x+y) + f(x-y) = 2f(x)g(y) \tag{12.7}$$

for all $x, y \in G$, where g is defined in (12.6).

Proof. Let x and y be two arbitrarily fixed points of G. Then, using (a) in the proof of Lemma 12.5 and (12.5) yields

$$|f(z)||f(x+y) + f(x-y) - 2f(x)g(y)|$$
$$= |f(z)f(x+y) + f(z)f(x-y) - 2f(x)f(z)g(y)|$$

$$\leq \left| f(z) f(x+y) - f\left(\frac{z+x+y}{2}\right)^2 + f\left(\frac{z-x-y}{2}\right)^2 \right|$$

$$+ \left| f(z) f(x-y) - f\left(\frac{z+x-y}{2}\right)^2 + f\left(\frac{z-x+y}{2}\right)^2 \right|$$

$$+ \left| f\left(\frac{z+x+y}{2}\right)^2 - f\left(\frac{z-x+y}{2}\right)^2 - f(z+y) f(x) \right|$$

$$+ \left| f\left(\frac{z+x-y}{2}\right)^2 - f\left(\frac{z-x-y}{2}\right)^2 - f(z-y) f(x) \right|$$

$$+ \left| \left(f(z+y) + f(z-y) \right) f(x) - 2 f(z) g(y) f(x) \right|$$

$$\leq 4\delta + \delta |f(x)|$$

for each $z \in G$. Hence,

$$|f(x+y) + f(x-y) - 2f(x)g(y)| \leq (4 + |f(x)|)\delta / |f(z)|.$$

Since f is assumed to be an unbounded function and x is a fixed element, the right-hand side of the last inequality can be as small as possible. Hence, the equation (12.7) holds true for all $x, y \in G$. □

Now, we are able to prove the theorem of Cholewa (ref. [69]).

Theorem 12.7 (Cholewa). *Let $(G, +)$ be an abelian group in which division by 2 is uniquely performable. Every unbounded function $f : G \rightarrow \mathbb{C}$ satisfying the inequality (12.4) for some $\delta > 0$ and for all $x, y \in G$ is a solution of the sine functional equation (12.3).*

Proof. Let f be an unbounded solution of the inequality (12.4). If we put $x = 0$ in (12.7), it then follows from Lemma 12.4 that

$$f(y) = -f(-y) \tag{a}$$

for any $y \in G$. Now, put $x = (u + v)/2$ and $y = (u - v)/2$ in (12.7). Then, we get

$$f(u) + f(v) = 2f\left((u+v)/2\right)g\left((u-v)/2\right) \tag{b}$$

for every $u, v \in G$. From Lemma 12.4 and (b), it follows that

$$f(x+y) = f(x+y) + f(0) = 2f\left((x+y)/2\right)g\left((x+y)/2\right) \tag{c}$$

for all $x, y \in G$, and

$$f(x-y) = f(x-y) + f(0) = 2f\left((x-y)/2\right)g\left((x-y)/2\right) \tag{d}$$

for any $x, y \in G$. Using (a) and (b) yields

$$f(x) - f(y) = f(x) + f(-y) = 2f\big((x-y)/2\big)g\big((x+y)/2\big) \qquad (e)$$

for every $x, y \in G$. Now, using (b), (c), (d), and (e) yields

$$
\begin{aligned}
f(x+y)&f(x-y) \\
&= \big(2f\big((x+y)/2\big)g\big((x+y)/2\big)\big)\big(2f\big((x-y)/2\big)g\big((x-y)/2\big)\big) \\
&= \big(2f\big((x+y)/2\big)g\big((x-y)/2\big)\big)\big(2f\big((x-y)/2\big)g\big((x+y)/2\big)\big) \\
&= \big(f(x)+f(y)\big)\big(f(x)-f(y)\big) \\
&= f(x)^2 - f(y)^2
\end{aligned}
$$

for any $x, y \in G$. $\square$

J. Baker [16] and P. Găvruta [114] proved in Theorem 12.2 that, concerning the superstability of the cosine equation, a function f satisfying the inequality (12.2) is either a solution of the cosine equation (12.1) or it is bounded by a constant depending on δ only. It is not the case for the sine equation. Indeed, the bounded functions

$$f_n(x) = n \sin x + 1/n$$

satisfy the inequality (12.4) with $\delta = 3$, for all $x, y \in \mathbb{R}$ and for all $n \in \mathbb{N}$. However, for each $\varepsilon > 0$, the inequality $|f_n(x)| \le \varepsilon$ fails to hold true for certain x and n.

The sine functional equation (12.3) can be rewritten as

$$f(x)f(y) = f\left(\frac{x+y}{2}\right)^2 - f\left(\frac{x-y}{2}\right)^2.$$

For an endomorphism $\sigma : G \to G$ of order 2 of the uniquely 2-divisible abelian group G, G.-H. Kim [216] proved the stability of the following generalized sine functional equations:

$$
\begin{aligned}
g(x)f(y) &= f\left(\frac{x+y}{2}\right)^2 - f\left(\frac{x+\sigma y}{2}\right)^2, \\
f(x)g(y) &= f\left(\frac{x+y}{2}\right)^2 - f\left(\frac{x+\sigma y}{2}\right)^2, \\
g(x)g(y) &= f\left(\frac{x+y}{2}\right)^2 - f\left(\frac{x+\sigma y}{2}\right)^2.
\end{aligned}
$$

12.3   Trigonometric Equations with Two Unknowns

So far, we have seen the superstability results concerning the cosine and sine functional equations (12.1) and (12.3). Obviously, there are other equations which are satisfied by the cosine and sine functions.

L. Székelyhidi [345] introduced the functional equations

$$f(xy) = f(x)f(y) - g(x)g(y) \tag{12.8}$$

and

$$f(xy) = f(x)g(y) + f(y)g(x) \tag{12.9}$$

for complex-valued functions defined on a semigroup $(G, \cdot)$. We see that the equations (12.8) and (12.9) describe addition theorems for cosine and sine.

If G is a semigroup and $\mathcal{F}$ is a vector space of complex-valued functions on G, then we say that the functions $f, g : G \to \mathbb{C}$ are *linearly independent modulo* $\mathcal{F}$ if $\lambda f + \mu g \in \mathcal{F}$ implies $\lambda = \mu = 0$ for any $\lambda, \mu \in \mathbb{C}$. The vector space $\mathcal{F}$ is said to be *invariant* if $f \in \mathcal{F}$ implies that the functions $f(xy)$ and $f(yx)$ belong to $\mathcal{F}$ for each fixed $y \in G$.

L. Székelyhidi [345] presented the following lemma.

Lemma 12.8. *Let $(G, \cdot)$ be a semigroup, $f, g : G \to \mathbb{C}$ be functions, and $\mathcal{F}$ be an invariant vector space of complex-valued functions on G. Suppose that f and g are linearly independent modulo $\mathcal{F}$. If the functions*

$$x \mapsto f(xy) - f(x)f(y) + g(x)g(y)$$

and

$$x \mapsto f(xy) - f(yx)$$

belong to $\mathcal{F}$ for each fixed $y \in G$, then f and g satisfy the functional equation (12.8) *for all $x, y \in G$.*

Proof. Let us define

$$F(x, y) = f(xy) - f(x)f(y) + g(x)g(y) \tag{a}$$

for every $x, y \in G$. Then there are constants $\lambda_0, \lambda_1, \lambda_2 \in \mathbb{C}$ and $y_1 \in G$ with

$$g(x) = \lambda_0 f(x) + \lambda_1 f(xy_1) + \lambda_2 F(x, y_1) \tag{b}$$

for each $x \in G$. Using (a) and (b) yields

$$
\begin{aligned}
f((xy)z) &= f(xy)f(z) - g(z)g(xy) + F(xy, z) \\
&= f(x)f(y)f(z) - g(x)g(y)f(z) + F(x, y)f(z) \\
&\quad - \lambda_0 f(xy)g(z) - \lambda_1 f(xyy_1)g(z) - \lambda_2 F(xy, y_1)g(z) + F(xy, z) \\
&= f(x)f(y)f(z) - g(x)g(y)f(z) + F(x, y)f(z) \\
&\quad - \lambda_0 f(x)f(y)g(z) + \lambda_0 g(x)g(y)g(z) - \lambda_0 F(x, y)g(z) \\
&\quad - \lambda_1 f(x)f(yy_1)g(z) + \lambda_1 g(x)g(yy_1)g(z) - \lambda_1 F(x, yy_1)g(z) \\
&\quad - \lambda_2 F(xy, y_1)g(z) + F(xy, z)
\end{aligned}
$$

for all $x, y, z \in G$.

On the other hand, it follows from (a) that

$$f\big((xy)z\big) = f\big(x(yz)\big) = f(x)f(yz) - g(x)g(yz) + F(x, yz),$$

and equating the last two equalities yields

$$\begin{aligned}
f(x)\big(f(y)f(z) &- \lambda_0 f(y)g(z) - \lambda_1 f(yy_1)g(z) - f(yz)\big) \\
- g(x)\big(g(y)f(z) &- \lambda_0 g(y)g(z) - \lambda_1 g(yy_1)g(z) - g(yz)\big) \\
= F(x, yz) &- F(xy, z) - F(x, y)f(z) + \lambda_0 F(x, y)g(z) \\
&+ \lambda_1 F(x, yy_1)g(z) + \lambda_2 F(xy, y_1)g(z).
\end{aligned}$$

Using the linear independence of f and g modulo $\mathcal{F}$ yields

$$\begin{aligned}
F(x, yz) - F(xy, z) = F(x, y)f(z) &- \lambda_0 F(x, y)g(z) \\
&- \lambda_1 F(x, yy_1)g(z) - \lambda_2 F(xy, y_1)g(z).
\end{aligned}$$

By the hypotheses, it follows that the left-hand side belongs to $\mathcal{F}$ as a function of z for all fixed $x, y \in G$. Again using the linear independence of f and g modulo $\mathcal{F}$ and the fact that $\mathcal{F}$ is an invariant vector space yield $F(x, y) = 0$ for all $x, y \in G$, which ends the proof. $\square$

If $(G, \cdot)$ is a semigroup, then the function $a : G \to \mathbb{C}$ satisfying $a(xy) = a(x) + a(y)$, for all $x, y \in G$, will be called *additive*. If a function $m : G \to \mathbb{C}$ satisfies $m(xy) = m(x)m(y)$ for every $x, y \in G$, then m is said to be an *exponential* (see also Section 9.1).

Székelyhidi [345] also contributed to the following lemma.

Lemma 12.9. *Let $(G, \cdot)$ be a semigroup, let $f, g : G \to \mathbb{C}$ be functions, and let $\mathcal{F}$ be an invariant vector space of complex-valued functions on G. If the functions*

$$x \mapsto f(xy) - f(x)f(y) + g(x)g(y)$$

and

$$x \mapsto f(xy) - f(yx)$$

belong to $\mathcal{F}$ for each fixed $y \in G$, then we have the following possibilities:

 (i) *$f, g \in \mathcal{F}$;*
 (ii) *f is an exponential, and g belongs to $\mathcal{F}$;*
 (iii) *$f + g$ or $f - g$ is an exponential in $\mathcal{F}$;*
 (iv) *$f = \lambda^2(\lambda^2 - 1)^{-1}m - (\lambda^2 - 1)^{-1}b$, $g = \lambda(\lambda^2 - 1)^{-1}m - \lambda(\lambda^2 - 1)^{-1}b$, where $m : G \to \mathbb{C}$ is an exponential, $b : G \to \mathbb{C}$ belongs to $\mathcal{F}$, and $\lambda \in \mathbb{C}$ is a constant with $\lambda^2 \neq 1$;*
 (v) *f and g satisfy the functional equation (12.8) for all $x, y \in G$.*

Proof. If f and g are linearly independent modulo $\mathcal{F}$, then (v) follows from Lemma 12.8. If $g \in \mathcal{F}$, then (i) or (ii) follows from Theorem 9.2. If $f \in \mathcal{F}$, then $g \in \mathcal{F}$, hence (i) follows.

Now, we suppose that f and g are linearly independent modulo $\mathcal{F}$, but $f, g \notin \mathcal{F}$. Then there exists a constant $\lambda \neq 0$ with $f = \lambda g + b$ and $b \in \mathcal{F}$; hence, by the assumption, the function

$$x \mapsto g(xy) - (1/\lambda)\big((\lambda^2 - 1)g(y) + \lambda b(y)\big)g(x)$$

belongs to $\mathcal{F}$. By Theorem 9.2 again, we have that

$$(1/\lambda)(\lambda^2 - 1)g + b = m$$

is an exponential, which gives (iii) for $\lambda^2 = 1$ and (iv) for $\lambda^2 \neq 1$. $\qquad\square$

Székelyhidi [345] has proved the stability of the functional equation (12.8) for complex-valued functions on an amenable group, whereas we have seen the super-stability phenomena for the cosine equation (12.1) and the sine equation (12.3) (see Theorems 12.2 and 12.7).

Theorem 12.10. *Let $(G, \cdot)$ be an amenable group and let $f, g : G \to \mathbb{C}$ be functions. The function*

$$(x, y) \mapsto f(xy) - f(x)f(y) + g(x)g(y)$$

is bounded if and only if we have one of the following possibilities:

(i) *f and g are bounded;*
(ii) *f is an exponential, and g is bounded;*
(iii) *$f = (1 + a)m + b$, $g = am + b$ or $f = am + b$, $g = (1 - a)m - b$, where $a : G \to \mathbb{C}$ is additive, $m : G \to \mathbb{C}$ is a bounded exponential, and $b : G \to \mathbb{C}$ is bounded;*
(iv) *$f = \lambda^2(\lambda^2 - 1)^{-1}m - (\lambda^2 - 1)^{-1}b$, $g = \lambda(\lambda^2 - 1)^{-1}m - \lambda(\lambda^2 - 1)^{-1}b$, where $m : G \to \mathbb{C}$ is an exponential, $b : G \to \mathbb{C}$ is bounded, and $\lambda \in \mathbb{C}$ is a constant with $\lambda^2 \neq 1$;*
(v) *f and g satisfy the functional equation (12.8) for all $x, y \in G$.*

Proof. Let $\mathcal{F}$ be the set of all bounded complex-valued functions on G. First, we must prove the necessity. If g is bounded, then we have (i) or (ii) by Theorem 9.2. If $f + g$ or $f - g$ is a bounded exponential corresponding to Lemma 12.9 (iii), then we have (iii) by using Hyers's theorem (see [344]). Finally, the remaining follows from Lemma 12.9. The sufficiency follows by direct calculations. $\qquad\square$

We now discuss the stability of the functional equation (12.9). The following lemma is a modified version of Lemma 12.8 which is applicable to the functional equation (12.9). The proof of the lemma can be provided by the reader (or see [345, Lemma 2.1]).

Lemma 12.11. *Let $(G, \cdot)$ be a semigroup, let $f, g : G \to \mathbb{C}$ be functions, and let $\mathcal{F}$ be an invariant vector space of complex-valued functions on G. Suppose that f and g are linearly independent modulo $\mathcal{F}$. If the function*

$$x \mapsto f(xy) - f(x)g(y) - f(y)g(x)$$

belongs to $\mathcal{F}$ for each fixed $y \in G$, then f and g satisfy the functional equation (12.9) for all $x, y \in G$.

Lemma 12.12. *Let $(G, \cdot)$ be a semigroup, let $f, g : G \to \mathbb{C}$ be functions, and let $\mathcal{F}$ be an invariant vector space of complex-valued functions on G. If the function*

$$x \mapsto f(xy) - f(x)g(y) - f(y)g(x)$$

belongs to $\mathcal{F}$ for each fixed $y \in G$, then we have the following possibilities:

  (i) *$f = 0$ and g is arbitrary;*
 (ii) *$f, g \in \mathcal{F}$;*
(iii) *$g \in \mathcal{F}$ is an exponential;*
 (iv) *$f = \lambda m - \lambda b$, $g = (1/2)m + (1/2)b$, where $m : G \to \mathbb{C}$ is an exponential, $b : G \to \mathbb{C}$ belongs to $\mathcal{F}$, and $\lambda \in \mathbb{C}$ is a constant;*
  (v) *f and g satisfy the functional equation (12.9) for all $x, y \in G$.*

Proof. If f and g are linearly independent modulo $\mathcal{F}$, then (v) follows from Lemma 12.11.

Now, we suppose that there are constants $\mu, \nu \in \mathbb{C}$ (at least one of them is different from zero) such that $\mu f + \nu g \in \mathcal{F}$ but $f, g \notin \mathcal{F}$. Then we have $g = (2\lambda)^{-1} f + b$ with $b \in \mathcal{F}$ and $\lambda \neq 0$. Hence, our hypothesis implies that the function

$$x \mapsto f(xy) - \big((1/\lambda)f(y) + b(y)\big)f(x)$$

belongs to $\mathcal{F}$ for each fixed $y \in G$. From Theorem 9.2 it follows that

$$(1/\lambda)f(y) + b(y) = m(y),$$

where $m : G \to \mathbb{C}$ is an exponential, which implies (iv).

If $g \in \mathcal{F}$ and $f \notin \mathcal{F}$, then

$$x \mapsto f(xy) - f(x)g(y)$$

belongs to $\mathcal{F}$ for all fixed $y \in G$, and (iii) follows from Theorem 9.2.

If $f \in \mathcal{F}$ and $f \neq 0$, then $g \in \mathcal{F}$. If $f = 0$, then g is arbitrary.   □

Székelyhidi [345] proved the following theorem concerning the stability of the functional equation (12.9).

Theorem 12.13. *Let $(G, \cdot)$ be an amenable group and let $f, g : G \to \mathbb{C}$ be given functions. The function*

$$(x, y) \mapsto f(xy) - f(x)g(y) - f(y)g(x)$$

is bounded if and only if we have one of the following possibilities:

(i) $f = 0$ *and* g *is arbitrary;*

(ii) f *and* g *are bounded;*

(iii) $f = am + b$, $g = m$, *where* $a : G \to \mathbb{C}$ *is additive,* $m : G \to \mathbb{C}$ *is a bounded exponential, and* $b : G \to \mathbb{C}$ *is bounded;*

(iv) $f = \lambda m - \lambda b$, $g = (1/2)m + (1/2)b$, *where* $m : G \to \mathbb{C}$ *is an exponential,* $b : G \to \mathbb{C}$ *is bounded, and* $\lambda \in \mathbb{C}$ *is a constant;*

(v) f *and* g *satisfy the functional equation* (12.9) *for all* $x, y \in G$.

Proof. Applying Lemma 12.11 with $\mathcal{F}$ denoting the set of all bounded complex-valued functions on G, we see that either one of the above conditions (i), (ii), (iv), (v) is satisfied, or $g = m$ is a bounded exponential. In the latter case, the function

$$(x, y) \mapsto f(xy)m\big((xy)^{-1}\big) - f(x)m\big(x^{-1}\big) - f(y)m\big(y^{-1}\big)$$

is bounded. Hence, by Hyers's theorem (see [344])

$$f(x)m\big(x^{-1}\big) = a(x) + b_0(x)$$

holds true for all $x \in G$, where $a : G \to \mathbb{C}$ is additive and $b_0 : G \to \mathbb{C}$ is bounded, and our statement follows. (We have excluded the trivial case $m = 0$, and we have used the obvious identity $m(x)m\big(x^{-1}\big) = 1$, which holds true for any nonzero exponential.) The sufficiency follows by direct calculations. $\square$

12.4 Butler–Rassias Functional Equation

In 2003, S. Butler [53] posed the following problem:

Problem (Butler). *Show that for* $c < -1$ *there are exactly two solutions* $f : \mathbb{R} \to \mathbb{R}$ *of the functional equation,* $f(x + y) = f(x)f(y) + c \sin x \sin y$.

M. Th. Rassias has excellently answered this problem by proving the following theorem (see [283]):

Theorem 12.14 (Rassias). *Let* $c < -1$ *be a constant. The functional equation*

$$f(x + y) - f(x)f(y) - c \sin x \sin y = 0 \qquad (12.10)$$

has exactly two solutions in the class of functions $f : \mathbb{R} \to \mathbb{R}$. *More precisely,*

$$f(x) = a \sin x + \cos x \quad and \quad f(x) = -a \sin x + \cos x,$$

where $a = \sqrt{|c| - 1}$.

Proof. Replacing x with $x + z$ in (12.10) yields

$$f(x + y + z) - f(x + z)f(y) - c\sin(x + z)\sin y = 0 \qquad (a)$$

for all $x, y, z \in \mathbb{R}$. Similarly, if we replace y with $y + z$ in (12.10), then we get

$$f(x + y + z) - f(x)f(y + z) - c\sin x \sin(y + z) = 0 \qquad (b)$$

for any $x, y, z \in \mathbb{R}$.

It follows from (a) and (b) that

$$f(x)f(y + z) - f(x + z)f(y) + c\sin x \sin(y + z) - c\sin(x + z)\sin y = 0,$$

and hence

$$
\begin{aligned}
f(x)\big(f(y + z) &- f(y)f(z) - c\sin y \sin z\big) \\
&+ f(x)f(y)f(z) + cf(x)\sin y \sin z \\
&- \big(f(x + z) - f(x)f(z) - c\sin x \sin z\big)f(y) \\
&- f(x)f(y)f(z) - cf(y)\sin x \sin z \\
&+ c\sin x \sin(y + z) - c\sin(x + z)\sin y \\
= f(x)f(y + z) &- f(x + z)f(y) \\
&+ c\sin x \sin(y + z) - c\sin(x + z)\sin y \\
= 0 &\qquad (c)
\end{aligned}
$$

for every $x, y, z \in \mathbb{R}$.

Hence, it follows from (12.10) and (c) that

$$
\begin{aligned}
cf(x)\sin y \sin z &+ c\sin x \sin(y + z) \\
&- cf(y)\sin x \sin z - c\sin(x + z)\sin y \\
= f(x)\big(f(y + z) &- f(y)f(z) - c\sin y \sin z\big) \\
&+ f(x)f(y)f(z) + cf(x)\sin y \sin z \\
&- \big(f(x + z) - f(x)f(z) - c\sin x \sin z\big)f(y) \\
&- f(x)f(y)f(z) - cf(y)\sin x \sin z \\
&+ c\sin x \sin(y + z) - c\sin(x + z)\sin y \\
&+ f(x)\big(-f(y + z) + f(y)f(z) + c\sin y \sin z\big) \\
&+ f(y)\big(f(x + z) - f(x)f(z) - c\sin x \sin z\big) \\
= 0 &\qquad (d)
\end{aligned}
$$

for all $x, y, z \in \mathbb{R}$.

If we set $y = z = \pi/2$ in the last equality, then

$$f(x) - f(\pi/2)\sin x - \cos x = 0 \qquad (e)$$

for each $x \in \mathbb{R}$. Substituting π for x in (d) yields $f(\pi) = -1$. If we put $x = y = \pi/2$ in (12.10), then we obtain

$$f(\pi/2)^2 = f(\pi) - c = |c| - 1$$

and hence

$$f(\pi/2) = \sqrt{|c| - 1} = a \quad \text{or} \quad f(\pi/2) = -\sqrt{|c| - 1} = -a.$$

Consequently, by (e), we have

$$f(x) = a \sin x + \cos x \quad \text{or} \quad f(x) = -a \sin x + \cos x$$

for all $x \in \mathbb{R}$. □

Based on these historical facts, the equation (12.10) is called the *Butler–Rassias functional equation*. It is astonishing that M.Th. Rassias solved the Butler–Rassias functional equation (12.10) when he was a 16-year-old high school student.

In the following lemma, we will prove that every function $f : \mathbb{R} \to \mathbb{R}$ satisfying the inequality (12.11) is bounded.

Lemma 12.15. *Let c and δ be real constants with $c < -1$ and $0 < \delta < |c|$. If a function $f : \mathbb{R} \to \mathbb{R}$ satisfies the functional inequality*

$$|f(x + y) - f(x)f(y) - c \sin x \sin y| \leq \delta \tag{12.11}$$

for all $x, y \in \mathbb{R}$, then

$$\sup_{x \in \mathbb{R}} |f(x)| \leq (1/2)\left(1 + \sqrt{1 + 8|c|}\right)$$

holds true.

Proof. As $0 < \delta < |c|$, it follows from (12.11) that

$$-2|c| \leq f(x + y) - f(x)f(y) \leq 2|c|$$

for all $x, y \in \mathbb{R}$, which is equivalent to the inequality

$$|f(x + y) - f(x)f(y)| \leq 2|c|$$

for any $x, y \in \mathbb{R}$.

According to Theorem 9.1, f is either an exponential function or bounded. If f were an exponential function, then it would follow from (12.11) that $|c \sin x \sin y| \leq \delta$ for any $x, y \in \mathbb{R}$, which is contrary to our hypothesis, $\delta < |c|$. Indeed, f satisfies

$$|f(x)| \leq (1/2)\left(1 + \sqrt{1 + 8|c|}\right)$$

for each $x \in \mathbb{R}$. □

Based on Lemma 12.15, we can prove the following lemma.

Lemma 12.16. *Let c and δ be real constants with $c < -1$ and $0 < \delta < |c|$. If a function $f : \mathbb{R} \to \mathbb{R}$ satisfies the inequality (12.11) for all $x, y \in \mathbb{R}$, then*

$$|f(x) - f(\pi/2) \sin x - \cos x| \le \frac{3 + \sqrt{1 + 8|c|}}{|c|} \delta$$

for all $x \in \mathbb{R}$.

Proof. Following the steps from (a) to (d) in the proof of Theorem 12.14 yields

$$\big| cf(x) \sin y \sin z + c \sin x \sin(y + z)$$
$$-cf(y) \sin x \sin z - c \sin(x + z) \sin y \big|$$
$$\le \big(2 + |f(x)| + |f(y)|\big)\delta$$

for all $x, y, z \in \mathbb{R}$. Setting $y = z = \pi/2$ and considering Lemma 12.15 end the proof.   □

In the following theorem, we prove the Hyers–Ulam stability of the Butler–Rassias functional equation (see [179, 186]).

Theorem 12.17. *Let c and δ be real constants with $c < -1$ and $0 < \delta < |c|$. If a function $f : \mathbb{R} \to \mathbb{R}$ satisfies the inequality (12.11) for all $x, y \in \mathbb{R}$, then there exists a solution function $f_0 : \mathbb{R} \to \mathbb{R}$ of the Butler–Rassias functional equation (12.10) such that*

$$|f(x) - f_0(x)| \le \frac{\left(3 + \sqrt{1 + 8|c|}\right)\left(1 + \sqrt{|c| - 1}\right) + |c|}{|c|\sqrt{|c| - 1}} \delta$$

for all $x \in \mathbb{R}$.

Proof. Let $0 < \delta < |c|$ and let $f : \mathbb{R} \to \mathbb{R}$ satisfy the inequality (12.11) for all $x, y \in \mathbb{R}$. It follows from Lemma 12.16 that

$$|f(x) - f(\pi/2) \sin x - \cos x| \le \frac{3 + \sqrt{1 + 8|c|}}{|c|} \delta \qquad (a)$$

for all $x \in \mathbb{R}$. Put $x = \pi$ in (a) to get

$$|f(\pi) + 1| \le \frac{3 + \sqrt{1 + 8|c|}}{|c|} \delta. \qquad (b)$$

Moreover, setting $x = y = \pi/2$ in (12.11) yields

$$\left| f(\pi) - f(\pi/2)^2 - c \right| \leq \delta. \qquad (c)$$

Combining (b) and (c) yields

$$\left| f(\pi/2)^2 + c + 1 \right| \leq \frac{3 + |c| + \sqrt{1 + 8|c|}}{|c|} \delta.$$

With $a = \sqrt{|c| - 1} > 0$, we have

$$\left| f(\pi/2)^2 - a^2 \right| \leq \frac{3 + |c| + \sqrt{1 + 8|c|}}{|c|} \delta. \qquad (d)$$

Assume that $f(\pi/2) \geq 0$. It then follows from (d) that

$$\left| f(\pi/2) - a \right| \leq \frac{3 + |c| + \sqrt{1 + 8|c|}}{|c||f(\pi/2) + a|} \delta \leq \frac{3 + |c| + \sqrt{1 + 8|c|}}{a|c|} \delta. \qquad (e)$$

Hence, (a) and (e) imply that

$$|f(x) - a \sin x - \cos x|$$
$$\leq |f(x) - f(\pi/2) \sin x - \cos x| + |(f(\pi/2) - a) \sin x|$$
$$\leq \frac{\left(3 + \sqrt{1 + 8|c|}\right)\left(1 + \sqrt{|c| - 1}\right) + |c|}{|c|\sqrt{|c| - 1}} \delta$$

for all $x \in \mathbb{R}$.

Assume now that $f(\pi/2) < 0$. It then follows from (d) that

$$\left| f(\pi/2) + a \right| \leq \frac{3 + |c| + \sqrt{1 + 8|c|}}{|c||f(\pi/2) - a|} \delta \leq \frac{3 + |c| + \sqrt{1 + 8|c|}}{a|c|} \delta. \qquad (f)$$

Thus, combining (a) and (f) yields

$$|f(x) + a \sin x - \cos x|$$
$$\leq |f(x) - f(\pi/2) \sin x - \cos x| + |(f(\pi/2) + a) \sin x|$$
$$\leq \frac{\left(3 + \sqrt{1 + 8|c|}\right)\left(1 + \sqrt{|c| - 1}\right) + |c|}{|c|\sqrt{|c| - 1}} \delta$$

for any $x \in \mathbb{R}$.

We define

$$f_0(x) = \begin{cases} a \sin x + \cos x & \text{(for } f(\pi/2) \geq 0), \\ -a \sin x + \cos x & \text{(for } f(\pi/2) < 0) \end{cases}$$

for all $x \in \mathbb{R}$. In view of Theorem 12.14, both $a \sin x + \cos x$ and $-a \sin x + \cos x$ are solutions of the Butler–Rassias functional equation (12.10). Hence, f_0 is a solution function of the Butler–Rassias functional equation. $\square$

12.5   Remarks

R. Ger [121] considered the functional equations (12.8) and (12.9) simultaneously – more precisely, he considered the system

$$\begin{cases} f(xy) = f(x)g(y) + f(y)g(x), \\ g(xy) = g(x)g(y) - f(x)f(y) \end{cases} \tag{12.12}$$

and proved that the system is not superstable, but that it is stable in the sense of Hyers, Ulam, and Rassias.

Theorem 12.18. *Let $(G, \cdot)$ be a semigroup and let $h, k : G^2 \to [0, \infty)$ be functions such that the sections $h(\cdot, x)$ and $k(x, \cdot)$ are bounded for each fixed $x \in G$. Assume that $f, g : G \to \mathbb{C}$ satisfy the system of inequalities*

$$\begin{cases} |f(xy) - f(x)g(y) - f(y)g(x)| \leq h(x, y), \\ |g(xy) - g(x)g(y) + f(x)f(y)| \leq k(x, y) \end{cases}$$

for all $x, y \in G$. Then there exists a solution (s, c) of the system (12.12) such that $f - s$ and $g - c$ are bounded.

Chapter 13
Isometric Functional Equation

An *isometry* is a *distance-preserving map* between metric spaces. For normed spaces E_1 and E_2, a function $f : E_1 \to E_2$ is called an *isometry* if f satisfies the *isometric functional equation* $\| f(x) - f(y) \| = \| x - y \|$ for all $x, y \in E_1$. The historical background for Hyers–Ulam stability of isometries will be introduced in Section 13.1. The Hyers–Ulam–Rassias stability of isometries on a restricted domain will be surveyed in Section 13.2. Section 13.3 will be devoted to the fixed point method for studying the stability problem of isometries. In the final section, the Hyers–Ulam–Rassias stability of Wigner equation $|\langle f(x), f(y) \rangle| = |\langle x, y \rangle|$ on a restricted domain will be discussed.

13.1 Hyers–Ulam Stability

Throughout this section, assume that δ is a positive constant. For normed spaces $(E_1, \| \cdot \|_1)$ and $(E_2, \| \cdot \|_2)$, a function $f : E_1 \to E_2$ is called a δ-*isometry* if f changes distances at most δ, i.e.,

$$\big| \| f(x) - f(y) \|_2 - \| x - y \|_1 \big| \leq \delta \tag{13.1}$$

for all $x, y \in E_1$.

In 1945, D. H. Hyers and S. M. Ulam [139] proved the Hyers–Ulam stability of isometries by making use of a direct method. First, we will introduce their lemmas.

Lemma 13.1. *Let E be a complete abstract Euclidean space. Assume that $f : E \to E$ is a δ-isometry and $f(0) = 0$. The limit*

$$I(x) = \lim_{n \to \infty} 2^{-n} f(2^n x) \tag{13.2}$$

exists for any $x \in E$ and $I : E \to E$ is an isometry.

Proof. Assume that x is an arbitrary point of E and let $r = \| x \|$. Since f is a δ-isometry, it follows from (13.1) that

S.-M. Jung, *Hyers–Ulam–Rassias Stability of Functional Equations in Nonlinear Analysis*, Springer Optimization and Its Applications 48, DOI 10.1007/978-1-4419-9637-4_13, © Springer Science+Business Media, LLC 2011

$$\big|\,\|f(x)\| - r\,\big| \le \delta \quad \text{and} \quad \big|\,\|f(x) - f(2x)\| - r\,\big| \le \delta. \tag{a}$$

If we replace x and y in (13.1) with 0 and $2x$, respectively, and set $y_0 = (1/2)\,f(2x)$, then we have

$$\big|\,\|y_0\| - r\,\big| \le (1/2)\delta. \tag{b}$$

We consider the intersection of two balls:

$$S_1 = \{y \in E \mid \|y\| \le r + \delta\} \quad \text{and} \quad S_2 = \{y \in E \mid \|y - 2y_0\| \le r + \delta\}.$$

Then, it holds true that $f(x) \in S_1 \cap S_2$. For any $y \in S_1 \cap S_2$, we get

$$\begin{aligned}
\|y\|^2 &\le (r + \delta)^2, \\
\|y - 2y_0\|^2 &= \|y\|^2 + 4\|y_0\|^2 - 4\langle y, y_0 \rangle \le (r + \delta)^2,
\end{aligned} \tag{c}$$

where $\langle \cdot, \cdot \rangle$ denotes the inner product on E. It follows from (a), (b), and (c) that

$$\begin{aligned}
2\|y - y_0\|^2 &= \|y\|^2 + 4\|y_0\|^2 - 4\langle y, y_0 \rangle + \|y\|^2 - 2\|y_0\|^2 \\
&\le (r + \delta)^2 + \|y\|^2 - 2\|y_0\|^2 \\
&\le 2(r + \delta)^2 - 2\|y_0\|^2 \\
&\le \begin{cases} 2(r + \delta)^2 - 2(r - \delta/2)^2 & \text{(for } \|x\| \ge \delta), \\ 2(r + \delta)^2 & \text{(for } \|x\| < \delta) \end{cases} \\
&= \begin{cases} 6\delta r + (3/2)\delta^2 & \text{(for } \|x\| \ge \delta), \\ 2(r + \delta)^2 & \text{(for } \|x\| < \delta). \end{cases}
\end{aligned} \tag{d}$$

Since $f(x) \in S_1 \cap S_2$, substituting $f(x)$ for y in (d) yields

$$\|f(x) - (1/2)f(2x)\| \le \begin{cases} 2\big(\delta\|x\|\big)^{1/2} & \text{(for } \|x\| \ge \delta), \\ 2\delta & \text{(for } \|x\| < \delta). \end{cases}$$

Therefore, we have

$$\|f(x/2) - (1/2)f(x)\| \le 2^{-1/2}k\|x\|^{1/2} + 2\delta \tag{e}$$

for all $x \in E$, where we set $k = 2\delta^{1/2}$. By applying the induction on n, we will prove that

$$\|f(2^{-n}x) - 2^{-n}f(x)\| \le 2^{-n/2}k\|x\|^{1/2}\sum_{i=0}^{n-1} 2^{-i/2} + (1 - 2^{-n})4\delta \tag{f}$$

for any $x \in E$ and $n \in \mathbb{N}$. In view of (e), the inequality (f) holds true for $n = 1$. Assume that (f) is true for some integer $n > 0$. By dividing the inequality (f) by 2, we have

$$\left\| (1/2) f(2^{-n} x) - 2^{-(n+1)} f(x) \right\|$$

$$\leq 2^{-(n+1)/2} k \|x\|^{1/2} \sum_{i=1}^{n} 2^{-i/2} + (1 - 2^{-n}) 2\delta.$$

If we replace x with $2^{-n} x$ in (e), we get

$$\left\| f\left(2^{-(n+1)} x\right) - (1/2) f(2^{-n} x) \right\| \leq 2^{-(n+1)/2} k \|x\|^{1/2} + 2\delta.$$

By adding the last two inequalities, we obtain

$$\left\| f\left(2^{-(n+1)} x\right) - 2^{-(n+1)} f(x) \right\|$$

$$\leq 2^{-(n+1)/2} k \|x\|^{1/2} \sum_{i=0}^{n} 2^{-i/2} + \left(1 - 2^{-(n+1)}\right) 4\delta,$$

which proves the validity of (f) for all $x \in E$ and $n \in \mathbb{N}$.

We will now present that $\{2^{-n} f(2^n x)\}$ is a Cauchy sequence. If n and p are any positive integers, then it follows from (f) that

$$\| 2^{-n} f(2^n x) - 2^{-(n+p)} f(2^{n+p} x) \|$$

$$\leq 2^{-n} \| f(2^{-p} 2^{n+p} x) - 2^{-p} f(2^{n+p} x) \|$$

$$\leq k \|x\|^{1/2} \sum_{i=n}^{n+p-1} 2^{-i/2} + 2^{-n} (1 - 2^{-p}) 4\delta$$

$$\to 0 \quad \text{as } n \to \infty$$

for any $x \in E$, which implies that $\{2^{-n} f(2^n x)\}$ is a Cauchy sequence. Since E is complete, we can define a function $I : E \to E$ by (13.2).

Finally, it remains to prove that I is an isometry. Let x and y be arbitrary points of E. Replace x and y in (13.1) with $2^n x$ and $2^n y$, respectively, divide the inequality

$$\left| \| f(2^n x) - f(2^n y) \| - 2^n \|x - y\| \right| \leq \delta$$

by 2^n, and take the limit as $n \to \infty$. The result is $\| I(x) - I(y) \| = \|x - y\|$, which ends the proof.   □

Lemma 13.2. *Let E be a complete abstract Euclidean space. Assume that $f : E \to E$ is a δ-isometry and $f(0) = 0$. If u and x are any points of E with $\|u\| = 1$ and $\langle x, u \rangle = 0$, then $|\langle f(x), I(u) \rangle| \leq 3\delta$, where $\langle \cdot, \cdot \rangle$ is the inner product on E and I is defined by (13.2).*

Proof. For any integer n, we set $z = 2^n u$. Let y denote an arbitrary point of the sphere S_n of radius 2^n and center at z. Then, it holds true that $\|y - z\|^2 = \|z\|^2$ and hence we have $\langle y, u \rangle = 2^{-n-1} \langle y, y \rangle$.

Since f is a δ-isometry and $\|y - z\| = \|z\|$, we get

$$\left| \|f(y) - f(z)\| - \|y - z\| \right| \le \delta \quad \text{or} \quad \left| \|f(y) - f(z)\| - \|z\| \right| \le \delta$$

and

$$\left| \|z\| - \|f(z)\| \right| \le \delta.$$

By adding the last two inequalities, we obtain

$$\left| \|f(y) - f(z)\| - \|f(z)\| \right| \le 2\delta,$$

i.e., $\|f(y) - f(z)\| = \eta(y, z) + \|f(z)\|$, where $|\eta(y, z)| \le 2\delta$.

The last equality may be expressed as

$$2\langle f(y), f(z) \rangle = \langle f(y), f(y) \rangle - 2\eta \|f(z)\| - \eta^2.$$

If we divide the last equality by 2^{n+1}, then we obtain

$$\langle f(y), 2^{-n} f(2^n u) \rangle = 2^{-(n+1)} \left(\langle f(y), f(y) \rangle - \eta^2 \right) - \eta \|2^{-n} f(2^n u)\|. \quad (a)$$

Now, let x be any point of the hyperplane $\langle x, u \rangle = 0$. If we set $y = x + ru$ with $r = 2^n - \left(2^{2n} - \|x\|^2 \right)^{1/2}$, then y is a point of the sphere S_n:

$$\begin{aligned}
\|y - z\|^2 &= \langle y, y \rangle - 2\langle y, z \rangle + \langle z, z \rangle \\
&= \langle x, x \rangle + r^2 - 2\langle x, z \rangle - 2r\langle u, z \rangle + \langle z, z \rangle \\
&= r^2 - 2^{n+1} r + \|x\|^2 + \|z\|^2 \\
&= \|z\|^2.
\end{aligned}$$

Moreover, $\|y - x\| = r \to 0$ as $n \to \infty$. In view of Lemma 13.1, we see that

$$t = \lim_{n \to \infty} 2^{-n} f(2^n u)$$

exists and is a unit vector.

Finally, for any $\varepsilon > 0$ and n sufficiently large, it follows from (a) that

$$\begin{aligned}
|\langle f(x), t \rangle| \le \; &|\langle f(x), t - 2^{-n} f(2^n u) \rangle| + |\langle f(y), 2^{-n} f(2^n u) \rangle| \\
&+ |\langle f(x) - f(y), 2^{-n} f(2^n u) \rangle|
\end{aligned}$$

$$\leq \|f(x)\|\|t - 2^{-n} f(2^n u)\| + (1/2)\varepsilon$$
$$+ 2\delta \|2^{-n} f(2^n u)\| + \|f(x) - f(y)\| \|2^{-n} f(2^n u)\|$$
$$\leq \varepsilon + 3\delta(1 + \varepsilon)$$

and hence we have

$$|\langle f(x), I(u) \rangle| = |\langle f(x), t \rangle| \leq 3\delta,$$

which ends the proof. $\qquad \square$

Lemma 13.3. *Let E be a complete abstract Euclidean space. Assume that $f : E \to E$ is a surjective δ-isometry and $f(0) = 0$. Then the $I : E \to E$ defined by (13.2) is also surjective.*

Proof. For any $z \in E$, let $f^{-1}(z)$ denote any point whose f-image is z. Then, $f^{-1} : E \to E$ is a δ-isometry. According to Lemma 13.1, the limit

$$I^*(z) = \lim_{n \to \infty} 2^{-n} f^{-1}(2^n z)$$

exists, and I^* is an isometry of E.

Now, it is obvious that

$$\left\|2^n z - f\left(2^n I^*(z)\right)\right\| = \left\|f\left(2^n 2^{-n} f^{-1}(2^n z)\right) - f\left(2^n I^*(z)\right)\right\|$$
$$\leq 2^n \left\|2^{-n} f^{-1}(2^n z) - I^*(z)\right\| + \delta.$$

If we divide the last inequality by 2^n and let $n \to \infty$, then we see that $z = I\left(I^*(z)\right)$ for each $z \in E$. Therefore, we conclude that $I(E) = E$. $\qquad \square$

Using the preceding lemmas, Hyers and Ulam proved in [139] that the surjective isometries of a complete Euclidean space are stable in the sense of Hyers and Ulam.

Theorem 13.4 (Hyers and Ulam). *Let E be a complete abstract Euclidean space. Assume that $f : E \to E$ is a surjective δ-isometry and $f(0) = 0$. Then the $I : E \to E$ defined by (13.2) is a surjective isometry, and the inequality*

$$\|f(x) - I(x)\| \leq 10\delta$$

holds true for all $x \in E$.

Proof. For a given $x \in E \setminus \{0\}$, let M denote the linear manifold orthogonal to x. In view of Lemma 13.3, $I : E \to E$ is a surjective isometric transformation. Hence, $I(M)$ is the linear manifold orthogonal to $I(x)$. Let w be the projection of $f(x)$ on $I(M)$. Let us define

$$t = \begin{cases} 0 & (\text{for } w = 0), \\ (1/\|w\|)w & (\text{for } w \neq 0). \end{cases}$$

According to Lemma 13.2, we see that $|\langle f(x), t \rangle| \leq 3\delta$. Put $v = (1/\|x\|)I(x)$. Then, v is a unit vector orthogonal to t and is coplanar with $f(x)$ and t. Thus, by the Pythagorean theorem, we have

$$\|f(x) - I(x)\|^2 = \langle f(x), t \rangle^2 + (\|x\| - \langle f(x), v \rangle)^2. \qquad (a)$$

Let $z_n = 2^n x$ and let w_n denote the projection of $f(z_n)$ on $I(M)$. Let us define

$$t_n = \begin{cases} 0 & (\text{for } w_n = 0), \\ (1/\|w_n\|)w_n & (\text{for } w_n \neq 0). \end{cases}$$

In either case, we see that $\langle t_n, v \rangle = 0$ and $|\langle f(z_n), t_n \rangle| \leq 3\delta$. If $\|f(z_n)\| < 3\delta$, it is obvious that $\|f(z_n)\| - |\langle f(z_n), v \rangle| \leq 3\delta$. Assume now that $\|f(z_n)\| \geq 3\delta$. Since $(1 - \alpha^2)^{1/2} \geq 1 - \alpha^2$ for any $-1 \leq \alpha \leq 1$ and $|\langle f(z_n), t_n \rangle| \leq \|f(z_n)\|$, we have

$$
\begin{aligned}
0 &\leq \|f(z_n)\| - |\langle f(z_n), v \rangle| \\
&= \|f(z_n)\| - (\|f(z_n)\|^2 - \langle f(z_n), t_n \rangle^2)^{1/2} \\
&= \|f(z_n)\|\left(1 - (1 - \langle f(z_n), t_n \rangle^2/\|f(z_n)\|^2)^{1/2}\right) \\
&\leq \|f(z_n)\|\langle f(z_n), t_n \rangle^2/\|f(z_n)\|^2 \\
&\leq |\langle f(z_n), t_n \rangle| \\
&\leq 3\delta.
\end{aligned}
$$

Hence, the inequality

$$\big|\|z_n\| - |\langle f(z_n), v \rangle|\big| \leq 4\delta \qquad (b)$$

holds true since $\|z_n\| \leq \|f(z_n)\| + \delta$.

Two cases arise: If $\langle f(x), v \rangle \geq 0$, we put $n = 0$ in (b) and use the identity (a) to obtain the inequality $\|f(x) - I(x)\| \leq 5\delta$. If $\langle f(x), v \rangle < 0$, then for some integer $m \geq 0$ we must have $\langle f(z_m), v \rangle < 0$ and $\langle f(2z_m), v \rangle \geq 0$, since $\langle I(x), v \rangle$ is positive and $I(x) = \lim_{n \to \infty} 2^{-n} f(z_n)$. Hence, by (b), we get

$$\|f(2z_m) - f(z_m)\| \geq \langle f(2z_m), v \rangle - \langle f(z_m), v \rangle \geq 3\|z_m\| - 8\delta.$$

However, we know that $\|f(2z_m) - f(z_m)\| \leq \|z_m\| + \delta$. Therefore, we obtain $\|x\| \leq \|z_m\| \leq (9/2)\delta$ and hence

$$\|f(x) - I(x)\| \leq \|f(x)\| + \|I(x)\| \leq \|x\| + \delta + \|x\| \leq 10\delta,$$

which ends the proof. $\square$

Hyers and Ulam noticed that Theorem 13.4 is not always true for non-surjective δ-isometric transformations of one Euclidean space into another: We consider a function $f : \mathbb{R} \to \mathbb{R}^2$ defined by

$$f(x) = \begin{cases} (x, 0) & \text{(for } x \leq 1), \\ (x, c \ln x) & \text{(for } x > 1). \end{cases}$$

It is easy to present that f will be a δ-isometry if we choose c in such a way that $\delta > c^2 \max\limits_{x>1} (\ln x)^2/(2x - 2)$. On the other hand, f cannot be approximated by any isometry in the sense of Theorem 13.4.

D. G. Bourgin [25] further generalized this result of Hyers and Ulam and proved the following theorem:

Assume that E_1 is a real Banach space and E_2 belongs to a class of uniformly convex real Banach spaces which includes the spaces $L_p(0, 1)$ for $p \in (1, 2) \cup (2, \infty)$. For each δ-isometry $f : E_1 \to E_2$ with $f(0) = 0$, there exists a linear isometry $I : E_1 \to E_2$ such that $\| f(x) - I(x) \| \leq 12\delta$ for all $x \in E_1$.

Subsequently, Hyers and Ulam [140] studied a stability problem for spaces of continuous functions:

Let S_1 and S_2 be compact metric spaces and let $\big(C(S_i), \| \cdot \|_\infty\big)$ denote the space of real-valued continuous functions on S_i equipped with the metric topology. If a homeomorphism $T : C(S_1) \to C(S_2)$ satisfies the inequality

$$\big| \|T(f) - T(g)\|_\infty - \| f - g\|_\infty \big| \leq \delta \tag{13.3}$$

for all $f, g \in C(S_1)$, then there exists an isometry $I : C(S_1) \to C(S_2)$ such that $\|T(f) - I(f)\|_\infty \leq 21\delta$ for each $f \in C(S_1)$.

Bourgin significantly generalized this result of Hyers and Ulam again (see [26]):

Let S_1 and S_2 be completely regular Hausdorff spaces and let $T : C(S_1) \to C(S_2)$ be a surjective function satisfying the inequality (13.3) for all $f, g \in C(S_1)$. Then there exists a linear isometry $I : C(S_1) \to C(S_2)$ such that $\|T(f) - I(f)\|_\infty \leq 10\delta$ for any $f \in C(S_1)$.

R. D. Bourgin [27] continued the study of stability problems for isometries on finite-dimensional Banach spaces under the additional assumption that the set of extreme points of the unit ball in E_1 is totally disconnected. In 1978, P. M. Gruber [128] obtained an elegant result as follows:

Theorem 13.5 (Gruber). *Let E_1 and E_2 be real normed spaces. Assume that $f : E_1 \to E_2$ is a surjective δ-isometry and $I : E_1 \to E_2$ is an isometry with $f(p) = I(p)$ for some $p \in E_1$. If $\| f(x) - I(x) \| = o(\|x\|)$ as $\|x\| \to \infty$ uniformly, then I is a surjective linear isometry and $\| f(x) - I(x) \| \leq 5\delta$ for all $x \in E_1$. If in addition f is continuous, then $\| f(x) - I(x) \| \leq 3\delta$ for all $x \in E_1$.*

Using Theorem 13.5 and an idea from [356], J. Gevirtz [124] established the Hyers–Ulam stability of isometries between arbitrary Banach spaces.

Theorem 13.6 (Gevirtz). *Assume that E_1 and E_2 are real Banach spaces. For each surjective δ-isometry $f : E_1 \to E_2$, there exists a surjective linear isometry $I : E_1 \to E_2$ such that $\|f(x) - I(x)\| \le 5\delta$ for all $x \in E_1$.*

Proof. We introduce a lemma which will be proved in the last part of this proof:

> If $f : E_1 \to E_2$ is a surjective δ-isometry, then there exist constants A and B such that
>
> $$\left\| 2f\left(\frac{x_0 + x_1}{2}\right) - f(x_0) - f(x_1) \right\| \le 2A\big(\delta\|x_0 - x_1\|\big)^{1/2} + 2B\delta \qquad (a)$$
>
> for all $x_0, x_1 \in E_1$.

Without loss of generality, we assume that $f(0) = 0$. Applying (a) with $x_0 = 2^{n+1}x$ and $x_1 = 0$ and dividing by 2^{n+1} yield

$$\left\| 2^{-n} f(2^n x) - 2^{-(n+1)} f\big(2^{n+1}x\big) \right\| \le 2^{-n/2} A\big(2\delta\|x\|\big)^{1/2} + 2^{-n} B\delta \qquad (b)$$

for any $x \in E_1$ and $n \in \mathbb{N}_0$. It follows from (b) that for all $m, n \in \mathbb{N}$,

$$\left\| 2^{-n} f(2^n x) - 2^{-(n+m)} f\big(2^{n+m}x\big) \right\|$$
$$\le \sum_{i=n}^{n+m-1} \left\| 2^{-i} f(2^i x) - 2^{-(i+1)} f\big(2^{i+1}x\big) \right\|$$
$$\le 2\big(\sqrt{2} - 1\big)^{-1} A 2^{-n/2}\big(\delta\|x\|\big)^{1/2} + 2^{-(n-1)} B\delta$$
$$\to 0 \quad \text{as } n \to \infty.$$

Thus, $\{2^{-n} f(2^n x)\}$ is a Cauchy sequence. Since E_2 is complete, we can define a function $I : E_1 \to E_2$ by (13.2). Then, we see that $I(0) = 0$ and

$$\|f(x) - I(x)\| \le 2\big(\sqrt{2} - 1\big)^{-1} A\big(\delta\|x\|\big)^{1/2} + 2B\delta \qquad (c)$$

for any $x \in E_1$. Obviously, $\|f(x) - I(x)\|/\|x\| \to 0$ uniformly as $\|x\| \to \infty$.

Since f is a δ-isometry, we get

$$\Big| \|2^{-n} f(2^n x_0) - 2^{-n} f(2^n x_1)\| - \|x_0 - x_1\| \Big| \le 2^{-n}\delta$$

for any $x_0, x_1 \in E_1$ and $n \in \mathbb{N}$. If we let $n \to \infty$ in the preceding inequality, then we see that I is an isometry. Moreover, $f(0) = 0 = I(0)$. In view of Theorem 13.5, I is a surjective linear isometry and $\|f(x) - I(x)\| \le 5\delta$ for all $x \in E_1$.

We now introduce some terminologies which will be used in the proof of Lemma (a). A function $F : E_2 \to E_1$ is called an *ε-inverse* of f if

$$\|(f \circ F)(y) - y\| \le \varepsilon \qquad (d)$$

for all $y \in E_2$. The function f is called ε-*onto* if it has an ε-inverse. Each ε-onto δ-isometry will be called an (ε, δ)-*isometry*.

We assert that

> If $f : E_1 \to E_2$ is an (ε, δ)-isometry and F is an ε-inverse of f, $\qquad$ (e)
> then F is an $(\varepsilon + \delta, 2\varepsilon + \delta)$-isometry.

It follows from (d) that

$$\|(F \circ f)(x) - x\| \leq \|(f \circ F \circ f)(x) - f(x)\| + \delta \leq \varepsilon + \delta$$

for all $x \in E_1$, which proves that F is $(\varepsilon + \delta)$-onto. To show that F is a $(2\varepsilon + \delta)$-isometry, let $y_0, y_1 \in E_2$. Then it follows from (d) that $\|(f \circ F)(y_i) - y_i\| \leq \varepsilon$ for $i \in \{0, 1\}$. Since f is a δ-isometry, we get

$$\big| \|(f \circ F)(y_0) - (f \circ F)(y_1)\| - \|F(y_0) - F(y_1)\| \big| \leq \delta.$$

Hence, we obtain

$$\begin{aligned}
\big| \|F(y_0) &- F(y_1)\| - \|y_0 - y_1\| \big| \\
&\leq \big| \|F(y_0) - F(y_1)\| - \|(f \circ F)(y_0) - (f \circ F)(y_1)\| \big| \\
&\quad + \big| \|(f \circ F)(y_0) - (f \circ F)(y_1)\| - \|y_0 - y_1\| \big| \\
&\leq \delta + \|(f \circ F)(y_0) - y_0\| + \|(f \circ F)(y_1) - y_1\| \\
&\leq \delta + 2\varepsilon.
\end{aligned}$$

We assert that

> Let $f_1 : E_1 \to E_2$ be an $(\varepsilon_1, \delta_1)$-isometry and $f_2 : E_2 \to E_3$ be
> an $(\varepsilon_2, \delta_2)$-isometry. Then $f_2 \circ f_1$ is an $(\varepsilon_1 + \varepsilon_2 + \delta_2, \delta_1 + \delta_2)$- $\qquad$ (f)
> isometry.

It is immediate that $f_2 \circ f_1$ is a $(\delta_1 + \delta_2)$-isometry. To prove that $f_2 \circ f_1$ is $(\varepsilon_1 + \varepsilon_2 + \delta_2)$-onto, let F_i be an ε_i-inverse of f_i ($i \in \{0, 1\}$) and let $z \in E_3$, where E_3 is a real Banach space. Then we have

$$\begin{aligned}
\|(f_2 \circ f_1 &\circ F_1 \circ F_2)(z) - z\| \\
&\leq \|(f_2 \circ f_1 \circ F_1 \circ F_2)(z) - (f_2 \circ F_2)(z)\| + \|(f_2 \circ F_2)(z) - z\| \\
&\leq \|(f_2 \circ f_1 \circ F_1 \circ F_2)(z) - (f_2 \circ F_2)(z)\| + \varepsilon_2 \\
&\leq \|(f_1 \circ F_1 \circ F_2)(z) - F_2(z)\| + \delta_2 + \varepsilon_2 \\
&\leq \varepsilon_1 + \delta_2 + \varepsilon_2.
\end{aligned}$$

We will now prove (a). Let $x_0, x_1 \in E_1$, $y_i = f(x_i)$ $(i \in \{0, 1\})$, $p = (1/2)(x_0 + x_1)$, and $q = (1/2)(y_0 + y_1)$. First, we will assume that $y_0 \neq y_1$. Since f is a $(0, \delta)$-isometry, due to (e), f has a 0-inverse F which is a (δ, δ)-isometry and for which $F(y_i) = x_i$ $(i \in \{0, 1\})$. We define sequences $\{g_k\}$ and $\{G_k\}$ of functions of E_2 into E_2 with the following properties:

$$g_k \text{ is a } \left(4^{k+1}\delta, 4^{k+1}\delta\right)\text{-isometry and } g_k(y_i) = y_{1-i}, \tag{g}$$

$$G_k \text{ is a } 4^{k+1}\delta\text{-inverse of } g_k \text{ and } G_k(y_i) = y_{1-i} \tag{h}$$

for $i \in \{0, 1\}$. Set $g_0(y) = f\left(2p - F(y)\right)$ for all $y \in E_2$. Then, $g_0 = f_1 \circ f_2$, where $f_1 = f$ is a $(0, \delta)$-isometry and $f_2(y) = 2p - F(y)$ is a (δ, δ)-isometry. In view of (f), g_0 is a $(2\delta, 2\delta)$-isometry and it permutes y_0 and y_1. Thus, (g) is true for $k = 0$. Let G_0 be any function satisfying (h) for $k = 0$. Now, let $g_1(y) = G_1(y) = 2q - y$ for any $y \in E_2$. Obviously, (g) and (h) are satisfied for $k = 1$. Finally, assuming that we have $g_0, \ldots, g_n$ and $G_0, \ldots, G_n$ which satisfy the stipulated conditions, we define $g_{n+1} = g_{n-1} \circ g_n \circ G_{n-1}$. Then, g_{n-1} is a $(4^n\delta, 4^n\delta)$-isometry, g_n is a $(4^{n+1}\delta, 4^{n+1}\delta)$-isometry, and G_{n-1} is a $4^n\delta$-inverse of g_{n-1}. In view of (e), we know that G_{n-1} is a $(2 \cdot 4^n\delta, 3 \cdot 4^n\delta)$-isometry. Hence, it follows from (f) that g_{n+1} is a $(3 \cdot 4^{n+1}\delta, 2 \cdot 4^{n+1}\delta)$-isometry. Moreover, g_{n+1} permutes y_0 and y_1. Therefore, g_{n+1} satisfies (g) with $k = n + 1$. Now, G_{n+1} is taken to be any function satisfying (h) with $k = n + 1$.

We next define a sequence $\{a_n\}$ of points of E_2 recursively by $a_1 = q$ and $a_{n+1} = g_{n-1}(a_n)$ for $n \in \mathbb{N}$. Let $d = (1/2)\|y_0 - y_1\|$. Denoting by $B(y, r)$ the closed ball of radius r and center y yields

$$g_k\left(B(y_i, r)\right) \subset B\left(y_{1-i}, r + 4^{k+1}\delta\right).$$

Since $a_1 \in B(y_0, d) \cap B(y_1, d)$ and $a_n = (g_{n-2} \circ g_{n-3} \circ \cdots \circ g_0)(a_1)$, successive application of this inclusion with $k \in \{0, 1, \ldots, n - 2\}$ yields

$$a_n \in B(y_0, d + 4^n\delta) \cap B(y_1, d + 4^n\delta) \subset B(q, d + 4^n\delta).$$

Since the diameter of the last ball is $2(d + 4^n\delta)$, we conclude that

$$\|a_n - a_{n-1}\| \leq 2(d + 4^n\delta) \tag{i}$$

for all integers $n \geq 2$.

We now prove that

$$\|g_n(y) - y\| \geq 2\|a_n - y\| - 2(4^n - 1)\delta \tag{j}$$

for any $y \in E_2$ and $n \in \mathbb{N}$. Since $g_1(y) = 2q - y$ and $a_1 = q$, (j) is true for $n = 1$. Assuming that (j) is true for some $n \in \mathbb{N}$, it follows from (g), (h), and (j) that

$$\|g_{n+1}(y) - y\|$$
$$= \|(g_{n-1} \circ g_n \circ G_{n-1})(y) - y\|$$
$$\geq \|(g_{n-1} \circ g_n \circ G_{n-1})(y) - (g_{n-1} \circ G_{n-1})(y)\|$$
$$\quad - \|(g_{n-1} \circ G_{n-1})(y) - y\|$$
$$\geq \|(g_{n-1} \circ g_n \circ G_{n-1})(y) - (g_{n-1} \circ G_{n-1})(y)\| - 4^n \delta$$
$$\geq \|(g_n \circ G_{n-1})(y) - G_{n-1}(y)\| - 2 \cdot 4^n \delta$$
$$\geq 2\|a_n - G_{n-1}(y)\| - (4^{n+1} - 2)\delta$$
$$\geq 2(\|g_{n-1}(a_n) - (g_{n-1} \circ G_{n-1})(y)\| - 4^n \delta) - (4^{n+1} - 2)\delta$$
$$\geq 2(\|a_{n+1} - y\| - 2 \cdot 4^n \delta) - (4^{n+1} - 2)\delta$$
$$= 2\|a_{n+1} - y\| - 2(4^{n+1} - 1)\delta,$$

which proves the validity of (j) for all $y \in E_2$ and $n \in \mathbb{N}$.

It follows from (j) that

$$\|a_{n+1} - a_n\| = \|g_{n-1}(a_n) - a_n\| \geq 2\|a_n - a_{n-1}\| - 2 \cdot 4^{n-1}\delta,$$

which by induction gives

$$\|a_n - a_{n-1}\| \geq 2^{n-2}\|a_2 - a_1\| - 4^{n-1}\delta.$$

Together with (i), this implies that $\|a_2 - a_1\|$ is bounded above by

$$\|a_2 - a_1\| \leq 2(2^{-(n-2)}d + 18\delta 2^{n-2}) \tag{k}$$

for all integers $n \geq 2$.

On the other hand, since F is a 0-inverse of f, we have

$$\|a_2 - a_1\| = \|f(2p - F(q)) - q\|$$
$$= \|f(2p - F(q)) - (f \circ F)(q)\|$$
$$\geq 2\|p - F(q)\| - \delta$$
$$\geq 2(\|f(p) - (f \circ F)(q)\| - \delta) - \delta$$
$$= 2\|f(p) - q\| - 3\delta.$$

Hence, it follows from (k) that

$$\|f(p) - q\| \leq 2^{-(n-2)}d + 18\delta 2^{n-2} + 2\delta \tag{l}$$

for every integer $n \geq 2$.

For the moment, we assume that $d > 18\delta$ and let t satisfy $2^{-t}d = 18\delta 2^t$, i.e., $t = (\ln 4)^{-1} \ln(d/18\delta) > 0$. If we let $n - 2 = [t]$, the greatest integer less than or equal to t, then it follows from (l) that

$$\|f(p) - q\| \le 2d2^{-t} + 18\delta 2^t + 2\delta$$
$$= 3d2^{-t} + 2\delta$$
$$= 3(18\delta d)^{1/2} + 2\delta$$
$$\le 10\big(\delta\|x_0 - x_1\|\big)^{1/2} + 2\delta,$$

since $\|x_0 - x_1\| \ge \|y_0 - y_1\| - \delta = 2d - \delta \ge (35/18)d$.

On the other hand, if $d \le 18\delta$ (which covers the case $y_0 = y_1$ that was excluded at the beginning of the proof), then $\|y_0 - y_1\| \le 36\delta$ and so $\|x_0 - x_1\| \le 37\delta$. Thus, $\|x_i - p\| \le 19\delta$ and, consequently, $\|y_i - f(p)\| \le 20\delta$ ($i \in \{0, 1\}$). Since $q = (1/2)(y_0 + y_1)$, we have

$$\|f(p) - q\| \le (1/2)\|f(p) - y_0\| + (1/2)\|f(p) - y_1\| \le 20\delta.$$

Therefore, in either case, we conclude that

$$\|f(p) - q\| \le 10\big(\delta\|x_0 - x_1\|\big)^{1/2} + 20\delta,$$

which proves the validity of (a). $\qquad\qquad\qquad\qquad\qquad\qquad\qquad\qquad\qquad$ $\square$

Moreover, M. Omladič and P. Šemrl [259] obtained a sharp stability result for δ-isometries. The proof of the theorem of Omladič and Šemrl depends on the following lemma which is an extension of [237, Lemma 3].

Lemma 13.7. *Let E_1 and E_2 be real Banach spaces. Assume that $f : E_1 \to E_2$ is a surjective δ-isometry and that n is a positive integer. If there exist $x_1, \dots, x_n \in E_1$ such that $f(x_i) \ne f(x_j)$ for $i \ne j$, then for any sufficiently small $\eta > 0$ there exists a bijective function $g : E_1 \to E_2$ such that $\|f(x) - g(x)\| \le \eta$ for all $x \in E_1$ and $f(x_i) = g(x_i)$ for $i \in \{1, \dots, n\}$.*

Proof. First, we prove that card E_1 = card E_2. Because f is surjective, it holds true that card $E_1 \ge$ card E_2. To show the converse inequality, it suffices to prove that the *density character* of E_1 is not larger than that of E_2, where the density character of E_1 is the least cardinality of a dense subset of E_1. Let τ be such that $\phi(t) \le t/2$ for all $t \ge \tau$, where we define

$$\phi(t) = \sup\big\{\big|\|f(x) - f(y)\| - \|x - y\|\big| \,\big|\, \|x - y\| \le t \ \text{ or } \ \|f(x) - f(y)\| \le t\big\}$$

for all $t \ge 0$. It then follows that $\|f(x) - f(y)\| \ge \tau/2$ whenever $\|x - y\| \ge \tau$. Let A be a maximal subset of E_1 such that $\|x - y\| \ge \tau$ for every $x, y \in A$ with $x \ne y$. The density character of E_1 is equal to the cardinality of A and, since $\|f(x) - f(y)\| \ge \tau/2$ for all $x, y \in A$ with $x \ne y$, this cardinal is not larger than the density character of E_2.

Now, let

$$E_2 = E_{21} \cup \cdots \cup E_{2n} \cup \bigcup_\alpha E_{2\alpha},$$

where the sets $E_{21}, \ldots, E_{2n}, E_{2\alpha}$ are pairwise disjoint and such that

(*i*) the diameter of each of them is not larger than the given real number $\eta > 0$,
(*ii*) their cardinalities are equal to the cardinality of E_2,
(*iii*) $f(x_i) \in E_{2i}$ for all $i \in \{1, \ldots, n\}$.

Define $E_{1i} = f^{-1}(E_{2i})$ and $E_{1\alpha} = f^{-1}(E_{2\alpha})$ for all $i \in \{1, \ldots, n\}$ and α. Then, it follows from (*ii*) that card $E_{1i} =$ card E_{2i} for any $i \in \{1, \ldots, n\}$ and card $E_{1\alpha} =$ card $E_{2\alpha}$ for all α. Consequently, we can find bijective functions $g_i :$ $E_{1i} \rightarrow E_{2i}$ and $g_\alpha : E_{1\alpha} \rightarrow E_{2\alpha}$ such that $g_i(x_i) = f(x_i)$. If we define a function $g : E_1 \rightarrow E_2$ by

$$g(x) = \begin{cases} g_i(x) & \text{(for } x \in E_{1i}), \\ g_\alpha(x) & \text{(for } x \in E_{1\alpha}), \end{cases}$$

then (*i*) and (*iii*) yield the desired properties for g. $\square$

Using ideas from [124], Omladič and Šemrl [259] proved that the upper bound 5δ in Theorem 13.6 can be replaced with 2δ.

Theorem 13.8 (Omladič and Šemrl). *Let E_1 and E_2 be real Banach spaces. If $f : E_1 \rightarrow E_2$ is a surjective δ-isometry and $f(0) = 0$, then there exists a unique surjective linear isometry $I : E_1 \rightarrow E_2$ such that*

$$\| f(x) - I(x) \| \le 2\delta \tag{13.4}$$

for each $x \in E_1$.

Proof. Choose $x, y \in E_1$ and assume that $f(x)$, $f(y)$, and $f\big((x + y)/2\big)$ are distinct. If ε is a constant larger than δ, then Lemma 13.7 implies that there exists a bijective function $g : E_1 \rightarrow E_2$ such that $g(x) = f(x)$, $g(y) = f(y)$, $g\big((x + y)/2\big) = f\big((x+y)/2\big)$, and $\| f(u) - g(u) \| \le (1/2)(\varepsilon - \delta)$ for all $u \in E_1$. Obviously, g is an ε-isometry.

Let us define a sequence of bijective functions $h_n : E_2 \rightarrow E_2$ by

$$h_0(u) = g\big(x + y - g^{-1}(u)\big),$$
$$h_1(u) = g(x) + g(y) - u,$$
$$h_n = h_{n-2} \circ h_{n-1} \circ h_{n-2}^{-1}$$

for all integers $n \ge 2$. We also define a sequence of bijective functions

$$k_n = h_n \circ h_{n-1} \circ \cdots \circ h_0$$

for any $n \in \mathbb{N}_0$. Moreover, we define a sequence $\{a_n\}$ recursively by

$$a_1 = (1/2)\big(f(x) + f(y)\big) \quad \text{and} \quad a_{n+1} = h_{n-1}(a_n) = k_{n-1}(a_1)$$

for every $n \in \mathbb{N}$.

According to the second part of the proof of Theorem 13.6, we can prove that there exist sequences $\{p_n\}$, $\{q_n\}$, $\{r_n\}$, and $\{s_n\}$ of nonnegative real numbers such that:

(i) h_n is a $(p_n\varepsilon)$-isometry for all $n \in \mathbb{N}_0$;
(ii) k_n is a $(q_n\varepsilon)$-isometry for all $n \in \mathbb{N}_0$;
(iii) $\|h_n(u) - u\| \geq 2\|a_n - u\| - r_n\varepsilon$ for any $u \in E_2$ and $n \in \mathbb{N}$;

and moreover

$$\|a_2 - a_1\| \leq 2^{-(n-2)}\|a_n - a_{n-1}\| + s_n\varepsilon \qquad (a)$$

for any integer $n \geq 2$.

As in [124], we can apply (i), (ii), (iii), and (a) to prove that

$$\left\|2f\left(\frac{x+y}{2}\right) - f(x) - f(y)\right\| \leq 2^{-(n-2)}\|x - y\| + 2(n+2)\delta \qquad (b)$$

for all $n \in \mathbb{N}$. It is not difficult to see that (b) is true also when $f(x)$, $f(y)$, and $f\big((x+y)/2\big)$ are not distinct.

It follows from (b) that

$$I(x) = \lim_{n\to\infty} 2^{-n} f(2^n x)$$

exists for all $x \in E_1$ and that I is a linear isometry. It also implies that

$$\|f(x) - I(x)\| \leq 2(r+3)\delta + 2^{-r+3}\|x\| \qquad (c)$$

for any $x \in E_1$ and $r \in \mathbb{N}$.

We will now prove that I is surjective. Suppose to the contrary that the range of I is a proper subspace of E_2. As it is closed, there exists a $z \in E_2$ of norm one such that its distance to the range of I is larger than $1/2$. The surjectivity of f implies the existence of $x_t \in E_1$ with $f(x_t) = tz$ for any $t > 0$. Recall that f is a δ-isometry with $f(0) = 0$ and observe that $\big|\|f(x_t)\| - \|x_t\|\big| \leq \delta$ yields $t - \delta \leq \|x_t\| \leq t + \delta$. Thus, we have

$$\|f(x_t) - I(x_t)\| = t\|z - (1/t)I(x_t)\| > t/2 \qquad (d)$$

for every $t > 0$ since $(1/t)I(x_t)$ belongs to the range of I. On the other hand, (c) implies that

$$\|f(x_t) - I(x_t)\| \leq 2(r+3)\delta + 2^{-r+3}\|x_t\|$$
$$\leq 2(r+3)\delta + 2^{-r+3}(t+\delta) \qquad (e)$$

for all $t > 0$ and $r \in \mathbb{N}$. Combining (d) and (e) yields

$$t/2 < 2(r+3)\delta + 2^{-r+3}(t+\delta)$$

for any $t > 0$ and $r \in \mathbb{N}$. If we set $t = r^2$ and then let $r \to \infty$, then we get a contradiction. Hence, I has to be surjective.

Now, we define a function $T : E_1 \to E_1$ by $T = I^{-1} \circ f$. Obviously, T is a surjective δ-isometry with $T(0) = 0$. Thus, we obtain

$$\left| \|T(x)\| - \|x\| \right| \leq \delta \qquad (f)$$

for all $x \in E_1$. By (c), we have

$$\|T(x) - x\| = \left\| I^{-1}\big(f(x)\big) - x \right\| = \|f(x) - I(x)\| \leq 2(r+3)\delta + 2^{-r+3}\|x\| \quad (g)$$

for all $x \in E_1$ and $r \in \mathbb{N}$.

We now assert that

$$\|T(x) - x\| \leq 2\delta \qquad (h)$$

for each $x \in E_1$. Choose an $x \in E_1$ satisfying $\|T(x) - x\| = a > 0$ and set $y = (1/a)\big(T(x) - x\big)$. Then, we have $\|y\| = 1$. As T is surjective, we can find a $z_n \in E_1$ for every $n \in \mathbb{N}$ such that $T(x + z_n) = x + ay + ny$. Since $n = \|T(x + z_n) - T(x)\|$ and T is a δ-isometry, we conclude that

$$n - \delta \leq \|z_n\| \leq n + \delta \qquad (i)$$

for any $n \in \mathbb{N}$.

If we replace x in (g) with $x + z_n$ and divide the resulting inequality by n, then we get

$$\|(1/n)ay + y - (1/n)z_n\| \leq (1/n)\big(2(r+3)\delta + 2^{-r+3}(\|x\| + n + \delta)\big).$$

We put $r = \big[\sqrt{n}\big]$ and let $n \to \infty$ in the last inequality, where $[w]$ denotes the largest integer not exceeding w, then (i) gives

$$\lim_{n \to \infty} (1/\|z_n\|)z_n = y. \qquad (j)$$

Let q be a real number and define

$$\lambda = \limsup_{n \to \infty} \big(\|x + z_n\| - \|z_n\|\big).$$

It now follows from (i) that $1 - q/\|z_n\| > 0$ if n is large enough. For such integers n, it holds true that

$$\|x + z_n\| - \|z_n\| + q = \left\| x + (q/\|z_n\|)z_n + \big(1 - q/\|z_n\|\big)z_n \right\|$$
$$- \left\| \big(1 - q/\|z_n\|\big)z_n \right\|$$
$$\leq \left\| x + (q/\|z_n\|)z_n \right\|.$$

This implies together with (j) that

$$\lambda + q \le \|x + qy\|$$

for any $q \in \mathbb{R}$. Similarly, we obtain

$$\lambda + q \ge \limsup_{n \to \infty} \left(- \|x + (q/\|z_n\|)z_n\| \right) = -\|x + qy\|$$

for each $q \in \mathbb{R}$. Consequently, we have

$$|\lambda + q| \le \|x + qy\| \qquad (k)$$

for every $q \in \mathbb{R}$.

We will have to consider two special cases. First, we deal with the situation when x and y are linearly independent. Let us define a functional φ on the linear span of x and y by $\varphi(x) = \lambda$ and $\varphi(y) = 1$. It then follows from (k) that $\|\varphi\| = 1$. If we substitute $x + z_n$ for x in (f), then it follows from (k) that

$$
\begin{aligned}
\delta &\ge \|x + ay + ny\| - \|x + z_n\| \\
&\ge |\lambda + a + n| - \|x + z_n\| \\
&\ge \lambda + n + a - \|x + z_n\|
\end{aligned}
$$

for each $n \in \mathbb{N}$. Applying (i) to the last inequality, we see that

$$\delta \ge \lambda - \|x + z_n\| + (n + \delta) + a - \delta \ge \lambda - \left(\|x + z_n\| - \|z_n\| \right) + a - \delta$$

which together with the definition of λ implies that $2\delta - a \ge 0$. This ends the proof of (h) in the case that x and y are linearly independent.

Assume now that $x = \mu y$ for some $\mu \in \mathbb{R}$. In view of (k), we have $|\lambda + q| \le |\mu + q|$ for any $q \in \mathbb{R}$. If we put $q = -\mu$ in the last inequality, then we get $\lambda = \mu$. If an integer n is large enough, then $\|x + ay + ny\| = \lambda + a + n$. For such integers n, substituting $x + z_n$ for x in (f) and considering (i) and the definition of λ yield

$$
\begin{aligned}
\delta &\ge \|x + ay + ny\| - \|x + z_n\| \\
&= a + \lambda - \left(\|x + z_n\| - \|z_n\| \right) + n - \|z_n\| \\
&\ge a - \delta,
\end{aligned}
$$

i.e., $a \le 2\delta$, which proves the validity of (h) in the case that x and y are linearly dependent.

Since I is a surjective isometry with $I(0) = 0$, it follows from (h) that

$$\|f(x) - I(x)\| = \|T(x) - x\| \le 2\delta$$

for all $x \in E_1$. If $I' : E_1 \to E_2$ is another surjective linear isometry satisfying $I'(0) = 0$ as well as (13.4), then we obtain

$$\|2^{-m} f(2^m x) - I'(x)\| \le 2^{-(m-1)} \delta$$

for all $x \in E_1$ and $m \in \mathbb{N}$. If we let $m \to \infty$, then we get $I = I'$. $\qquad\square$

Moreover, Omladič and Šemrl presented that the inequality (13.4) is sharp by constructing the following example. Define a surjective function $f : \mathbb{R} \to \mathbb{R}$ by

$$f(x) = \begin{cases} -3x & (\text{for } x \in [0, 1/2]), \\ x - 1 & (\text{for } x \notin [0, 1/2]). \end{cases}$$

Obviously, f is a 1-isometry. In view of Theorem 13.8, f can be approximated by a linear isometry $I : \mathbb{R} \to \mathbb{R}$. There are only two linear isometries, namely, $I(x) = x$ and $I(x) = -x$. However, the second one does not approximate f uniformly and we can easily show that

$$\max_{x \in \mathbb{R}} |f(x) - x| = 2,$$

which proves that the inequality (13.4) is sharp in the one-dimensional case.

On the other hand, G. Dolinar [94] investigated the stability of isometries in connection with (ε, p)-*isometry*, where a function $f : E_1 \to E_2$ is called an (ε, p)-*isometry* if f satisfies the inequality

$$\big| \|f(x) - f(y)\| - \|x - y\| \big| \le \varepsilon \|x - y\|^p$$

for some $\varepsilon \ge 0$ and for all $x, y \in E_1$:

Let E_1 and E_2 be real Banach spaces and let $0 \le p < 1$. There exists a constant $N(p)$, independent of E_1 and E_2, such that for each surjective (ε, p)-isometry $f : E_1 \to E_2$ with $f(0) = 0$ there exists a surjective isometry $I : E_1 \to E_2$ satisfying

$$\|f(x) - I(x)\| \le \varepsilon N(p) \|x\|^p$$

for all $x \in E_1$.

He also proved the superstability of isometries. Indeed, it was proved that for $p > 1$ every surjective (ε, p)-isometry $f : E_1 \to E_2$ from a finite-dimensional real Banach space E_1 onto a finite-dimensional real Banach space E_2 is an isometry.

For more general information on the Hyers–Ulam stability of isometries and related topics, refer to [21, 154, 237, 297, 305, 307, 328, 338, 350, 352, 355].

13.2 Stability on a Restricted Domain

Let $m \ge 1$ be an integer. For each $E \subset \mathbb{R}^m$, $H(E)$ denotes the smallest *flat* containing E, i.e.,

$$H(E) = \{\lambda_0 p_0 + \cdots + \lambda_m p_m \mid p_0, \ldots, p_m \in E; \ \lambda_0, \ldots, \lambda_m \in \mathbb{R};$$
$$\lambda_0 + \cdots + \lambda_m = 1\}.$$

For any integer $k \in \{1, \ldots, m\}$, the points $p_0, \ldots, p_k \in \mathbb{R}^m$ are called *independent* if $H(p_0, \ldots, p_k)$ is k-dimensional. We use the notation $d(x, E)$ for the Euclidean distance between the point x and the subset E of $\mathbb{R}^m$.

In 1981, J. W. Fickett [103] proved the following lemma and applied it to the proof of the Hyers–Ulam stability of isometries defined on a bounded domain.

Lemma 13.9. *Let $p_0, \ldots, p_{m-1}$ be independent points in $\mathbb{R}^m$, where $m \geq 2$ is an integer. Define $H = H(p_0, \ldots, p_{m-2})$ and let p_m be any point of $\mathbb{R}^m$ satisfying $d(p_m, H) \leq d(p_{m-1}, H)$. Define $d_i = |p_m - p_i|$ for every $i \in \{0, \ldots, m-2\}$. Suppose that $0 \leq d_i \leq 1$ for $i \in \{0, \ldots, m-2\}$, and that d_{m-1} and ε are given with $\left|d_{m-1} - |p_m - p_{m-1}|\right| \leq \varepsilon$ and $0 < \varepsilon \leq 1$. If there exists a point q with $|q - p_i| = d_i$ for $i \in \{0, \ldots, m-1\}$, then there exists such a q with $|p_m - q| \leq 2\sqrt{\varepsilon}$.*

For a given integer $n \geq 2$, let S be a bounded subset of $\mathbb{R}^n$ and let $f : S \to \mathbb{R}^n$ be a δ-isometry. Fickett [103] introduced a construction of isometry $I : S \to \mathbb{R}^n$ which is quite different from the (direct) method of Hyers and Ulam:

(i) We extend the domain S of f to $\overline{S}$, the closure of S. For any $s \in \overline{S} \setminus S$, we define $f_1(s)$ to be any element of the set

$$\bigcap_{m=1}^{\infty} \overline{f(\{y \in S \mid d(s, y) < 1/m\})},$$

and let $f_1(s) = f(s)$ for $s \in S$. Then it is easy to show that f_1 is again a δ-isometry.

(ii) Let k be the dimension of $H(S)$ and choose $s_0, \ldots, s_k \in \overline{S}$ to satisfy

$$\begin{aligned} |s_0 - s_1| &= \text{diameter of } \overline{S}, \\ d\left(s_i, H(s_0, \ldots, s_{i-1})\right) &= \sup \left\{d\left(s, H(s_0, \ldots, s_{i-1})\right) \mid s \in S\right\} \end{aligned} \tag{13.5}$$

for each $i \in \{2, \ldots, k\}$. Note that $s_0, \ldots, s_k$ are independent.

(iii) We define

$$\begin{aligned} I(s_0) &= f_1(s_0), \\ I(s_i) &= \text{a point as close as possible to } f_1(s_i) \text{ satisfying} \\ & \quad |I(s_i) - I(s_j)| = |s_i - s_j| \text{ for } 0 \leq j \leq i \leq k, \\ I(s) &= \text{the unique point in } H\left(I(s_0), \ldots, I(s_k)\right) \text{ satisfying} \\ & \quad |I(s) - I(s_i)| = |s - s_i| \text{ for } i \in \{0, \ldots, k\}. \end{aligned} \tag{13.6}$$

For an $S \subset \mathbb{R}^n$, diam S denotes the diameter of S. Moreover, we define

$$K_0(\delta) = K_1(\delta) = \delta, \quad K_2(\delta) = 3\sqrt{3\delta}, \quad K_i(\delta) = 27\delta^{2^{1-i}}$$

for all integers $i \geq 3$.

J. W. Fickett [103] proved the Hyers–Ulam stability of isometries on a bounded domain.

Theorem 13.10 (Fickett). *Let S be a bounded subset of $\mathbb{R}^n$ and let $f : S \to \mathbb{R}^n$ be a δ-isometry, where $0 \le K_n(\delta/\operatorname{diam} S) \le 1/3$. Then the isometry $I : S \to \mathbb{R}^n$ constructed by the preceding steps satisfies*

$$|f(s) - I(s)| \le K_{n+1}(\delta/\operatorname{diam} S) \cdot \operatorname{diam} S$$

for all $s \in S$.

Proof. By a homothety argument, we can reduce the general case to the case where $\operatorname{diam} S = 1$. By induction on m, we prove that if $t_0, \ldots, t_m \in \overline{S}$, with $t_0, \ldots, t_{m-1}$ independent, satisfy

$$d\left(t_i, H(t_0, \ldots, t_{i-1})\right) \ge d\left(t_j, H(t_0, \ldots, t_{i-1})\right) \tag{a}$$

for all integers i, j with $1 \le i \le j \le m$, and if $h : \{t_0, \ldots, t_m\} \to \mathbb{R}^n$ is defined inductively by

$$h(t_i) = \text{a point as close as possible to } f_1(t_i) \text{ satisfying}$$
$$|h(t_i) - h(t_j)| = |t_i - t_j| \text{ for } 0 \le j \le i \le m,$$

then

$$|h(t_i) - f_1(t_i)| \le K_m(\delta)$$

for each integer $i \in \{0, \ldots, m\}$.

This claim is true for $m \in \{0, 1\}$. Let $m \ge 2$. We assume that the claim is true for m-point sets, and let $t_0, \ldots, t_m$, and h be as described. Then $t_0, \ldots, t_{m-2}, t_{m-1}$ and $t_0, \ldots, t_{m-2}, t_m$ each satisfy an m-point version of (a).

For $i \in \{0, \ldots, m-1\}$, we define $h_1(t_i) = h(t_i)$ and

$$h_1(t_m) = \text{a point as near as possible to } f_1(t_m) \text{ satisfying}$$
$$|h_1(t_m) - h_1(t_i)| = |t_m - t_i| \text{ for } i \in \{0, \ldots, m-2\}.$$

Note that $h_1(t_m)$ is in $H\left(h(t_0), \ldots, h(t_m)\right)$. By the induction hypothesis, we have

$$|h_1(t_i) - f_1(t_i)| \le K_{m-1}(\delta)$$

for any $i \in \{0, \ldots, m\}$. Hence, we get

$$\left||h_1(t_{m-1}) - h_1(t_m)| - |t_{m-1} - t_m|\right| \le \delta + 2K_{m-1}(\delta) \le 3K_{m-1}(\delta).$$

We now apply Lemma 13.9 with $p_0, \ldots, p_m$, d_{m-1}, ε, and $\mathbb{R}^m$ there equal to $h_1(t_0), \ldots, h_1(t_m)$, $|t_m - t_{m-1}|$, $3K_{m-1}(\delta)$, and $H\left(h(t_0), \ldots, h(t_m)\right)$ to get a $q \in H\left(h(t_0), \ldots, h(t_m)\right)$ with $|q - h(t_i)| = |t_m - t_i|$ for each $i \in \{0, \ldots, m-2\}$ and $|q - h_1(t_m)| \le 2\left(3K_{m-1}(\delta)\right)^{1/2}$. Thus,

$$|f_1(t_m) - h(t_m)| \le |f_1(t_m) - q|$$
$$\le |f_1(t_m) - h_1(t_m)| + |h_1(t_m) - q|$$
$$\le K_{m-1}(\delta) + 2\left(3K_{m-1}(\delta)\right)^{1/2}$$
$$\le K_m(\delta),$$

which ends the proof of our first claim.

Now, let $s \in S$ be arbitrary. By (13.5) and (13.7), we may apply the preceding steps with $m = k + 1, t_0 = s_0, \ldots, t_k = s_k, t_m = s$, and $h = I$ to conclude that

$$|I(s) - f_1(s)| \le K_{k+1}(\delta) \le K_{n+1}(\delta)$$

for all $s \in S$.   □

Let $(G, +)$ be an abelian metric group with a metric $d(\cdot, \cdot)$ satisfying

$$d(x + z, y + z) = d(x, y) \quad \text{and} \quad d(2x, 2y) = 2d(x, y) \tag{13.7}$$

for all $x, y, z \in G$. Moreover, we assume that for any given $y \in G$ the equation

$$x + x = y$$

is uniquely solvable. We here promise that $2^{-1}y$ or $y/2$ stands for the unique solution of the above equation and we inductively define $2^{-(n+1)}y = 2^{-1}\left(2^{-n}y\right)$ for each $y \in G$ and $n \in \mathbb{N}$. We may usually write $2^{-n}x$ instead of $x/2^n$ for each $x \in G$ and $n \in \mathbb{N}$. The second condition in (13.7) also implies that

$$d(x/2, y/2) = (1/2)d(x, y)$$

for all $x, y \in G$.

S.-M. Jung [178] applied a direct method and proved the Hyers–Ulam stability of isometries on a restricted domain.

Theorem 13.11. *Let E be a subset of G with the property that*

$$0 \in E \quad \text{and} \quad 2^k x \in E \quad \text{for all } x \in E \text{ and } k \in \mathbb{N}$$

and let F be a real Hilbert space with the associated inner product $\langle \cdot, \cdot \rangle$. If a function $f : E \to F$ satisfies the inequality

$$\left| \|f(x) - f(y)\| - d(x, y) \right| \le \varepsilon d(x, y)^p$$

for some $\varepsilon \ge 0, 0 \le p < 1$ and for all $x, y \in E$, then there exists an isometry $I : E \to F$ that satisfies

$$\|f(x) - I(x) - f(0)\|$$
$$\le \frac{2^{(1-p)/2}}{2^{(1-p)/2} - 1} \max\left\{ \sqrt{4.5\varepsilon}, 2\varepsilon \right\} \max\left\{ d(x, 0)^p, d(x, 0)^{(1+p)/2} \right\} \tag{13.8}$$

for all $x \in E$. For $0 < p < 1$, the isometry I is uniquely determined.

Proof. If we define a function $g : E \to F$ by $g(x) = f(x) - f(0)$, then we have

$$\big| \|g(x) - g(y)\| - d(x,y) \big| \leq \varepsilon d(x,y)^p \tag{a}$$

for any $x, y \in E$. With $y = 0$ and $y = 2x$ separately, the inequality (a) together with (13.7) yields

$$\big| \|g(x)\| - d(x,0) \big| \leq \varepsilon d(x,0)^p,$$
$$\big| \|g(x) - g(2x)\| - d(x,0) \big| \leq \varepsilon d(x,0)^p, \tag{b}$$

respectively.

It follows from (b) that

$$A(x)^2 \leq \|g(x)\|^2 \leq \big(d(x,0) + \varepsilon d(x,0)^p\big)^2 \tag{c}$$

and

$$\|g(x) - g(2x)\|^2 = \|g(x)\|^2 + \|g(2x)\|^2 - 2\langle g(x), g(2x)\rangle$$
$$\leq \big(d(x,0) + \varepsilon d(x,0)^p\big)^2, \tag{d}$$

where we set

$$A(x) = \begin{cases} 0 & (\text{for } d(x,0) \leq \varepsilon^{1/(1-p)}), \\ d(x,0) - \varepsilon d(x,0)^p & (\text{for } d(x,0) > \varepsilon^{1/(1-p)}). \end{cases}$$

If $d(x,0) > (1/2)\varepsilon^{1/(1-p)}$, then $A(2x) = d(2x,0) - \varepsilon d(2x,0)^p$ and $A(2x)^2 \leq \|g(2x)\|^2$. Hence, it follows from (13.7), (c), and (d) that

$$2\|g(x) - (1/2)g(2x)\|^2$$
$$= 2\|g(x)\|^2 + (1/2)\|g(2x)\|^2 - 2\langle g(x), g(2x)\rangle$$
$$= \|g(x)\|^2 + \big(\|g(x)\|^2 + \|g(2x)\|^2 - 2\langle g(x), g(2x)\rangle\big) - (1/2)\|g(2x)\|^2$$
$$\leq 2\big(d(x,0) + \varepsilon d(x,0)^p\big)^2 - (1/2)\big(d(2x,0) - \varepsilon d(2x,0)^p\big)^2$$
$$= 4\varepsilon\big(1 + 2^{p-1}\big)d(x,0)^{1+p} + 2\varepsilon^2\big(1 - 2^{2(p-1)}\big)d(x,0)^{2p}$$
$$= \big(4 + 2^{p+1} + 2\varepsilon d(x,0)^{p-1}\big(1 - 2^{2(p-1)}\big)\big)\varepsilon d(x,0)^{1+p}$$
$$< \big(4 + 2^{p+1} + 2^{2-p}\big(1 - 2^{2(p-1)}\big)\big)\varepsilon d(x,0)^{1+p}$$
$$\leq 9\varepsilon d(x,0)^{1+p}.$$

On the other hand, for $d(x,0) \leq (1/2)\varepsilon^{1/(1-p)}$, it analogously follows from (c) and (d) that

$$2\|g(x) - (1/2)g(2x)\|^2 \leq 2\big(d(x,0) + \varepsilon d(x,0)^p\big)^2 \leq 8\varepsilon^2 d(x,0)^{2p}.$$

Hence, we have

$$\|g(x) - (1/2)g(2x)\| \leq C \max\big\{d(x,0)^p, \, d(x,0)^{(1+p)/2}\big\} \qquad (e)$$

for all $x \in E$, where we set $C = \max\big\{\sqrt{4.5\varepsilon}, \, 2\varepsilon\big\}$.

The last inequality implies the validity of the inequality

$$\|g(x) - 2^{-n}g(2^n x)\| \leq C \max\big\{d(x,0)^p, \, d(x,0)^{(1+p)/2}\big\} \sum_{i=0}^{n-1} 2^{-i(1-p)/2} \quad (f)$$

for $n = 1$. Assume now that the inequality (f) is true for some $n \in \mathbb{N}$. It then follows from (13.7), (e), and (f) that

$$\begin{aligned}
&\|g(x) - 2^{-(n+1)}g(2^{n+1}x)\| \\
&\quad \leq \|g(x) - 2^{-n}g(2^n x)\| + \|2^{-n}g(2^n x) - 2^{-(n+1)}g(2^{n+1}x)\| \\
&\quad \leq C \max\big\{d(x,0)^p, \, d(x,0)^{(1+p)/2}\big\} \sum_{i=0}^{n-1} 2^{-i(1-p)/2} \\
&\qquad + 2^{-n} C \max\big\{d(2^n x,0)^p, \, d(2^n x,0)^{(1+p)/2}\big\} \\
&\quad \leq C \max\big\{d(x,0)^p, \, d(x,0)^{(1+p)/2}\big\} \sum_{i=0}^{n} 2^{-i(1-p)/2},
\end{aligned}$$

which implies the validity of (f) for all $x \in E$ and $n \in \mathbb{N}$.

For given $m, n \in \mathbb{N}$ with $n > m$, we use (13.7) and (f) to verify

$$\begin{aligned}
&\|2^{-m}g(2^m x) - 2^{-n}g(2^n x)\| \\
&\quad = 2^{-m}\|g(2^m x) - 2^{-(n-m)}g(2^{n-m} \cdot 2^m x)\| \\
&\quad \leq C \max\big\{d(x,0)^p, \, d(x,0)^{(1+p)/2}\big\} \sum_{i=m}^{n-1} 2^{-i(1-p)/2} \\
&\quad \to 0 \quad \text{as } m \to \infty.
\end{aligned}$$

Thus, $\{2^{-n}g(2^n x)\}$ is a Cauchy sequence for any $x \in E$. Let us define a function $I : E \to F$ by

$$I(x) = \lim_{n \to \infty} 2^{-n}g(2^n x). \qquad (g)$$

If we substitute $2^n x$ and $2^n y$ for x and y in (a), divide the resulting inequality by 2^n, and consider the case that n goes to infinity, then we see that I is an isometry. The inequality (f), together with (g), presents that the inequality (13.8) holds true for any $x \in E$.

Assume now that $0 < p < 1$ and $I' : E \to F$ is an isometry satisfying the inequality (13.8). Based on our assumption $p > 0$, it then follows from (13.8) that $I'(0) = 0$. Since

$$\|I'(x) - I'(y)\| = d(x, y)$$

for all $x, y \in E$, it follows from (13.7) that

$$\|I'(2x) - I'(x)\| = d(2x, x) = d(x, 0) = \|I'(x)\|$$

and

$$\|I'(2x)\| = d(2x, 0) = 2d(x, 0) = 2\|I'(x)\|.$$

Hence, we have

$$\|I'(2x) - I'(x)\|^2 = \|I'(2x)\|^2 - 2\langle I'(2x), I'(x)\rangle + \|I'(x)\|^2 = \|I'(x)\|^2.$$

Thus, we get

$$\|I'(2x)\|\|I'(x)\| = \langle I'(2x), I'(x)\rangle,$$

i.e.,

$$I'(2x) = 2I'(x) \qquad (h)$$

for any $x \in E$. Assume that

$$2^{-k} I'(2^k x) = I'(x) \qquad (i)$$

for all $x \in E$ and some $k \in \mathbb{N}$. Then, by (h) and (i), we obtain

$$2^{-(k+1)} I'(2^{k+1} x) = 2^{-k} I'(2^k x) = I'(x),$$

which implies that, for $0 < p < 1$, the equality (i) is true for all $x \in E$ and all $k \in \mathbb{N}$.

For some $0 < p < 1$ and for an arbitrary $x \in E$, it follows from (13.8) and (i) that

$$\|I(x) - I'(x)\| = 2^{-k}\|I(2^k x) - I'(2^k x)\|$$

$$\leq \frac{1}{2^k} \frac{2 \cdot 2^{(1-p)/2}}{2^{(1-p)/2} - 1} C \max\{d(2^k x, 0)^p, d(2^k x, 0)^{(1+p)/2}\}$$

$$\leq 2^{-k(1-p)/2}\frac{2^{(3-p)/2}}{2^{(1-p)/2}-1}C \max\left\{d(x,0)^p, d(x,0)^{(1+p)/2}\right\}$$
$$\to 0 \quad \text{as } k \to \infty.$$

This implies the uniqueness of I for the case $0 < p < 1$. □

It would be interesting to compare the previous theorem with [335, Theorem 1] because the function f involved in Theorem 13.11 is a kind of asymptotic isometry. The following corollary may be proven by use of Theorem 13.11 or in a straight forward manner. Indeed, it is an immediate consequence of [335, Proposition 4].

Corollary 13.12. *Let G and F be a real normed space and a real Hilbert space, respectively. Assume that a function $f : G \to F$ satisfies $f(0) = 0$ and $f(2^k x) = 2^k f(x)$ for all $x \in G$ and $k \in \mathbb{N}$. The function f is a linear isometry if and only if there exists a $0 < p < 1$ such that*

$$\left|\|f(x) - f(y)\| - \|x - y\|\right| = O\left(\|x - y\|^p\right)$$

as $\|x\| \to \infty$ and $\|y\| \to \infty$.

Jung [178] also proved the Hyers–Ulam stability of isometries on a restricted domain for the case $p > 1$.

Theorem 13.13. *Let E be a subset of G with the property that*

$$0 \in E \quad \text{and} \quad 2^{-k}x \in E \quad \text{for all } x \in E \text{ and } k \in \mathbb{N}$$

and let F be a real Hilbert space with the associated inner product $\langle \cdot, \cdot \rangle$. If a function $f : E \to F$ satisfies the inequality

$$\left|\|f(x) - f(y)\| - d(x, y)\right| \leq \varepsilon d(x, y)^p$$

for some $\varepsilon \geq 0$, $p > 1$, and for all $x, y \in E$, then there exists a unique isometry $I : E \to F$ such that

$$\|f(x) - I(x) - f(0)\|$$
$$\leq \frac{2^{(p-1)/2}}{2^{(p-1)/2}-1}\max\left\{2\sqrt{\varepsilon}, 2\varepsilon\right\}\max\left\{d(x,0)^p, d(x,0)^{(1+p)/2}\right\}$$

for any $x \in E$.

Corollary 13.14. *Let G and F be a real normed space and a real Hilbert space, respectively. Suppose a function $f : G \to F$ satisfies $f(0) = 0$ and $f(2^k x) = 2^k f(x)$ for all $x \in G$ and $k \in \mathbb{N}$. The function f is a linear isometry if and only if there exists a $p > 1$ such that*

$$\left| \|f(x) - f(y)\| - \|x - y\| \right| = O\left(\|x - y\|^p\right)$$

as $\|x\| \to 0$ and $\|y\| \to 0$.

For more general information on this subject, we refer the reader to [176, 188].

13.3 Fixed Point Method

Recently, Cădariu and Radu [57] applied the fixed point method to the proof of the Hyers–Ulam stability of the Cauchy additive functional equation. Using their idea, S.-M. Jung could present a short and simple proof for the Hyers–Ulam–Rassias stability of isometries whose domain is a normed space and range is a Banach space in which the parallelogram law holds true (see [182]).

Theorem 13.15 (Jung). *Let E_1 be a normed space over $\mathbb{K}$ and let E_2 be a Banach space over $\mathbb{K}$ in which the parallelogram law holds true. Assume that $\delta : E_1 \to [0, \infty)$ is an even function such that there exists a constant L, $0 < L < 1$, with*

$$\begin{cases} \delta(2x) \le 2L^2\delta(x) & \text{(for all } x \in E_1), \\ \|x\| \le L^2\delta(x) & \text{(for all } x \in E_1 \text{ with } \|x\| < 1), \\ \|x\| \ge L^2\delta(x) & \text{(for all } x \in E_1 \text{ with } \|x\| \ge 1), \end{cases} \tag{13.9}$$

and that

$$\lim_{n \to \infty} 2^{-n}\delta(2^n x) = 0 \tag{13.10}$$

for all $x \in E_1$. If a function $f : E_1 \to E_2$ satisfies

$$\left| \|f(x) - f(y)\| - \|x - y\| \right| \le \delta(x - y) \tag{13.11}$$

for all $x, y \in E_1$, then there exists an isometry $I : E_1 \to E_2$ such that

$$\|f(x) - f(0) - I(x)\| \le \frac{1 + L^2}{1 - L}\psi(x) \tag{13.12}$$

for all $x \in E_1$, where

$$\psi(x) = \begin{cases} \left(1/\sqrt{1 + 2L^2}\right)\left(\|x\| + \delta(x)\right) & \text{(for } \|x\| < 1), \\ (1/L)\sqrt{\|x\|\delta(x)} & \text{(for } \|x\| \ge 1). \end{cases}$$

The I is the unique isometry satisfying (13.12) *and $I(2x) = 2I(x)$ for any $x \in E_1$.*

Proof. First, we will prove that

$$\psi(2x) \le 2L\psi(x) \tag{a}$$

for all $x \in E_1$. If $\|x\| \geq 1$, then (a) immediately follows from the first condition of (13.9). If $\|x\| < 1/2$, then the first and second conditions of (13.9) imply that

$$2L\psi(x) - \psi(2x) \geq \frac{2(1-L)}{\sqrt{1+2L^2}}\left(L\delta(x) - \|x\|\right) \geq \frac{2(1-L)}{\sqrt{1+2L^2}}\left(L^2\delta(x) - \|x\|\right) \geq 0.$$

Now, let $1/2 \leq \|x\| < 1$, i.e., $\|2x\| \geq 1$. It then follows from the first two conditions of (13.9) that

$$\begin{aligned}
\frac{1}{4L^2}\psi(2x)^2 &= \frac{1}{4L^2}\frac{1}{L^2}\|2x\|\delta(2x) \\
&\leq \frac{1}{L^2}\|x\|\delta(x) \\
&= \frac{1}{L^2(1+2L^2)}\|x\|\delta(x) + \frac{2}{1+2L^2}\|x\|\delta(x) \\
&\leq \frac{1}{1+2L^2}\delta(x)^2 + \frac{2}{1+2L^2}\|x\|\delta(x) \\
&\leq \frac{1}{1+2L^2}\left(\|x\| + \delta(x)\right)^2 \\
&= \psi(x)^2
\end{aligned}$$

for all $x \in E_1$.

Let us define

$$X = \{h : E_1 \to E_2 \mid h(0) = 0\}$$

and introduce a generalized metric on X as follows:

$$d(h_1, h_2) = \inf\{C \in [0, \infty] \mid \|h_1(x) - h_2(x)\| \leq C\psi(x) \text{ for all } x \in E_1\}.$$

Then, it is easy to prove that (X, d) is a generalized complete metric space (see the proof of [189, Theorem 3.1] or [57, Theorem 2.5]). We now define an operator $\Lambda : X \to X$ by

$$(\Lambda h)(x) = (1/2)h(2x)$$

for all $h \in X$ and $x \in E_1$.

We assert that Λ is a strictly contractive operator. Given $h_1, h_2 \in X$, let $C \in [0, \infty]$ be an arbitrary constant with $d(h_1, h_2) \leq C$. From the definition of d, it follows that

$$\|h_1(x) - h_2(x)\| \leq C\psi(x)$$

for each $x \in E_1$. By the last inequality and (a), we have

$$\|(\Lambda h_1)(x) - (\Lambda h_2)(x)\| = (1/2)\|h_1(2x) - h_2(2x)\| \leq (1/2)C\psi(2x) \leq CL\psi(x)$$

for all $x \in E_1$. Hence, it holds true that $d(\Lambda h_1, \Lambda h_2) \leq CL$, i.e., $d(\Lambda h_1, \Lambda h_2) \leq Ld(h_1, h_2)$ for any $h_1, h_2 \in X$.

If we set $g(x) = f(x) - f(0)$ for any $x \in E_1$, then it follows from (13.11) that

$$\big|\|g(x) - g(y)\| - \|x - y\|\big| \leq \delta(x - y) \qquad (b)$$

for all $x, y \in E_1$.

Now, we can apply the parallelogram law to the parallelogram

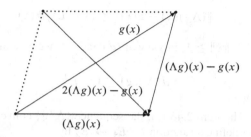

and we conclude that

$$2\|(\Lambda g)(x) - g(x)\|^2 + 2\|(\Lambda g)(x)\|^2 = \|2(\Lambda g)(x) - g(x)\|^2 + \|g(x)\|^2$$

for any $x \in E_1$. Since $(\Lambda g)(x) = (1/2)g(2x)$, it follows from the last equality and (b) that

$$2\|(\Lambda g)(x) - g(x)\|^2 = \|g(2x) - g(x)\|^2 + \|g(x)\|^2 - (1/2)\|g(2x)\|^2$$
$$\leq 2\big(\|x\| + \delta(x)\big)^2 - (1/2)\|g(2x)\|^2 \qquad (c)$$

for each $x \in E_1$.

If $\|x\| < 1$, then it follows from (c) that

$$2\|(\Lambda g)(x) - g(x)\|^2 \leq 2\big(\|x\| + \delta(x)\big)^2$$

or equivalently

$$\|(\Lambda g)(x) - g(x)\| \leq \sqrt{1 + 2L^2}\, \psi(x) \qquad (d)$$

for all $x \in E_1$ with $\|x\| < 1$. If $\|x\| \geq 1$, then it follows from (13.9), (b), and (c) that

$$2\|(\Lambda g)(x) - g(x)\|^2 \leq 2\big(\|x\| + \delta(x)\big)^2 - (1/2)\|g(2x)\|^2$$
$$\leq 2\big(\|x\| + \delta(x)\big)^2 - (1/2)\big(\|2x\| - \delta(2x)\big)^2$$
$$\leq 2\big(\|x\| + \delta(x)\big)^2 - (1/2)\big(2\|x\| - 2L^2\delta(x)\big)^2$$
$$= 4(1 + L^2)\|x\|\delta(x) + 2(1 - L^4)\delta(x)^2$$

$$\leq 4(1 + L^2)\|x\|\delta(x) + 2\frac{1 - L^4}{L^2}\|x\|\delta(x)$$

$$= 2\left(\frac{1 + L^2}{L}\right)^2 \|x\|\delta(x).$$

Hence, it follows that

$$\|(\Lambda g)(x) - g(x)\| \leq (1 + L^2)\psi(x) \tag{e}$$

for all $x \in E_1$ with $\|x\| \geq 1$. In view of (d) and (e), we conclude that

$$d(\Lambda g, g) \leq 1 + L^2. \tag{f}$$

According to Theorem 2.43 (i), the sequence $\{\Lambda^n g\}$ converges to a fixed point I of Λ, i.e., if we define a function $I : E_1 \to E_2$ by

$$I(x) = \lim_{n \to \infty} (\Lambda^n g)(x) = \lim_{n \to \infty} 2^{-n} g(2^n x) \tag{g}$$

for all $x \in E_1$, then I belongs to X and I satisfies

$$I(2x) = 2I(x) \tag{h}$$

for any $x \in E_1$. Moreover, it follows from Theorem 2.43 (iii) and (f) that

$$d(g, I) \leq \frac{1}{1 - L}d(\Lambda g, g) \leq \frac{1 + L^2}{1 - L},$$

i.e., inequality (13.12) holds true for every $x \in E_1$.

If we replace x with $2^n x$ and y with $2^n y$ in (b), divide by 2^n both sides of the resulting inequality and let n go to infinity, then it follows from (13.10) and (g) that I is an isometry.

Finally, it remains to prove the uniqueness of I. Let I' be another isometry satisfying (13.12) and (h) in place of I. If we substitute g, I, and 0 for x, x^*, and n_0 in Theorem 2.43, respectively, then (13.12) implies that

$$d(\Lambda^{n_0} g, I') = d(g, I') \leq \frac{1 + L^2}{1 - L} < \infty.$$

Hence, $I' \in X^*$ (see Theorem 2.43 for the definition of X^*). By (h), we further have $I'(x) = (1/2)I'(2x) = (\Lambda I')(x)$ for all $x \in E_1$, i.e., I' is a "fixed point" of Λ. Therefore, Theorem 2.43 (ii) implies that $I = I'$. $\square$

We notice that the parallelogram law is specifically true for norms derived from inner products. It is also known that every isometry from a real normed space into a real Hilbert space is affine (see [15]). Since the isometry I satisfies $I(0) = 0$ (see the proof of Theorem 13.15), the following corollary is a consequence of Theorem 13.15.

Corollary 13.16. *Let E_1 and E_2 be a real normed space and a real Hilbert space, respectively. Given any $0 \leq p < 1$, choose a constant ε with $1 < \varepsilon \leq 2^{1-p}$ and define a function $\delta : E_1 \to [0, \infty)$ by*

$$\delta(x) = \varepsilon \|x\|^p$$

for all $x \in E_1$. If a function $f : E_1 \to E_2$ satisfies the inequality (13.11) for all $x, y \in E_1$, then there exists a unique linear isometry $I : E_1 \to E_2$ such that

$$\|f(x) - f(0) - I(x)\| \leq \frac{1 + \varepsilon}{\varepsilon - \sqrt{\varepsilon}} \psi(x)$$

for all $x \in E_1$, where

$$\psi(x) = \begin{cases} \sqrt{\varepsilon/(\varepsilon + 2)}\big(\|x\| + \varepsilon\|x\|^p\big) & (\text{for } \|x\| < 1), \\ \varepsilon\|x\|^{(1+p)/2} & (\text{for } \|x\| \geq 1). \end{cases}$$

We simply need to put $L = \varepsilon^{-1/2}$ in Theorem 13.15 to prove Corollary 13.16. Similarly, Jung proved the following theorem and corollary.

Theorem 13.17. *Let E_1 be a normed space over $\mathbb{K}$ and let E_2 be a Banach space over $\mathbb{K}$ in which the parallelogram law holds true. Assume that $\delta : E_1 \to [0, \infty)$ is an even function such that there exists a constant L, $0 < L < 1$, with*

$$\begin{cases} 2\delta(x) \leq L^2\delta(2x) & (\text{for all } x \in E_1), \\ \|x\| \geq \delta(x) & (\text{for all } x \in E_1 \text{ with } \|x\| < 1), \\ \|x\| \leq L^2\delta(x) & (\text{for all } x \in E_1 \text{ with } \|x\| \geq 1), \end{cases}$$

and

$$\lim_{n \to \infty} 2^n \delta(2^{-n}x) = 0$$

for all $x \in E_1$. If a function $f : E_1 \to E_2$ satisfies the inequality (13.11) for all $x, y \in E_1$, then there exists an isometry $I : E_1 \to E_2$ such that the inequality (13.12) holds true for all $x \in E_1$, where

$$\psi(x) = \begin{cases} (1/L)\sqrt{\|x\|\delta(x)} & (\text{for } \|x\| < 1), \\ (1/\sqrt{1 + 2L^2})\big(\|x\| + \delta(x)\big) & (\text{for } \|x\| \geq 1). \end{cases}$$

The I is the unique isometry satisfying (13.12) and $I(2x) = 2I(x)$ for any $x \in E_1$.

Corollary 13.18. *Let E_1 and E_2 be a real normed space and a real Hilbert space, respectively. For a given $p > 1$, choose constants ε_1 and ε_2 with $0 < \varepsilon_1 \leq 1 < \varepsilon_2 \leq 2^{p-1}$ and define a function $\delta : E_1 \to [0, \infty)$ by*

$$\delta(x) = \begin{cases} \varepsilon_1 \|x\|^p & (for \ \|x\| < 1), \\ \varepsilon_2 \|x\|^p & (for \ \|x\| \geq 1) \end{cases}$$

for all $x \in E_1$. If a function $f : E_1 \to E_2$ satisfies the inequality (13.11) for all $x, y \in E_1$, then there exists a unique linear isometry $I : E_1 \to E_2$ such that

$$\|f(x) - f(0) - I(x)\| \leq \frac{1 + \varepsilon_2}{\varepsilon_2 - \sqrt{\varepsilon_2}} \psi(x)$$

for all $x \in E_1$, where

$$\psi(x) = \begin{cases} \sqrt{\varepsilon_1 \varepsilon_2} \, \|x\|^{(1+p)/2} & (for \ \|x\| < 1), \\ \sqrt{\varepsilon_2/(\varepsilon_2 + 2)} \big(\|x\| + \varepsilon_2 \|x\|^p \big) & (for \ \|x\| \geq 1). \end{cases}$$

13.4 Wigner Equation

Let E_1 and E_2 be real or complex Hilbert spaces (we denote the scalar field by $\mathbb{K}$) with the inner products and the associated norms denoted by $\langle \cdot, \cdot \rangle$ and $\| \cdot \|$, respectively.

A function $f : E_1 \to E_2$ is called *inner product preserving* if it is a solution of the *orthogonality equation*

$$\langle f(x), f(y) \rangle = \langle x, y \rangle \tag{13.13}$$

for all $x, y \in E_1$. We can present that f satisfies (13.13) if and only if it is a linear isometry. Similarly, $f : E_1 \to E_2$ is a solution of the functional equation

$$\langle f(x), f(y) \rangle = \langle y, x \rangle, \tag{13.14}$$

for every $x, y \in E_1$, if and only if f is a *conjugate-linear isometry*, i.e., f is an isometry and $f(\lambda x + \mu y) = \overline{\lambda} f(x) + \overline{\mu} f(y)$ for any $x, y \in E_1$ and $\lambda, \mu \in \mathbb{K}$.

Functions $f, g : E_1 \to E_2$ are called *phase-equivalent* if and only if there exists a function $\sigma : E_1 \to S$ such that $g(x) = \sigma(x) f(x)$ for each $x \in E_1$, where we set

$$S = \big\{ z \in \mathbb{K} \mid |z| = 1 \big\}.$$

A functional equation

$$|\langle f(x), f(y) \rangle| = |\langle x, y \rangle|, \tag{13.15}$$

for any $x, y \in E_1$, is called the *Wigner equation* (or the *generalized orthogonality equation*) because it first appeared in the book by E. P. Wigner [358]. The following theorem plays a crucial role in Wigner's time reversal operator theory.

Theorem 13.19 (Wigner). *If a function $f : E_1 \to E_2$ satisfies (13.15), then f is phase-equivalent to a linear or a conjugate-linear isometry.*

The "mathematical" proof of this theorem can be found in [317,329]. The Hyers–Ulam stability of the Wigner equation has been proved, in a more general setting, in [65]. For the real case, a more elementary proof was given in [64]:

Let E be a real Hilbert space with dim $E \geq 2$. If a function $f : E \to E$ satisfies the inequality

$$\left| |\langle f(x), f(y) \rangle| - |\langle x, y \rangle| \right| \leq \delta \tag{13.16}$$

for all $x, y \in E$ and some $\delta \geq 0$, then there exists a solution function $I : E \to E$ of the Wigner equation (13.15) such that

$$\| f(x) - I(x) \| \leq \sqrt{\delta}$$

for any $x \in E$.

Let c and d be given constants, where $c > 0$ ($c \neq 1$) and $d \geq 0$. From now on, we denote by D a subset of E_1 defined by

$$D = \begin{cases} \{x \in E_1 \mid \|x\| \geq d\} & \text{(for } 0 < c < 1\text{),} \\ \{x \in E_1 \mid \|x\| \leq d\} & \text{(for } c > 1\text{)} \end{cases}$$

if there is no specification. We will exclude the trivial case $D = \{0\}$. We consider a function $\varphi : E_1^2 \to [0, \infty)$ satisfying the property

$$\lim_{m+n \to \infty} c^{m+n} \varphi(c^{-m} x, c^{-n} y) = 0 \tag{13.17}$$

for all $x, y \in D$.

In order to define a class of approximate solutions of the Wigner equation, we introduce the functional inequality

$$\left| |\langle f(x), f(y) \rangle| - |\langle x, y \rangle| \right| \leq \varphi(x, y) \tag{13.18}$$

for all $x, y \in D$. Let us define a functional sequence $\{f_n\}$ by

$$f_n(x) = c^n f(c^{-n} x) \tag{13.19}$$

for any $x \in E_1$. Then, it easily follows from (13.18) that

$$\|x\|^2 - c^{m+n} \varphi(c^{-m} x, c^{-n} x)$$

$$\leq |\langle f_m(x), f_n(x) \rangle|$$

$$\leq \|x\|^2 + c^{m+n} \varphi(c^{-m} x, c^{-n} x) \tag{13.20}$$

for any $x \in D$ and $m, n \in \mathbb{N}_0$. In particular, if we set $m = n$ in (13.20), then we have

$$\|x\|^2 - c^{2n} \varphi(c^{-n} x, c^{-n} x) \leq \|f_n(x)\|^2 \leq \|x\|^2 + c^{2n} \varphi(c^{-n} x, c^{-n} x). \tag{13.21}$$

With these considerations, J. Chmieliński and S.-M. Jung [66] proved the Hyers–Ulam–Rassias stability of the Wigner equation on a restricted domain.

Theorem 13.20 (Chmieliński and Jung). *Let E_1 and E_2 be real (or complex) Hilbert spaces. If a function $f : E_1 \to E_2$ satisfies (13.18) with the function $\varphi : E_1^2 \to [0, \infty)$ satisfying the property (13.17), then there exists a solution function $I : E_1 \to E_2$ of the Wigner equation (13.15) such that*

$$\|f(x) - I(x)\| \leq \sqrt{\varphi(x, x)}$$

for all $x \in D$. The function I is unique up to a phase-equivalent function.

Proof. The right-hand-side inequality in (13.21) and property (13.17) imply that the sequence $\{f_n(x)\}$ is bounded for any $x \in D$. Thus, there exists a subsequence $\{f_{l_n(x)}(x)\}$ of $\{f_n(x)\}$ weakly convergent in E_2 (cf. [241, Theorem 2]). Next, we choose $\sigma_{l_n(x)} \in \mathcal{S}$ such that

$$\sigma_{l_n(x)} \langle f_{l_n(x)}(x), f(x) \rangle \geq 0.$$

As $\mathcal{S}$ is compact in $\mathbb{K}$, we can find a convergent subsequence $\{\sigma_{k_n(x)}\}$. We will now write $\{\sigma_n(x)\}$ instead of $\{\sigma_{k_n(x)}\}$. Then, we have

$$\langle \sigma_n(x) f_{k_n(x)}(x), f(x) \rangle \geq 0$$

which together with (13.20) gives us

$$\langle \sigma_n(x) f_{k_n(x)}(x), f(x) \rangle$$

$$= |\langle \sigma_n(x) f_{k_n(x)}(x), f(x) \rangle|$$

$$= |\langle f_{k_n(x)}(x), f(x) \rangle|$$

$$\geq \|x\|^2 - c^{k_n(x)} \varphi(c^{-k_n(x)} x, x). \tag{a}$$

Since the sequence $\{\sigma_n(x) f_{k_n(x)}(x)\}$ is weakly convergent, we can define

$$f_*(x) = \underset{n \to \infty}{\text{w-lim}}\, \sigma_n(x) f_{k_n(x)}(x)$$

for each $x \in D$. It follows from (13.18) that

$$\left| |\langle \sigma_m(x) f_{k_m(x)}(x),\, \sigma_n(y) f_{k_n(y)}(y)\rangle| - |\langle x, y\rangle| \right|$$
$$\leq c^{k_m(x)+k_n(y)} \varphi\left(c^{-k_m(x)}x,\, c^{-k_n(y)}y\right) \tag{b}$$

for any $x, y \in D$ and $m, n \in \mathbb{N}_0$. For a fixed $y \in D$ and $n \in \mathbb{N}_0$, we define

$$\alpha(z) = \langle z,\, \sigma_n(y) f_{k_n(y)}(y)\rangle$$

for every $z \in E_2$. Then, we have $\alpha \in E_2^*$ and we can write (b) as

$$\left| |\alpha(\sigma_m(x) f_{k_m(x)}(x))| - |\langle x, y\rangle| \right| \leq c^{k_m(x)+k_n(y)} \varphi\left(c^{-k_m(x)}x,\, c^{-k_n(y)}y\right).$$

Letting $m \to \infty$, we get

$$\left| |\alpha(f_*(x))| - |\langle x, y\rangle| \right| \leq 0$$

and then

$$\left| \langle \sigma_n(y) f_{k_n(y)}(y),\, f_*(x)\rangle \right| = |\langle x, y\rangle| \tag{c}$$

for all $x \in D$.

For a fixed $x \in D$, we define a functional β from the dual space E_2^* by

$$\beta(z) = \langle z,\, f_*(x)\rangle$$

for any $z \in E_2$, and then (c) can be expressed as

$$\left| \beta(\sigma_n(y) f_{k_n(y)}(y)) \right| = |\langle x, y\rangle|.$$

Letting $n \to \infty$, we get from the definition of f_* that

$$\left| \beta(f_*(y)) \right| = |\langle x, y\rangle|,$$

i.e.,

$$\left| \langle f_*(x),\, f_*(y)\rangle \right| = |\langle x, y\rangle| \tag{d}$$

for every $x, y \in D$.

From (13.21) and (a), we obtain

$$\|\sigma_n(x) f_{k_n(x)}(x) - f(x)\|^2$$
$$= \|\sigma_n(x) f_{k_n(x)}(x)\|^2 + \|f(x)\|^2 - 2\Re\left(\langle \sigma_n(x) f_{k_n(x)}(x),\, f(x)\rangle\right)$$
$$\le \|x\|^2 + c^{2k_n(x)} \varphi\left(c^{-k_n(x)}x, c^{-k_n(x)}x\right)$$
$$+ \|x\|^2 + \varphi(x, x) - 2\|x\|^2 + 2c^{k_n(x)} \varphi\left(c^{-k_n(x)}x, x\right)$$

for all $x \in D$. Let us fix $x \in D$. If $f_*(x) \neq f(x)$, then we define

$$\gamma(z) = \left\langle z,\, \frac{f_*(x) - f(x)}{\|f_*(x) - f(x)\|} \right\rangle$$

for all $z \in E_2$. Obviously, $\gamma \in E_2^*$ and $\|\gamma\| = 1$. Thus,

$$\left|\gamma\left(\sigma_n(x) f_{k_n(x)}(x) - f(x)\right)\right|$$
$$\le \|\sigma_n(x) f_{k_n(x)}(x) - f(x)\|$$
$$\le \sqrt{c^{2k_n(x)} \varphi\left(c^{-k_n(x)}x, c^{-k_n(x)}x\right) + \varphi(x, x) + 2c^{k_n(x)} \varphi\left(c^{-k_n(x)}x, x\right)}\,.$$

Letting $n \to \infty$ and using property (13.17) yield $|\gamma(f_*(x) - f(x))| \le \sqrt{\varphi(x, x)}$. Hence,

$$\|f_*(x) - f(x)\| \le \sqrt{\varphi(x, x)} \tag{e}$$

for any $x \in D$. (Obviously, the last inequality also holds true if $f_*(x) = f(x)$.)

Now, we have to extend f_* from D to the whole E_1. Putting

$$n(x) = \min\left\{n \in \mathbb{N}_0 \mid c^{-n}x \in D\right\}$$

for each $x \in E_1 \setminus \{0\}$, we define

$$I(x) = \begin{cases} c^{n(x)} f_*(c^{-n(x)}x) & (\text{for } x \in E_1 \setminus \{0\}), \\ 0 & (\text{for } x = 0). \end{cases} \tag{f}$$

If $x = 0$ or $y = 0$, then $\langle I(x), I(y)\rangle = \langle x, y\rangle = 0$. For $x, y \in E_1 \setminus \{0\}$, we have

$$|\langle I(x), I(y)\rangle| = \left|\langle c^{n(x)} f_*(c^{-n(x)}x),\, c^{n(y)} f_*(c^{-n(y)}y)\rangle\right|$$
$$= c^{n(x)+n(y)} \left|\langle f_*(c^{-n(x)}x),\, f_*(c^{-n(y)}y)\rangle\right|$$
$$= c^{n(x)+n(y)} \left|\langle c^{-n(x)}x,\, c^{-n(y)}y\rangle\right|$$
$$= |\langle x, y\rangle|,$$

i.e., I satisfies the Wigner equation (13.15). Moreover, if $x \in D$, then $n(x) = 0$ and hence $I(x) = f_*(x)$. It then follows from (e) that

$$\|I(x) - f(x)\| \leq \sqrt{\varphi(x, x)}$$

for each $x \in D$.

It now suffices to prove the uniqueness of I. Suppose that $I_1, I_2 : E_1 \to E_2$ satisfy the assertion of the theorem, i.e., that both are solutions of (13.15) such that $\|I_i(x) - f(x)\| \leq \sqrt{\varphi(x, x)}$ for all $x \in D$ and $i \in \{1, 2\}$. Then

$$\|I_1(x) - I_2(x)\| \leq 2\sqrt{\varphi(x, x)}$$

for every $x \in D$. In view of Theorem 13.19, there exist two linear or conjugate-linear isometries $T_1, T_2 : E_1 \to E_2$ and functions $\sigma_1, \sigma_2 : E_1 \to S$ such that

$$I_1(x) = \sigma_1(x)T_1(x) \quad \text{and} \quad I_2(x) = \sigma_2(x)T_2(x)$$

for any $x \in E_1$. Let us fix $x \in E_1$. As before, let $n(x)$ be the smallest nonnegative integer such that $c^{-n(x)}x \in D$. For $n \geq n(x)$, we have $c^{-n}x \in D$ and hence

$$
\begin{aligned}
4\varphi(c^{-n}x, c^{-n}x) &\geq \|I_1(c^{-n}x) - I_2(c^{-n}x)\|^2 \\
&= \|\sigma_1(c^{-n}x)T_1(c^{-n}x) - \sigma_2(c^{-n}x)T_2(c^{-n}x)\|^2 \\
&= \|T_1(c^{-n}x)\|^2 + \|T_2(c^{-n}x)\|^2 \\
&\quad - 2\Re\big(\langle \sigma_1(c^{-n}x)T_1(c^{-n}x),\ \sigma_2(c^{-n}x)T_2(c^{-n}x)\rangle\big) \\
&\geq 2\|c^{-n}x\|^2 - 2\big|\langle T_1(c^{-n}x), T_2(c^{-n}x)\rangle\big| \\
&= 2c^{-2n}\|x\|^2 - 2c^{-2n}\big|\langle T_1(x), T_2(x)\rangle\big|.
\end{aligned}
$$

Therefore, we have

$$\|x\|^2 - 2c^{2n}\varphi(c^{-n}x, c^{-n}x) \leq |\langle T_1(x), T_2(x)\rangle| \qquad (g)$$

for all $x \in E_1$ and $n \geq n(x)$.

Assume that $T_1(x)$ and $T_2(x)$ were linearly independent for some $x \in E_1 \setminus \{0\}$. Then, we would have

$$|\langle T_1(x), T_2(x)\rangle| < \|T_1(x)\| \cdot \|T_2(x)\| = \|x\|^2,$$

i.e., there would exist $\xi \in [0, 1)$ such that $|\langle T_1(x), T_2(x)\rangle| = \xi \|x\|^2$. This together with (g) would give

$$\|x\|^2 - 2c^{2n}\varphi(c^{-n}x, c^{-n}x) \leq \xi \|x\|^2$$

and consequently

$$(1 - \xi)\|x\|^2 \le 2c^{2n}\varphi(c^{-n}x, c^{-n}x)$$

for any $n \ge n(x)$. Due to (13.17), the right-hand side of the above inequality tends to zero as $n \to \infty$, a contradiction. Thus, the vectors $T_1(x)$ and $T_2(x)$ are linearly dependent for any $x \in E_1$.

Assume now that $T_1(x) = \sigma T_2(x)$ and $|\sigma| \ne 1$ for some $x \in E_1 \setminus \{0\}$. Without loss of generality, we can assume that $|\sigma| < 1$. Once again, it would follow from (g) that

$$
\begin{aligned}
\|x\|^2 - 2c^{2n}\varphi(c^{-n}x, c^{-n}x) &\le |\langle T_1(x), T_2(x)\rangle| \\
&= |\langle \sigma T_2(x), T_2(x)\rangle| \\
&= |\sigma| \cdot |\langle T_2(x), T_2(x)\rangle| \\
&= |\sigma| \cdot \|x\|^2
\end{aligned}
$$

for all $n \ge n(x)$. Therefore,

$$\left(1 - |\sigma|\right)\|x\|^2 \le 2c^{2n}\varphi(c^{-n}x, c^{-n}x)$$

for any $n \ge n(x)$. Letting $n \to \infty$, we obtain

$$\left(1 - |\sigma|\right)\|x\|^2 \le 0,$$

a contradiction. Thus (defining $\sigma(0) = 1$) we have proved that for any $x \in E_1$ there exists $\sigma(x) \in \mathbb{K}$ such that $|\sigma(x)| = 1$ and $T_1(x) = \sigma(x)T_2(x)$. This implies

$$I_1(x) = \sigma_1(x)T_1(x) = \sigma_1(x)\sigma(x)T_2(x) = \frac{\sigma_1(x)\sigma(x)}{\sigma_2(x)}I_2(x)$$

for any $x \in E_1$, which means that I_1 and I_2 are phase-equivalent and ends the proof.   □

Let $\varphi(x, y) = \varepsilon\|x\|^p\|y\|^p$ with $\varepsilon \ge 0$. Then

$$c^{m+n}\varphi(c^{-m}x, c^{-n}y) = \varepsilon c^{(m+n)(1-p)}\|x\|^p\|y\|^p$$

for each $m, n \in \mathbb{N}$. For either $0 < c < 1$ and $p < 1$ or $c > 1$ and $p > 1$, the right-hand side of the above equality tends to zero as $m + n \to \infty$; i.e., φ satisfies (13.17). Thus, by applying Theorem 13.20, we can easily prove the following corollary.

Corollary 13.21. *Assume either that $p < 1$ and $D = \{x \in E_1 \mid \|x\| \ge d\}$ $(d \ge 0)$ or that $p > 1$ and $D = \{x \in E_1 \mid \|x\| \le d\}$ $(d > 0)$. If a function $f : E_1 \to E_2$ satisfies the inequality*

$$\left|\,|\langle f(x), f(y)\rangle| - |\langle x, y\rangle|\,\right| \le \varepsilon\|x\|^p\|y\|^p$$

for all $x, y \in D$, then there exists a solution function $I : E_1 \to E_2$ of the Wigner equation (13.15) *such that*

$$\|f(x) - I(x)\| \leq \sqrt{\varepsilon}\,\|x\|^p$$

for each $x \in D$. The function I is unique up to a phase-equivalent function.

If we put $p = 0$ in Corollary 13.21, then we obtain the Hyers–Ulam stability of the Wigner equation.

Corollary 13.22. *If a function $f : E_1 \to E_2$ satisfies the inequality* (13.16) *for all $x, y \in D = \{x \in E_1 \mid \|x\| \geq d\}$ ($d \geq 0$), then there exists a solution function $I : E_1 \to E_2$ of the Wigner equation* (13.15) *such that*

$$\|f(x) - I(x)\| \leq \sqrt{\delta}$$

for any $x \in D$. The function I is unique up to a phase-equivalent function.

In the following theorem, we prove the Hyers–Ulam–Rassias stability of the orthogonality equation.

Theorem 13.23. *Let E_1 and E_2 be real (or complex) Hilbert spaces. If a function $f : E_1 \to E_2$ satisfies the inequality*

$$|\langle f(x), f(y)\rangle - \langle x, y\rangle| \leq \varphi(x, y) \tag{13.22}$$

for any $x, y \in D$ and $\varphi : E_1^2 \to [0, \infty)$ satisfies (13.17), *then there exists a unique linear isometry $I : E_1 \to E_2$ such that*

$$\|f(x) - I(x)\| \leq \sqrt{\varphi(x, x)} \tag{13.23}$$

for all $x \in D$.

Proof. We define $f_n(x) = c^n f(c^{-n} x)$ and observe that

$$\|x\|^2 - c^{m+n}\varphi(c^{-m}x, c^{-n}x) \leq \Re\big(\langle f_m(x), f_n(x)\rangle\big)$$
$$\leq \|x\|^2 + c^{m+n}\varphi(c^{-m}x, c^{-n}x)$$

for all $x \in D$. Using these inequalities, we present that

$$\|f_m(x) - f_n(x)\|$$
$$\leq \sqrt{c^{2m}\varphi(c^{-m}x, c^{-m}x) + c^{2n}\varphi(c^{-n}x, c^{-n}x) + 2c^{m+n}\varphi(c^{-m}x, c^{-n}x)}$$

for all $x \in D$. The right-hand side tends to zero as $m, n \to \infty$. Since E_2 is complete, we may define

$$f_*(x) = \lim_{n \to \infty} f_n(x)$$

for any $x \in D$. Now, using (f) in the proof of Theorem 13.20, we extend f_* to I defined on E_1 and prove that I satisfies (13.13) and (13.23).

If I' also satisfies the assertion of the theorem, then (due to linearity of I' and (13.23))

$$\|I'(x) - f_n(x)\| \leq c^n \sqrt{\varphi(c^{-n}x, c^{-n}x)} \to 0 \quad \text{as } n \to \infty$$

for all $x \in D$. Therefore,

$$I'(x) = \lim_{n \to \infty} f_n(x) = f_*(x)$$

for each $x \in D$, which implies that $I = I'$ on E_1. $\square$

On the other hand, S.-M. Jung and P. K. Sahoo [207] proved the superstability of the Wigner equation on restricted domains under some strong conditions imposed on the control function:

Given an integer $n \geq 2$ and positive real numbers $c \neq 1$ and d, we define

$$D_n = \begin{cases} \{x \in \mathbb{R}^n \mid \|x\| \geq d\} & \text{(for } 0 < c < 1), \\ \{x \in \mathbb{R}^n \mid \|x\| < d\} & \text{(for } c > 1), \end{cases}$$

where $\| \cdot \|$ denotes the usual norm on $\mathbb{R}^n$ defined by

$$\|x\| = \sqrt{\langle x, x \rangle}$$

with the usual inner product $\langle \cdot, \cdot \rangle$ defined by

$$\langle x, y \rangle = x_1 y_1 + x_2 y_2 + \cdots + x_n y_n$$

for all points $x = (x_1, \ldots, x_n)$ and $y = (y_1, \ldots, y_n)$ of $\mathbb{R}^n$.

Suppose $\varphi : \mathbb{R}^n \times \mathbb{R}^n \to [0, \infty)$ is a symmetric function which satisfies the following conditions:

(i) There exists a function $\phi : [0, \infty)^2 \to [0, \infty)$ such that $\varphi(x, y) = \phi(\|x\|, \|y\|)$ for all $x, y \in \mathbb{R}^n$;

(ii) For all $x, y \in \mathbb{R}^n$

$$(1/|\lambda|)\varphi(\lambda x, y) = O\left(-\frac{\ln c}{\ln |\lambda|}\right)$$

either as $|\lambda| \to \infty$ (for $0 < c < 1$) or as $|\lambda| \to 0$ (for $c > 1$);

(iii) If both $|\lambda|$ and $|\mu|$ are different from 1, then for all $x, y \in \mathbb{R}^n$,

$$(1/|\lambda\mu|)\varphi(\lambda x, \mu y) = O\left(\left|\frac{\ln c}{\ln|\lambda|}\frac{\ln c}{\ln|\mu|}\right|\right)$$

either as $|\lambda\mu| \to \infty$ (for $0 < c < 1$) or as $|\lambda\mu| \to 0$ (for $c > 1$).

Under these notations and conditions, Jung and Sahoo [207] proved the superstability of the Wigner equation on restricted domains.

Theorem 13.24. *If a function $f : D_n \to \mathbb{R}^n$ satisfies the inequality (13.18) for all $x, y \in D_n$, then f satisfies the Wigner equation (13.15) for all $x, y \in D_n$.*

They also proved the superstability of the orthogonality equation on restricted domains.

Theorem 13.25. *If a function $f : D_n \to \mathbb{R}^n$ satisfies the inequality (13.22) for all $x, y \in D_n$, then f satisfies the orthogonality equation (13.13) for all $x, y \in D_n$.*

Moreover, S.-M. Jung [177, 183] and Th. M. Rassias investigated the Hyers–Ulam stability of the orthogonality equation (13.13) on a closed ball in $\mathbb{R}^3$. Here, we introduce a result of Rassias:

If a function $f : D \to D$, defined on a closed ball $D \subset \mathbb{R}^3$ of radius $d > 0$ and with center at the origin, satisfies $f(0) = 0$ and

$$|\langle f(x), f(y)\rangle - \langle x, y\rangle| \le \delta$$

for some $0 \le \delta < \min\{1/4, d^2/17\}$ and for all $x, y \in D$, then there exists an isometry $I : D \to D$ such that

$$|f(x) - I(x)| < \begin{cases} 14\sqrt{\delta} & \text{(for } d < \sqrt{17}/2), \\ (5d + 3)\sqrt{\delta} & \text{(for } d \ge \sqrt{17}/2) \end{cases}$$

for any $x \in D$.

Chapter 14
Miscellaneous

One of the simplest functional equations is the associativity equation. This functional equation represents the famous associativity axiom $x \cdot (y \cdot z) = (x \cdot y) \cdot z$. Section 14.1 deals with the superstability of the associativity equation. In Section 14.2, an important functional equation defining multiplicative derivations in algebras will be introduced, and the Hyers–Ulam stability of the equation for functions on $(0, 1]$ will be proved. The gamma function Γ is very useful to develop other functions which have physical applications. In Section 14.3, the Hyers–Ulam–Rassias stability of the gamma functional equation and a generalized beta functional equation will be proved. The Hyers–Ulam stability of the Fibonacci functional equation will be proved in the last section.

14.1 Associativity Equation

The *associativity axiom* $x \cdot (y \cdot z) = (x \cdot y) \cdot z$ plays an important role in definitions of algebraic structures. The functional equation

$$F\big(x, F(y,z)\big) = F\big(F(x, y), z\big)$$

is called the *associativity equation*.

The aim of this section is to solve the functional inequality

$$\big|F\big(kx, kF(y,z)\big) - F\big(kF(x, y), kz\big)\big| \le \delta, \tag{14.1}$$

where δ is a given positive constant and the variables x, y, z, and k run over $[0, \infty)$.

C. Alsina [4] presented that the functional inequality (14.1) is superstable.

Theorem 14.1 (Alsina). *Let a function* $F : [0, \infty)^2 \to [0, \infty)$ *be given with* $F(0, x) = x$ *for all* $x \ge 0$. *If* F *satisfies the inequality* (14.1) *for all* $x, y, z, k \ge 0$, *then* F *satisfies the equation*

$$F\big(kx, kF(y,z)\big) = F\big(kF(x, y), kz\big) \tag{14.2}$$

for all $x, y, z, k \ge 0$.

S.-M. Jung, *Hyers–Ulam–Rassias Stability of Functional Equations in Nonlinear Analysis*, Springer Optimization and Its Applications 48, DOI 10.1007/978-1-4419-9637-4_14, © Springer Science+Business Media, LLC 2011

Proof. Let $u > 1$ be fixed. Putting $k = u^n$ $(n \in \mathbb{N})$, $x = 0$ in (14.1), and dividing the resulting inequality by u^n yield

$$|F(y,z) - u^{-n} F(u^n y, u^n z)| \le u^{-n} \delta.$$

Hence, we get

$$F(y,z) = \lim_{n \to \infty} u^{-n} F(u^n y, u^n z) \qquad (a)$$

for any $y, z \ge 0$ and any $u > 1$. It follows from (a) that

$$F(y,z) = \lim_{n \to \infty} u^n F(u^{-n} y, u^{-n} z) \qquad (b)$$

for $y, z \ge 0$ and $u < 1$.

Using (14.1) and (a) yields

$$\big| F\big(x, F(y,z)\big) - F\big(F(x,y), z\big) \big|$$
$$= \lim_{n \to \infty} 2^{-n} \big| F\big(2^n x, 2^n F(y,z)\big) - F\big(2^n F(x,y), 2^n z\big) \big|$$
$$\le \lim_{n \to \infty} 2^{-n} \delta$$
$$= 0$$

for all $x, y, z \ge 0$, where the associativity of F follows.

Using this and (a) yields

$$\big| F\big(kx, kF(y,z)\big) - F\big(kF(x,y), kz\big) \big|$$
$$= \lim_{n \to \infty} \big| k^{-n} F\big(k^{n+1} x, k^{n+1} F(y,z)\big) - k^{-n} F\big(k^{n+1} F(x,y), k^{n+1} z\big) \big|$$
$$= k \cdot \bigg| \lim_{n \to \infty} k^{-(n+1)} F\big(k^{n+1} x, k^{n+1} F(y,z)\big)$$
$$\qquad - \lim_{n \to \infty} k^{-(n+1)} F\big(k^{n+1} F(x,y), k^{n+1} z\big) \bigg|$$
$$= k \cdot \big| F\big(x, F(y,z)\big) - F\big(F(x,y), z\big) \big|$$
$$= 0$$

for every $x, y, z \ge 0$ and $k > 1$. Hence, (14.2) holds true whenever $k > 1$.

Similarly, using (b) and the associativity of F we show that (14.2) holds true for $0 \le k < 1$. $\qquad \square$

Moreover, Alsina [4] showed that if we fix $k = 1$, the superstability fails to hold true for the associativity equation. It still remains open whether the associativity equation is stable.

14.2 Equation of Multiplicative Derivation

Let X be an infinite-dimensional real (or complex) Banach space. By $B(X)$ we mean the algebra of all *bounded linear operators* on X. We denote by $F(X)$ the subalgebra of all bounded finite rank operators. A subalgebra $\mathcal{A}$ of $B(X)$ is said to be *standard* if $\mathcal{A}$ contains $F(X)$. Let $\mathcal{A}$ be a standard operator algebra on X. A function $f : \mathcal{A} \to B(X)$ is called a *multiplicative derivation* if $f(AB) = Af(B) + f(A)B$ for all $A, B \in \mathcal{A}$.

We introduce an important functional equation which defines multiplicative derivations in algebras:

$$f(xy) = xf(y) + f(x)y. \tag{14.3}$$

P. Šemrl [327] obtained the first result concerning the superstability of this equation for functions between operator algebras. Here, we introduce a Hyers–Ulam stability result presented by J. Tabor [349] as an answer to a question by G. Maksa [239].

Theorem 14.2 (Tabor). *Let E be a Banach space, and let a function $f : (0, 1] \to E$ satisfy the inequality*

$$\| f(xy) - xf(y) - f(x)y \| \leq \delta \tag{14.4}$$

for some $\delta > 0$ and for all $x, y \in (0, 1]$. Then there exists a solution $D : (0, 1] \to E$ of the equation (14.3) such that

$$\| f(x) - D(x) \| \leq (4e)\delta$$

for every $x \in (0, 1]$.

Proof. Let us define a function $g : (0, 1] \to E$ by

$$g(x) = f(x)/x$$

for each $x \in (0, 1]$. Then g satisfies the inequality

$$\| g(xy) - g(x) - g(y) \| \leq \delta/(xy)$$

for any $x, y \in (0, 1]$. Define a function $G : [0, \infty) \to E$ by $G(v) = g(e^{-v})$. Then

$$\| G(u + v) - G(u) - G(v) \| \leq \delta e^{u+v} \tag{a}$$

for every $u, v \geq 0$, which implies that

$$\| G(u + v) - G(u) - G(v) \| \leq \delta e^{c}$$

for all $u, v \in [0, c)$ with $u + v < c$, where $c > 1$ is an arbitrarily given constant.

According to Lemma 2.28, there exists an additive function $A : \mathbb{R} \to E$ such that

$$\|G(u) - A(u)\| \le 3e^c \delta$$

for every $u \in [0, c)$. If we let $c \to 1$ in the last inequality, we then get

$$\|G(u) - A(u)\| \le 3e\delta \qquad (b)$$

for every $u \in [0, 1]$. Moreover, it follows from (a) that

$$\|G(u + 1) - G(u) - G(1)\| \le \delta e^{u+1},$$
$$\|G(u + 2) - G(u + 1) - G(1)\| \le \delta e^{u+2},$$
$$\vdots$$
$$\|G(u + k) - G(u + k - 1) - G(1)\| \le \delta e^{u+k}$$

for each $u \in [0, 1]$ and $k \in \mathbb{N}$. Summing up these inequalities, we obtain

$$\|G(u + k) - G(u) - kG(1)\| \le \delta e \cdot e^{u+k} \qquad (c)$$

for $u \in [0, 1]$ and $k \in \mathbb{N}$.

Let $v \ge 0$ and let $k \in \mathbb{N}_0$ be given with $v - k \in [0, 1]$. Then, by (b) and (c), we have

$$\begin{aligned}
\|G(v) - A(v)\| &\le \|G(v) - G(v - k) - kG(1)\| \\
&\quad + \|G(v - k) - A(v - k)\| + \| -A(k) + kG(1)\| \\
&\le \delta e \cdot e^v + 3\delta e + k\|A(1) - G(1)\| \\
&\le \delta e \cdot e^v + 3\delta e + 3\delta ev \\
&\le \delta e(e^v + 3(1 + v)) \\
&\le 4\delta e \cdot e^v.
\end{aligned}$$

This and the definition of G imply that

$$\|g(x) - A(-\ln x)\| \le 4\delta e \cdot e^{-\ln x} = 4\delta e / x$$

for $x \in (0, 1]$, i.e.,

$$\|f(x)/x - A(-\ln x)\| \le 4\delta e / x \qquad (d)$$

for all $x \in (0, 1]$. If we put $D(x) = xA(-\ln x)$ for $x \in (0, 1]$, we can easily check that D is a solution of (14.3). This and (d) yield that

$$\|f(x) - D(x)\| = \|f(x) - xA(-\ln x)\| \le (4e)\delta$$

for each $x \in (0, 1]$. $\square$

Similarly, Z. Páles [260] proved that the functional equation (14.3) for real-valued functions on $[1, \infty)$ is stable in the sense of Hyers and Ulam. Moreover, he remarked that the equation (14.3) has a stronger property, i.e., a superstability on $[1, \infty)$.

14.3 Gamma Functional Equation

The *gamma function*

$$\Gamma(x) = \int_0^\infty e^{-t} t^{x-1} dt,$$

for $x > 0$, appears occasionally in the physical applications. The gamma function is especially very useful to develop other functions which have physical applications. It is well-known that the gamma function satisfies the functional equations

$$f(x + 1) = x f(x) \tag{14.5}$$

for all $x > 0$,

$$\prod_{k=0}^{p-1} f\big((x + k)/p\big) = (2\pi)^{(1/2)(p-1)} p^{1/2-x} f(x)$$

for any $x > 0$ and $p \in \mathbb{N}$, and

$$f(x) f(1 - x) = \pi (\sin \pi x)^{-1}$$

for each $x \in (0, 1)$.

The functional equation (14.5), however, is the most well-known among them because it is the simplest and the most remarkable. For convenience, we call the equation (14.5) the *gamma functional equation.*

Let $\mathbb{R}^\circ = \mathbb{R} \setminus \{0, -1, -2, \ldots\}$ and let a function $\varphi : \mathbb{R}^\circ \to [0, \infty)$ satisfy

$$\Phi(x) = \sum_{j=0}^\infty \varphi(x + j) \prod_{i=0}^j |x + i|^{-1} < \infty \tag{14.6}$$

for all $x \in \mathbb{R}^\circ$.

S.-M. Jung [160] proved the Hyers–Ulam–Rassias stability of the gamma functional equation (14.5) (cf. [5, 168]).

Theorem 14.3 (Jung). *Assume the function* $f : \mathbb{R}^\circ \to \mathbb{R}$ *satisfies the functional inequality*

$$|f(x + 1) - x f(x)| \le \varphi(x) \tag{14.7}$$

for all $x \in \mathbb{R}^\circ$. Then there exists a unique function $G : \mathbb{R}^\circ \to \mathbb{R}$ which satisfies the gamma functional equation (14.5) for all $x \in \mathbb{R}^\circ$ and the inequality

$$|f(x) - G(x)| \leq \Phi(x) \qquad (14.8)$$

for every $x \in \mathbb{R}^\circ$.

Proof. Let x be an arbitrary element of $\mathbb{R}^\circ$. First, we use induction on n to prove that

$$\left| f(x + n) - f(x) \prod_{i=0}^{n-1}(x + i) \right| \leq \sum_{j=0}^{n-1} \varphi(x + j) \prod_{i=1}^{n-1-j} |x + i + j|, \qquad (a)$$

for any $n \in \mathbb{N}$, where we define $\prod_{i=1}^{0} |x + i| = 1$ conventionally. The inequality (a) immediately follows from (14.7) for the case of $n = 1$. Assume that (a) holds true for some integer $n > 0$. Using (14.7) and (a), we obtain for $n + 1$

$$\left| f(x + n + 1) - f(x) \prod_{i=0}^{n}(x + i) \right|$$

$$\leq |f(x + n + 1) - (x + n)f(x + n)|$$

$$+ |x + n| \left| f(x + n) - f(x) \prod_{i=0}^{n-1}(x + i) \right|$$

$$\leq \varphi(x + n) + |x + n| \sum_{j=0}^{n-1} \varphi(x + j) \prod_{i=1}^{n-1-j} |x + i + j|$$

$$= \varphi(x + n) + \sum_{j=0}^{n-1} \varphi(x + j) \prod_{i=1}^{n-j} |x + i + j|$$

$$= \sum_{j=0}^{n} \varphi(x + j) \prod_{i=1}^{n-j} |x + i + j|,$$

which ends the proof of (a).

If we divide both sides in (a) by $|x(x + 1) \cdots (x + n - 1)|$, we get

$$\left| f(x + n) \prod_{i=0}^{n-1}(x + i)^{-1} - f(x) \right| \leq \sum_{j=0}^{n-1} \varphi(x + j) \prod_{i=0}^{j} |x + i|^{-1} \qquad (b)$$

for every $n \in \mathbb{N}$. By using (14.7) and (14.6), we have for $n > m > 0$

$$\left| f(x + m) \prod_{i=0}^{m-1}(x + i)^{-1} - f(x + n) \prod_{i=0}^{n-1}(x + i)^{-1} \right|$$

$$= \left| f(x+m) \prod_{i=0}^{m-1} (x+i)^{-1} - f(x+m+1) \prod_{i=0}^{m} (x+i)^{-1} \right.$$

$$+ f(x+m+1) \prod_{i=0}^{m} (x+i)^{-1} - f(x+m+2) \prod_{i=0}^{m+1} (x+i)^{-1}$$

$$+ - \cdots$$

$$\left. + f(x+n-1) \prod_{i=0}^{n-2} (x+i)^{-1} - f(x+n) \prod_{i=0}^{n-1} (x+i)^{-1} \right|$$

$$\leq \sum_{j=m}^{n-1} |(x+j)f(x+j) - f(x+j+1)| \prod_{i=0}^{j} |x+i|^{-1}$$

$$\leq \sum_{j=m}^{n-1} \varphi(x+j) \prod_{i=0}^{j} |x+i|^{-1}$$

$$\to 0 \quad \text{as} \quad m \to \infty.$$

Therefore, the sequence

$$\left\{ f(x+n) \prod_{i=0}^{n-1} (x+i)^{-1} \right\}$$

is a Cauchy sequence, and we can define a function $G : \mathbb{R}^{\circ} \to \mathbb{R}$ by

$$G(x) = \lim_{n \to \infty} f(x+n) \prod_{i=0}^{n-1} (x+i)^{-1}.$$

In view of (b) and (14.6), the inequality (14.8) is true. By the definition of G we can easily verify that G satisfies (14.5) for its domain of definition:

$$G(x+1) = \lim_{n \to \infty} f(x+n+1) \prod_{i=0}^{n-1} (x+i+1)^{-1}$$

$$= x \lim_{n \to \infty} f(x+n+1) \prod_{i=0}^{n} (x+i)^{-1}$$

$$= xG(x).$$

Assume now that $G' : \mathbb{R}^{\circ} \to \mathbb{R}$ is another function which satisfies (14.5) and (14.8). It follows from (14.5) that

$$G(x) = G(x+n) \prod_{i=0}^{n-1} (x+i)^{-1} \quad \text{and} \quad G'(x) = G'(x+n) \prod_{i=0}^{n-1} (x+i)^{-1} \quad (c)$$

for each $n \in \mathbb{N}$. By (c), (14.8), and (14.6), we obtain

$$
\begin{aligned}
\left| G(x) - G'(x) \right| &= \left| G(x + n) - G'(x + n) \right| \prod_{i=0}^{n-1} |x + i|^{-1} \\
&\leq 2\Phi(x + n) \prod_{i=0}^{n-1} |x + i|^{-1} \\
&= 2 \sum_{j=0}^{\infty} \varphi(x + n + j) \prod_{i=0}^{n+j} |x + i|^{-1} \\
&= 2 \sum_{j=n}^{\infty} \varphi(x + j) \prod_{i=0}^{j} |x + i|^{-1} \\
&\to 0 \quad \text{as } n \to \infty,
\end{aligned}
$$

which implies the uniqueness of G. $\square$

The condition in (14.6) for φ is not very strong. By the ratio test for convergence of infinite series, we can easily demonstrate that almost all functions $\varphi : \mathbb{R}^\circ \to [0, \infty)$ which are familiar to us, for example, $\varphi(x) = \delta, c|x|^p, c \ln |x|, c \exp(|x|)$, etc., satisfy the condition (14.6).

Even though a function $G : \mathbb{R}^\circ \to \mathbb{R}$ satisfies the gamma functional equation (14.5), G is not necessarily equal to the gamma function Γ on $(0, \infty)$. If G is logarithmically convex on $(0, \infty)$ and satisfies the gamma functional equation (14.5) for $x > 0$ and $G(1) = 1$, then G necessarily equals the gamma function Γ on $(0, \infty)$ (see [208]).

We will now introduce a new functional equation

$$
f(x + p, y + q) = \psi(x, y) f(x, y). \tag{14.9}
$$

The gamma functional equation and the beta functional equation are special cases of this functional equation (14.9).

In the rest of this section, we investigate the Hyers–Ulam stability of the functional equation (14.9). Indeed, K.-W. Jun, G.-H. Kim, and Y.-W. Lee [148] proved the following theorem.

Theorem 14.4 (Jun, Kim, and Lee). *Let δ, p, and q be positive real numbers and let n_0 be a nonnegative integer. If a function $f : (0, \infty)^2 \to \mathbb{R}$ satisfies the inequality*

$$
|f(x + p, y + q) - \psi(x, y) f(x, y)| \leq \delta \tag{14.10}
$$

for all $x, y > n_0$ and if the function $\psi : (0, \infty)^2 \to (0, \infty)$ satisfies

$$
\Psi(x, y) = \sum_{i=0}^{\infty} \prod_{j=0}^{i} \frac{1}{\psi(x + jp, y + jq)} < \infty \tag{14.11}
$$

for all $x, y > n_0$, *then there exists a unique solution* $G : (0, \infty)^2 \to \mathbb{R}$ *of the functional equation* (14.9) *with*

$$|f(x, y) - G(x, y)| \leq \Psi(x, y)\delta \tag{14.12}$$

for any $x, y > n_0$.

Proof. For any $x, y > 0$ and $n \in \mathbb{N}$, we define

$$P_n(x, y) = f(x + np, y + nq) \prod_{j=0}^{n-1} \frac{1}{\psi(x + jp, y + jq)}.$$

By (14.10), we have

$$\begin{aligned}
|P_{n+1}(x, y) &- P_n(x, y)| \\
&= \left| f(x + np + p, y + nq + q) \right. \\
&\quad \left. - \psi(x + np, y + nq) f(x + np, y + nq) \right| \prod_{j=0}^{n} \frac{1}{\psi(x + jp, y + jq)} \\
&\leq \delta \prod_{j=0}^{n} \frac{1}{\psi(x + jp, y + jq)}
\end{aligned} \tag{a}$$

for all $x, y > n_0$ and $n \in \mathbb{N}$.

We now use induction on n to prove

$$|P_n(x, y) - f(x, y)| \leq \delta \sum_{i=0}^{n-1} \prod_{j=0}^{i} \frac{1}{\psi(x + jp, y + jq)} \tag{b}$$

for all $x, y > n_0$ and $n \in \mathbb{N}$. In view of (14.10), the inequality (b) is true for $n = 1$. Assume that the inequality (b) is true for some integer $n > 0$. Then, it follows from (a) and (b) that

$$\begin{aligned}
|P_{n+1}(x, y) - f(x, y)| &\leq |P_{n+1}(x, y) - P_n(x, y)| + |P_n(x, y) - f(x, y)| \\
&\leq \delta \sum_{i=0}^{n} \prod_{j=0}^{i} \frac{1}{\psi(x + jp, y + jq)},
\end{aligned}$$

which ends the proof of (b).

We claim that $\{P_n(x, y)\}$ is a Cauchy sequence. Indeed, for any $m, n \in \mathbb{N}$ with $n > m$ and $x, y > n_0$, it follows from (14.11) and (a) that

$$|P_n(x, y) - P_m(x, y)| \leq \sum_{i=m}^{n-1} |P_{i+1}(x, y) - P_i(x, y)|$$

$$\leq \delta \sum_{i=m}^{n-1} \prod_{j=0}^{i} \frac{1}{\psi(x+jp, y+jq)}$$

$$\to 0 \quad \text{as} \quad m \to \infty.$$

Hence, we can define a function $G_0 : (n_0, \infty)^2 \to \mathbb{R}$ by

$$G_0(x, y) = \lim_{n \to \infty} P_n(x, y).$$

Since $P_n(x + p, y + q) = \psi(x, y) P_{n+1}(x, y)$, we have

$$G_0(x + p, y + q) = \psi(x, y) G_0(x, y) \tag{c}$$

for all $x, y > n_0$. In view of (14.11) and (b), we get

$$|G_0(x, y) - f(x, y)| = \lim_{n \to \infty} |P_n(x, y) - f(x, y)|$$

$$\leq \delta \sum_{i=0}^{\infty} \prod_{j=0}^{i} \frac{1}{\psi(x+jp, y+jq)}$$

$$= \Psi(x, y) \delta$$

for any $x, y > n_0$. If $G_1 : (n_0, \infty)^2 \to \mathbb{R}$ is another function which satisfies

$$G_1(x+p, y+q) = \psi(x, y) G_1(x, y) \quad \text{and} \quad |G_1(x, y) - f(x, y)| \leq \Psi(x, y)\delta \tag{d}$$

for any $x, y > n_0$, then it follows from (14.12), (c), and (d) that

$$|G_0(x, y) - G_1(x, y)|$$

$$= |G_0(x + np, y + nq) - G_1(x + np, y + nq)| \prod_{j=0}^{n-1} \frac{1}{\psi(x+jp, y+jq)}$$

$$\leq 2\delta \Psi(x + np, y + nq) \prod_{j=0}^{n-1} \frac{1}{\psi(x+jp, y+jq)}$$

$$= 2\delta \sum_{i=n}^{\infty} \prod_{j=0}^{i} \frac{1}{\psi(x+jp, y+jq)}$$

$$\to 0 \quad \text{as} \quad n \to \infty$$

for any $x, y > n_0$, which implies the uniqueness of G_0.

Finally, we extend the domain of G_0 to $(0, \infty)^2$. For $0 < x, y \leq n_0$, we define

$$G(x, y) = G_0(x + kp, y + kq) \prod_{n=0}^{k-1} \frac{1}{\psi(x + np, y + nq)},$$

where k is the smallest natural number satisfying both the inequalities $x + kp > n_0$ and $y + kq > n_0$. It then holds true that $G(x + p, y + q) = \psi(x, y)G(x, y)$ for all $x, y > 0$ and $G(x, y) = G_0(x, y)$ for any $x, y > n_0$. Thus, the inequality (14.12) holds true for all $x, y > n_0$.   □

In the following corollary, we prove the Hyers–Ulam stability of the gamma functional equation (14.5). We can compare this corollary with Theorem 14.3 or [5].

Corollary 14.5. *Let δ be a positive constant and let n_0 be a nonnegative integer. If a function $f : (0, \infty) \to \mathbb{R}$ satisfies the inequality*

$$|f(x + 1) - xf(x)| \leq \delta$$

for all $x > n_0$, then there exists a unique solution $G : (0, \infty) \to \mathbb{R}$ of the gamma functional equation (14.5) such that

$$|f(x) - G(x)| \leq e\delta/x$$

for all $x > n_0$.

Proof. We apply Theorem 14.4 (for a single-variable case) with $p = 1$ and $\psi(x) = x$. For any $x > 0$, we have

$$\sum_{i=0}^{\infty} \prod_{j=0}^{i} \frac{1}{x + j} = (1/x)\left(1 + \frac{1}{x + 1} + \frac{1}{(x + 1)(x + 2)} + \cdots\right)$$
$$\leq (1/x)\left(1 + 1 + \frac{1}{2!} + \frac{1}{3!} + \cdots\right)$$
$$= e/x.$$

Then, the sequence

$$\left\{\sum_{i=0}^{\infty} \prod_{j=0}^{i} \frac{1}{x + j}\right\}$$

converges to $\Psi(x)$ and $\Psi(x) \leq e/x$ for any $x > 0$. Hence, we complete the proof by using Theorem 14.4.   □

Jun, Kim, and Lee have also proved the stability of the functional equation (14.9) in the sense of Ger (see [148]).

Theorem 14.6. *Let p and q be positive real numbers and let n_0 be a nonnegative integer. Let $f : (0, \infty)^2 \to (0, \infty)$ be a function that satisfies the inequality*

$$\left| \frac{f(x + p, y + q)}{\psi(x, y) f(x, y)} - 1 \right| \leq \varphi(x, y)$$

for all $x, y > n_0$, where $\psi : (0, \infty)^2 \to (0, \infty)$ is a function such that

$$\Psi(x, y) = \sum_{i=0}^{\infty} \prod_{j=0}^{i} \frac{1}{\psi(x + jp, y + jq)} < \infty$$

for all $x, y > n_0$ and $\varphi : (0, \infty)^2 \to (0, 1)$ is a function such that

$$\alpha(x, y) = \sum_{i=0}^{\infty} \ln \left(1 - \varphi(x + ip, y + iq) \right) < \infty,$$
$$\beta(x, y) = \sum_{i=0}^{\infty} \ln \left(1 + \varphi(x + ip, y + iq) \right) < \infty$$

for all $x, y > n_0$. Then there exists a unique solution $G : (0, \infty)^2 \to (0, \infty)$ of the functional equation (14.9) *with*

$$e^{\alpha(x,y)} \leq G(x, y)/f(x, y) \leq e^{\beta(x,y)}$$

for any $x, y > n_0$.

For more detailed information on the stability results of the functional equation (14.9), we refer the reader to [217, 219, 234–236].

14.4 Fibonacci Functional Equation

The *Fibonacci sequence* is one of the most well-known number sequences. Let us denote by F_n the nth *Fibonacci number* for any $n \in \mathbb{N}$. In particular, we will define $F_0 := 0$. It is well-known that the Fibonacci numbers satisfy the equation

$$F_n = F_{n-1} + F_{n-2}$$

for all integers $n \geq 2$ (ref. [225]). From this famous formula, we may derive a functional equation

$$f(x) = f(x - 1) + f(x - 2), \tag{14.13}$$

which may be called the *Fibonacci functional equation*. A function $f : \mathbb{R} \to \mathbb{R}$ will be called a *Fibonacci function* if it is a solution of the Fibonacci equation (14.13).

The reader is referred to the recent book by M. Th. Rassias [284] providing a self-contained and rigorous presentation of some of the most important theorems and results from Number Theory with a wide selection of solved Olympiad-caliber problems.

In this section, for fixed real numbers p and q with $q \neq 0$ and $p^2 - 4q \neq 0$, we generalize the Fibonacci functional equation (14.13) into

$$f(x) = pf(x - 1) - qf(x - 2) \tag{14.14}$$

and prove its Hyers–Ulam stability in the class of functions $f : \mathbb{R} \to E$, where E is a real (or complex) Banach space.

By a and b we denote the distinct roots of the equation $x^2 - px + q = 0$. More precisely, we set

$$a = (1/2)\left(p + \sqrt{p^2 - 4q}\right) \quad \text{and} \quad b = (1/2)\left(p - \sqrt{p^2 - 4q}\right).$$

Moreover, for any $n \in \mathbb{Z}$, we define

$$U_n = U_n(p, q) = \frac{a^n - b^n}{a - b}.$$

If p and q are integers, then $\{U_n(p, q)\}$ is called the *Lucas sequence of the first kind*. It is not difficult to see that

$$U_{n+2} = pU_{n+1} - qU_n \tag{14.15}$$

for all integer n. For any $x \in \mathbb{R}$, $[x]$ stands for the largest integer that does not exceed x.

S.-M. Jung [185] investigated the general solution of the generalized Fibonacci functional equation (14.14).

Theorem 14.7. *Let E be either a real vector space if $p^2 - 4q > 0$ or a complex vector space if $p^2 - 4q < 0$. A function $f : \mathbb{R} \to E$ is a solution of the functional equation (14.14) if and only if there exists a function $h : [-1, 1) \to E$ such that*

$$f(x) = U_{[x]+1}h(x - [x]) - qU_{[x]}h(x - [x] - 1). \tag{14.16}$$

Proof. Since $a + b = p$ and $ab = q$, it follows from (14.14) that

$$\begin{aligned}
f(x) - af(x - 1) &= b(f(x - 1) - af(x - 2)), \\
f(x) - bf(x - 1) &= a(f(x - 1) - bf(x - 2)).
\end{aligned} \tag{a}$$

By mathematical induction, we can easily verify that

$$\begin{aligned}
f(x) - af(x - 1) &= b^n(f(x - n) - af(x - n - 1)), \\
f(x) - bf(x - 1) &= a^n(f(x - n) - bf(x - n - 1))
\end{aligned} \tag{b}$$

for all $x \in \mathbb{R}$ and $n \in \mathbb{N}_0$. If we substitute $x + n$ $(n \geq 0)$ for x in (b) and divide the resulting equations by b^n resp. a^n, and if we substitute $-m$ for n in the resulting equations, then we obtain the equations in (b) with m in place of n, where $m \in \{0, -1, -2, \ldots\}$. Therefore, the equations in (b) are true for all $x \in \mathbb{R}$ and $n \in \mathbb{Z}$.

We multiply the first and the second equation of (b) by b and a, respectively. If we subtract the first resulting equation from the second one, then we obtain

$$f(x) = U_{n+1} f(x - n) - q U_n f(x - n - 1) \qquad (c)$$

for any $x \in \mathbb{R}$ and $n \in \mathbb{Z}$.

If we put $n = [x]$ in (c), then

$$f(x) = U_{[x]+1} f\big(x - [x]\big) - q U_{[x]} f\big(x - [x] - 1\big)$$

for all $x \in \mathbb{R}$.

We know that $0 \leq x - [x] < 1$ and $-1 \leq x - [x] - 1 < 0$. If we define a function $h : [-1, 1) \to E$ by $h := f|_{[-1,1)}$, then we see that f is a function of the form (14.16).

Assume now that f is a function of the form (14.16), where $h : [-1, 1) \to E$ is an arbitrary function. Then, it follows from (14.16) that

$$f(x) = U_{[x]+1} h\big(x - [x]\big) - q U_{[x]} h\big(x - [x] - 1\big),$$
$$f(x - 1) = U_{[x]} h\big(x - [x]\big) - q U_{[x]-1} h\big(x - [x] - 1\big),$$
$$f(x - 2) = U_{[x]-1} h\big(x - [x]\big) - q U_{[x]-2} h\big(x - [x] - 1\big)$$

for any $x \in \mathbb{R}$. Thus, by (14.15), we obtain

$$f(x) - p f(x - 1) + q f(x - 2)$$
$$= \big(U_{[x]+1} - p U_{[x]} + q U_{[x]-1}\big) h\big(x - [x]\big)$$
$$\quad - q\big(U_{[x]} - p U_{[x]-1} + q U_{[x]-2}\big) h\big(x - [x] - 1\big)$$
$$= 0,$$

which ends the proof.   □

Remark 1. It should be remarked that the functional equation (14.14) is a particular case of the linear equation $\displaystyle\sum_{i=0}^{n} p_i f\big(g^i(x)\big) = 0$ with $g(x) = x - 1$ and $n = 2$. Moreover, a substantial part of the proof of Theorem 14.7 can be derived from theorems presented in the books [227, 229]. However, the theorems in [227, 229] deal with solutions of the linear equation under some regularity conditions, for example, the continuity, convexity, differentiability, analyticity, and so on, while Theorem 14.7 deals with the general solution of (14.14) without regularity conditions. Indeed, the proof of Theorem 14.7 is simple and straightforward.

We now denote by a and b the distinct roots of the equation $x^2 - px + q = 0$ satisfying $|a| > 1$ and $0 < |b| < 1$.

S.-M. Jung [185] proved the Hyers–Ulam stability of the generalized Fibonacci functional equation (14.14) and J. Brzdęk and S.-M. Jung have proved a more general theorem (ref. [44]).

Theorem 14.8 (Jung). *Let $(E, \| \cdot \|)$ be either a real Banach space if $p^2 - 4q > 0$ or a complex Banach space if $p^2 - 4q < 0$. If a function $f : \mathbb{R} \to E$ satisfies the inequality*

$$\| f(x) - pf(x - 1) + qf(x - 2) \| \leq \delta \qquad (14.17)$$

for all $x \in \mathbb{R}$ and for some $\delta \geq 0$, then there exists a unique solution function $F : \mathbb{R} \to E$ of the generalized Fibonacci functional equation (14.14) such that

$$\| f(x) - F(x) \| \leq \frac{|a| - |b|}{|a - b|} \frac{\delta}{(|a| - 1)(1 - |b|)} \qquad (14.18)$$

for all $x \in \mathbb{R}$.

Proof. Analogously to the first equation of (a) in the proof of Theorem 14.7, it follows from (14.17) that

$$\left\| f(x) - af(x - 1) - b\big(f(x - 1) - af(x - 2)\big) \right\| \leq \delta$$

for each $x \in \mathbb{R}$. If we replace x with $x - k$ in the last inequality, then we have

$$\left\| f(x - k) - af(x - k - 1) - b\big(f(x - k - 1) - af(x - k - 2)\big) \right\| \leq \delta$$

and further

$$\left\| b^k \big(f(x - k) - af(x - k - 1)\big) - b^{k+1}\big(f(x - k - 1) - af(x - k - 2)\big) \right\| \leq |b|^k \delta \qquad (a)$$

for all $x \in \mathbb{R}$ and $k \in \mathbb{Z}$. By (a), we obviously have

$$\left\| f(x) - af(x - 1) - b^n\big(f(x - n) - af(x - n - 1)\big) \right\|$$
$$\leq \sum_{k=0}^{n-1} \left\| b^k\big(f(x - k) - af(x - k - 1)\big) \right.$$
$$\left. - b^{k+1}\big(f(x - k - 1) - af(x - k - 2)\big) \right\|$$
$$\leq \sum_{k=0}^{n-1} |b|^k \delta \qquad (b)$$

for every $x \in \mathbb{R}$ and $n \in \mathbb{N}$.

For any $x \in \mathbb{R}$, (a) implies that the sequence $\{b^n(f(x-n) - af(x-n-1))\}$ is a Cauchy sequence. (Note that $0 < |b| < 1$.) Therefore, we can define a function $F_1 : \mathbb{R} \to E$ by

$$F_1(x) = \lim_{n \to \infty} b^n(f(x-n) - af(x-n-1)),$$

since E is complete. In view of the definition of F_1, we obtain

$$
\begin{aligned}
pF_1&(x-1) - qF_1(x-2) \\
&= (p/b) \lim_{n \to \infty} b^{n+1}(f(x - (n+1)) - af(x - (n+1) - 1)) \\
&\quad - (q/b^2) \lim_{n \to \infty} b^{n+2}(f(x - (n+2)) - af(x - (n+2) - 1)) \\
&= (p/b)F_1(x) - (q/b^2)F_1(x) \\
&= F_1(x)
\end{aligned}
\tag{c}
$$

for all $x \in \mathbb{R}$, since $b^2 = pb - q$. If n goes to infinity, then (b) yields

$$\|f(x) - af(x-1) - F_1(x)\| \le \frac{\delta}{1 - |b|} \tag{d}$$

for every $x \in \mathbb{R}$.

On the other hand, it also follows from (14.17) that

$$\|f(x) - bf(x-1) - a(f(x-1) - bf(x-2))\| \le \delta.$$

Analogously to (a), replacing x with $x+k$ in the above inequality and then dividing by $|a|^k$ both sides of the resulting inequality, we have

$$
\begin{aligned}
\|a^{-k}&(f(x+k) - bf(x+k-1)) \\
&- a^{-k+1}(f(x+k-1) - bf(x+k-2))\| \le |a|^{-k}\delta
\end{aligned}
\tag{e}
$$

for all $x \in \mathbb{R}$ and $k \in \mathbb{Z}$. By using (e), we further obtain

$$
\begin{aligned}
\|a^{-n}&(f(x+n) - bf(x+n-1)) - (f(x) - bf(x-1))\| \\
&\le \sum_{k=1}^{n} \|a^{-k}(f(x+k) - bf(x+k-1)) \\
&\qquad\qquad - a^{-k+1}(f(x+k-1) - bf(x+k-2))\| \\
&\le \sum_{k=1}^{n} |a|^{-k}\delta
\end{aligned}
\tag{f}
$$

for $x \in \mathbb{R}$ and $n \in \mathbb{N}$.

On account of (e), we see that the sequence $\{a^{-n}(f(x+n) - bf(x+n-1))\}$ is a Cauchy sequence for any fixed $x \in \mathbb{R}$. (Note that $|a| > 1$.) Hence, we can define a function $F_2 : \mathbb{R} \to E$ by

$$F_2(x) = \lim_{n \to \infty} a^{-n}\big(f(x+n) - bf(x+n-1)\big).$$

Using the definition of F_2 yields

$$
\begin{aligned}
pF_2&(x-1) - qF_2(x-2) \\
&= (p/a) \lim_{n \to \infty} a^{-(n-1)}\big(f(x+n-1) - bf(x+(n-1)-1)\big) \\
&\quad - (q/a^2) \lim_{n \to \infty} a^{-(n-2)}\big(f(x+n-2) - bf(x+(n-2)-1)\big) \\
&= (p/a)F_2(x) - (q/a^2)F_2(x) \\
&= F_2(x)
\end{aligned}
\tag{g}
$$

for any $x \in \mathbb{R}$. If we let n go to infinity, then it follows from (f) that

$$\|F_2(x) - f(x) + bf(x-1)\| \le \frac{\delta}{|a|-1} \tag{h}$$

for each $x \in \mathbb{R}$.

By (d) and (h), we have

$$
\begin{aligned}
\bigg\| f(x) &- \bigg(\frac{b}{b-a}F_1(x) - \frac{a}{b-a}F_2(x)\bigg)\bigg\| \\
&= \frac{1}{|b-a|}\big\| (b-a)f(x) - \big(bF_1(x) - aF_2(x)\big)\big\| \\
&\le \frac{1}{|a-b|}\|bf(x) - abf(x-1) - bF_1(x)\| \\
&\quad + \frac{1}{|a-b|}\|aF_2(x) - af(x) + abf(x-1)\| \\
&\le \frac{|a|-|b|}{|a-b|}\frac{\delta}{(|a|-1)(1-|b|)}
\end{aligned}
\tag{i}
$$

for all $x \in \mathbb{R}$. We now define a function $F : \mathbb{R} \to E$ by

$$F(x) = \frac{b}{b-a}F_1(x) - \frac{a}{b-a}F_2(x)$$

for all $x \in \mathbb{R}$. Then, it follows from (c) and (g) that

$$
\begin{aligned}
pF&(x-1) - qF(x-2) \\
&= \frac{pb}{b-a}F_1(x-1) - \frac{pa}{b-a}F_2(x-1) - \frac{qb}{b-a}F_1(x-2) + \frac{qa}{b-a}F_2(x-2) \\
&= \frac{b}{b-a}\big(pF_1(x-1) - qF_1(x-2)\big) - \frac{a}{b-a}\big(pF_2(x-1) - qF_2(x-2)\big) \\
&= \frac{b}{b-a}F_1(x) - \frac{a}{b-a}F_2(x) \\
&= F(x)
\end{aligned}
$$

for each $x \in \mathbb{R}$, i.e., F is a solution of (14.14). Moreover, by (i), we obtain the inequality (14.18).

Now, it only remains to prove the uniqueness of F. Assume that $F, F' : \mathbb{R} \to E$ are solutions of (14.14) and that there exist positive constants C_1 and C_2 with

$$\|f(x) - F(x)\| \leq C_1 \quad \text{and} \quad \|f(x) - F'(x)\| \leq C_2 \tag{j}$$

for all $x \in \mathbb{R}$. According to Theorem 14.7, there exist functions $h, h' : [-1, 1) \to E$ such that

$$
\begin{aligned}
F(x) &= U_{[x]+1}h\big(x - [x]\big) - qU_{[x]}h\big(x - [x] - 1\big), \\
F'(x) &= U_{[x]+1}h'\big(x - [x]\big) - qU_{[x]}h'\big(x - [x] - 1\big)
\end{aligned}
\tag{k}
$$

for any $x \in \mathbb{R}$, since F and F' are solutions of (14.14).

Fix a t with $0 \leq t < 1$. It then follows from (j) and (k) that

$$
\begin{aligned}
\big\|U_{n+1}\big(h(t) - h'(t)\big) &+ U_n\big(qh'(t - 1) - qh(t - 1)\big)\big\| \\
&= \big\|\big(U_{n+1}h(t) - qU_nh(t - 1)\big) - \big(U_{n+1}h'(t) - qU_nh'(t - 1)\big)\big\| \\
&= \|F(n + t) - F'(n + t)\| \\
&\leq \|F(n + t) - f(n + t)\| + \|f(n + t) - F'(n + t)\| \\
&\leq C_1 + C_2
\end{aligned}
$$

for each $n \in \mathbb{Z}$, i.e.,

$$
\begin{aligned}
\Big\|\frac{a^{n+1} - b^{n+1}}{a - b}\big(h(t) - h'(t)\big) &+ \frac{a^n - b^n}{a - b}\big(qh'(t - 1) - qh(t - 1)\big)\Big\| \\
&\leq C_1 + C_2
\end{aligned}
\tag{l}
$$

for every $n \in \mathbb{Z}$. Dividing both sides by $|a|^n$ yields

$$
\begin{aligned}
\Big\|\frac{a - (b/a)^n b}{a - b}\big(h(t) - h'(t)\big) &+ \frac{1 - (b/a)^n}{a - b}\big(qh'(t - 1) - qh(t - 1)\big)\Big\| \\
&\leq (C_1 + C_2)|a|^{-n}
\end{aligned}
$$

and letting $n \to \infty$, we obtain

$$a\big(h(t) - h'(t)\big) + q\big(h'(t - 1) - h(t - 1)\big) = 0. \tag{m}$$

Analogously, if we divide both sides of (l) by $|b|^n$ and let $n \to -\infty$, then we get

$$b\big(h(t) - h'(t)\big) + q\big(h'(t - 1) - h(t - 1)\big) = 0. \tag{n}$$

By (m) and (n), we have

$$
\begin{pmatrix} a & q \\ b & q \end{pmatrix} \begin{pmatrix} h(t) - h'(t) \\ h'(t - 1) - h(t - 1) \end{pmatrix} = \begin{pmatrix} 0 \\ 0 \end{pmatrix}.
$$

Since $aq - bq \neq 0$ (where both a and b are nonzero and so $q = ab \neq 0$), it should hold true that

$$h(t) - h'(t) = h'(t-1) - h(t-1) = 0$$

for any $t \in [0, 1)$, i.e., $h(t) = h'(t)$ for all $t \in [-1, 1)$. Therefore, we conclude that $F(x) = F'(x)$ for any $x \in \mathbb{R}$. (The presented proof of uniqueness of F is somewhat long and involved. Indeed, the uniqueness can be obtained directly from [48, Proposition 1].) $\qquad\square$

Remark 2. The functional equation (14.14) is a particular case of the linear equations of higher orders and the Hyers–Ulam stability of the linear equations has been proved in [48, Theorem 2]. Indeed, J. Brzdęk, D. Popa, and B. Xu have proved an interesting theorem, from which the following corollary follows (see also [46,353]):

Let a function $f : \mathbb{R} \to E$ satisfy the inequality (14.17) for all $x \in \mathbb{R}$ and for some $\delta \geq 0$ and let a, b be the distinct roots of the equation $x^2 - px + q = 0$. If $|a| > 1$, $0 < |b| < 1$, and $|b| \neq 1/2$, then there exists a solution function $F : \mathbb{R} \to E$ of (14.14) such that

$$\|f(x) - F(x)\| \leq \frac{4\delta}{|2|a| - 1||2|b| - 1|} \tag{14.19}$$

for all $x \in \mathbb{R}$.

If either $0 < |b| < 1/2$ and $|a| > 3/2 - |b|$ or $1/2 < |b| < 3/4$ and $|a| > (5 - 6|b|)/(6 - 8|b|)$, then

$$\frac{4\delta}{|2|a| - 1||2|b| - 1|} > \frac{\delta}{(|a| - 1)(1 - |b|)} \geq \frac{|a| - |b|}{|a - b|} \frac{\delta}{(|a| - 1)(1 - |b|)}.$$

Hence, the estimation (14.18) of Theorem 14.8 is better in these cases than the estimation (14.19).

Remark 3. As we know, $\{U_n(1, -1)\}_{n=1,2,\ldots}$ is the Fibonacci sequence. So if we set $p = 1$ and $q = -1$ in Theorems 14.7 and 14.8, then we obtain the same results as in [184, Theorems 2.1, 3.1, and 3.3].

Bibliography

1. J. Aczél, Lectures on Functional Equations and Their Applications, Academic Press, New York/London, 1966.
2. J. Aczél, A Short Course on Functional Equations, Reidel, Dordrecht, 1987.
3. J. Aczél and J. Dhombres, Functional Equations in Several Variables, Cambridge Univ. Press, Cambridge, 1989.
4. C. Alsina, On the stability of a functional equation related to associativity, Ann. Polon. Math. 53 (1991), 1–5.
5. H. Alzer, Remarks on the stability of the gamma functional equation, Results Math. 35 (1999), 199–200.
6. H. Amann, Fixed points of asymptotically linear maps in ordered Banach spaces, J. Funct. Anal. 14 (1973), 162–171.
7. T. Aoki, On the stability of the linear transformation in Banach spaces, J. Math. Soc. Japan 2 (1950), 64–66.
8. C. Badea, On the Hyers-Ulam stability of mappings: the direct method, In 'Stability of Mappings of Hyers-Ulam Type' (edited by Th. M. Rassias and J. Tabor), Hadronic Press, Palm Harbor, Florida, 1994, pp. 7–13.
9. R. Badora, On the stability of the cosine functional equation, Rocznik Nauk. Dydakt. Prace Mat. 15 (1998), 5–14.
10. R. Badora and R. Ger, On some trigonometric functional inequalities, In 'Functional Equations – Results and Advances, Volume 3' (edited by Z. Daróczy and Z. Páles), Kluwer, Dordrecht, 2002.
11. J.-H. Bae and K.-W. Jun, On the generalized Hyers-Ulam-Rassias stability of an n-dimensional quadratic functional equation, J. Math. Anal. Appl. 258 (2001), no. 1, 183–193.
12. J.-H. Bae, K.-W. Jun and S.-M. Jung, On the stability of a quadratic functional equation, Kyungpook Math. J. 48 (2003), no. 3, 415–423.
13. J.-H. Bae and W.-G. Park, A functional equation originating from quadratic forms, J. Math. Anal. Appl. 326 (2007), 1142–1148.
14. J.-H. Bae and W.-G. Park, Stability of a Cauchy-Jensen functional equation in quasi-Banach spaces, J. Inequal. Appl. 2010 (2010), Article ID 151547, 9 pages. (doi:10.1155/2010/151547)
15. J. Baker, Isometries in normed spaces, Amer. Math. Monthly 78 (1971), 655–658.
16. J. Baker, The stability of the cosine equation, Proc. Amer. Math. Soc. 80 (1980), 411–416.
17. J. Baker, J. Lawrence and F. Zorzitto, The stability of the equation $f(x+y) = f(x)f(y)$, Proc. Amer. Math. Soc. 74 (1979), 242–246.
18. B. Belaid, E. Elhoucien and Th. M. Rassias, On the generalized Hyers-Ulam stability of the quadratic functional equation with a general involution, Nonlinear Funct. Anal. Appl. 12 (2007), 247–262.
19. B. Belaid, E. Elhoucien and Th. M. Rassias, On the Hyers-Ulam stability of approximately Pexider mappings, Math. Inequal. Appl. 11 (2008), no. 4, 805–818.

S.-M. Jung, *Hyers–Ulam–Rassias Stability of Functional Equations in Nonlinear Analysis*, Springer Optimization and Its Applications 48, DOI 10.1007/978-1-4419-9637-4, © Springer Science+Business Media, LLC 2011

20. G. Berruti and F. Skof, *Risultati di equivalenza per un'equazione di Cauchy alternativa negli spazi normati*, Atti Accad. Sci. Torino Cl. Sci. Fis. Mat. Natur. 125 (1991), 154–167.

21. R. Bhatia and P. Šemrl, *Approximate isometries on Euclidean spaces*, Amer. Math. Monthly 104 (1997), 497–504.

22. F. F. Bonsall and J. Duncan, Numerical Ranges of Operators on Normed Spaces and Elements of Normed Algebras, London Math. Soc. Lecture Notes 2, Cambridge Univ. Press, London, 1971.

23. C. Borelli, *On Hyers-Ulam stability of Hosszú's functional equation*, Results Math. 26 (1994), 221–224.

24. C. Borelli and G. L. Forti, *On a general Hyers-Ulam stability result*, Int. J. Math. Math. Sci. 18 (1995), 229–236.

25. D. G. Bourgin, *Approximate isometries*, Bull. Amer. Math. Soc. 52 (1946), 704–714.

26. D. G. Bourgin, *Approximately isometric and multiplicative transformations on continuous function rings*, Duke Math. J. 16 (1949), 385–397.

27. R. D. Bourgin, *Approximate isometries on finite dimensional Banach spaces*, Trans. Amer. Math. Soc. 207 (1975), 309–328.

28. D. G. Bourgin, *Classes of transformations and bordering transformations*, Bull. Amer. Math. Soc. 57 (1951), 223–237.

29. J. Brzdęk, *On the Cauchy difference*, Glas. Mat. Ser. III 27 (47) (1992), 263–269.

30. J. Brzdęk, *A note on stability of additive mappings*, In 'Stability of Mappings of Hyers-Ulam Type' (edited by Th. M. Rassias and J. Tabor), Hadronic Press, Palm Harbor, Florida, 1994, pp. 19–22.

31. J. Brzdęk, *On functionals which are orthogonally additive modulo* $\mathbb{Z}$, Results Math. 30 (1996), 25–38.

32. J. Brzdęk, *The Cauchy and Jensen differences on semigroups*, Publ. Math. Debrecen 48 (1996), 117–136.

33. J. Brzdęk, *On the Cauchy difference on normed spaces*, Abh. Math. Sem. Univ. Hamburg 66 (1996), 143–150.

34. J. Brzdęk, *On orthogonally exponential and orthogonally additive mappings*, Proc. Amer. Math. Soc. 125 (1997), 2127–2132.

35. J. Brzdęk, *On orthogonally exponential functionals*, Pacific J. Math. 181 (1997), 247–267.

36. J. Brzdęk, *On measurable orthogonally exponential functions*, Arch. Math. (Basel) 72 (1999), 185–191.

37. J. Brzdęk, *On the isosceles orthogonally exponential mappings*, Acta Math. Hungar. 87 (2000), 147–152.

38. J. Brzdęk, *On functions which are almost additive modulo a subgroup*, Glas. Mat. Ser. III 36 (56) (2001), 1–9.

39. J. Brzdęk, *On the quotient stability of a family of functional equations*, Nonlinear Anal. 71 (2009), 4396–4404.

40. J. Brzdęk, *On a method of proving the Hyers-Ulam stability of functional equations on restricted domains*, Aust. J. Math. Anal. Appl. 6 (2009), Issue 1, Article 4, 10 pages.

41. J. Brzdęk, *On stability of a family of functional equations*, Acta Math. Hungar. 128 (2010), 139–149.

42. J. Brzdęk, *A note on stability of the Popoviciu functional equation on restricted domain*, Demonstratio Math. 43 (2010), 635–641.

43. J. Brzdęk and A. Grabiec, *Remarks to the Cauchy difference*, In 'Stability of Mappings of Hyers-Ulam Type' (edited by Th. M. Rassias and J. Tabor), Hadronic Press, Palm Harbor, Florida, 1994, pp. 23–30.

44. J. Brzdęk and S.-M. Jung, *A note on stability of a linear functional equation of second order connected with the Fibonacci numbers and Lucas sequences*, J. Inequal. Appl. 2010 (2010), Article ID 793947, 10 pages. (doi:10.1155/2010/793947)

45. J. Brzdęk and A. Pietrzyk, *A note on stability of the general linear equation*, Aequationes Math. 75 (2008), 267–270.

46. J. Brzdęk, D. Popa and B. Xu, *Note on nonstability of the linear recurrence*, Abh. Math. Sem. Univ. Hamburg 76 (2006), 183–189.

47. J. Brzdęk, D. Popa and B. Xu, *The Hyers-Ulam stability of nonlinear recurrences*, J. Math. Anal. Appl. 335 (2007), 443–449.
48. J. Brzdęk, D. Popa and B. Xu, *Hyers-Ulam stability for linear equations of higher orders*, Acta Math. Hungar. 120 (2008), 1–8.
49. J. Brzdęk, D. Popa and B. Xu, *On nonstability of the linear recurrence of order one*, J. Math. Anal. Appl. 367 (2010), 146–153.
50. J. Brzdęk, D. Popa and B. Xu, *Remarks on stability of the linear recurrence of higher order*, Appl. Math. Lett. 23 (2010), 1459–1463.
51. J. Brzdęk and J. Sikorska, *A conditional exponential functional equation and its stability*, Nonlinear Anal. 72 (2010), 2923–2934.
52. J. Brzdęk and J. Tabor, *Stability of the Cauchy congruence on restricted domain*, Arch. Math. (Basel) 82 (2004), 546–550.
53. S. Butler, *Problem no. 11030*, Amer. Math. Monthly 110 (2003), 637–637.
54. N. P. Câc and J. A. Gatica, *Fixed point theorems for mappings in ordered Banach spaces*, J. Math. Anal. Appl. 71 (1979), 547–557.
55. L. Cădariu and V. Radu, *Fixed points and the stability of Jensen's functional equation*, J. Inequal. Pure Appl. Math. 4 (2003), Issue 1, Article 4, 7 pages.
56. L. Cădariu and V. Radu, *Fixed points and the stability of quadratic functional equations*, An. Univ. Timişoara Ser. Mat.-Inform. 41 (2003), no. 1, 25–48.
57. L. Cădariu and V. Radu, *On the stability of the Cauchy functional equation: a fixed point approach*, Grazer Math. Ber. 346 (2004), 43–52.
58. L. Cădariu and V. Radu, *Fixed point methods for the generalized stability of functional equations in a single variable*, Fixed Point Theory Appl. 2008 (2008), Article ID 749392, 15 pages. (doi:10.1155/2008/749392)
59. E. Castillo and M. R. Ruiz-Cobo, Functional Equations and Modelling in Science and Engineering, Dekker, New York/Basel, 1992.
60. A. L. Cauchy, Cours d'analyse de l'École Polytechnique, Vol. I, Analyse algébrique, Debure, Paris, 1821.
61. I.-S. Chang and H.-M. Kim, *Hyers-Ulam-Rassias stability of a quadratic equation*, Kyungpook Math. J. 42 (2002), 71–86.
62. I.-S. Chang and H.-M. Kim, *On the Hyers-Ulam stability of quadratic functional equations*, J. Inequal. Pure Appl. Math. 3 (2002), 1–12.
63. J. Chang and J. Chung, *Stability of trigonometric functional equations in generalized functions*, J. Inequal. Appl. 2010 (2010), Article ID 801502, 12 pages. (doi:10.1155/2010/801502)
64. J. Chmieliński, *On the Hyers-Ulam stability of the generalized orthogonality equation in real Hilbert spaces*, In 'Stability of Mappings of Hyers-Ulam Type' (edited by Th. M. Rassias and J. Tabor), Hadronic Press, Palm Harbor, Florida, 1994, pp. 31–41.
65. J. Chmieliński, *The stability of the Wigner equation in complex Hilbert spaces*, Rocznik Nauk. Dydak. WSP Krakowie 196 (1998), 49–55.
66. J. Chmieliński and S.-M. Jung, *The stability of the Wigner equation on a restricted domain*, J. Math. Anal. Appl. 254 (2001), 309–320.
67. J. Chmieliński and J. Tabor, *On approximate solutions of the Pexider equation*, Aequationes Math. 46 (1993), 143–163.
68. Y.-S. Cho and H.-M. Kim, *Stability of functional inequalities with Cauchy-Jensen additive mappings*, Abstr. Appl. Anal. 2007 (2007), Article ID 89180, 13 pages. (doi:10.1155/2007/89180)
69. P. W. Cholewa, *The stability of the sine equation*, Proc. Amer. Math. Soc. 88 (1983), 631–634.
70. P. W. Cholewa, *Remarks on the stability of functional equations*, Aequationes Math. 27 (1984), 76–86.
71. H.-Y. Chu, D.-S. Kang and Th. M. Rassias, *On the stability of a mixed n-dimensional quadratic functional equation*, Bull. Belg. Math. Soc. Simon Stevin 15 (2008), no. 1, 9–24.
72. J. Chung, *A distributional version of functional equations and their stabilities*, Nonlinear Anal. 62 (2005), 1037–1051.
73. J. Chung, *Approximately additive Schwartz distributions*, J. Math. Anal. Appl. 324 (2006), 1449–1457.

74. J. Chung, *Distributional methods for a class of functional equations and their stabilities*, Acta Math. Sin. (Engl. Ser.) 23 (2007), no. 11, 2017–2026.

75. J. Chung, *Stability of approximately quadratic Schwartz distributions*, Nonlinear Anal. 67 (2007), 175–186.

76. J. Chung, *A functional equation of Aczél and Chung in generalized functions*, Adv. Difference Equ. 2008 (2008), Article ID 147979, 11 pages. (doi:10.1155/2008/147979)

77. J. Chung, *Stability of exponential equations in Schwartz distributions*, Nonlinear Anal. 69 (2008), 3503–3511.

78. J. Chung, *On a stability of Pexiderized exponential equation*, Bull. Korean Math. Soc. 46 (2009), no. 2, 295–301.

79. J. Chung, *Stability of a generalized quadratic functional equation in Schwartz distributions*, Acta Math. Sin. (Engl. Ser.) 25 (2009), no. 9, 1459–1468.

80. J. Chung, *A heat kernel approach to the stability of exponential equation in distributions and hyperfunctions*, J. Math. Phys. 51 (2010), Article 053523, 9 pages. (doi:10.1063/1.3376657)

81. J. Chung, *Stability of functional equations on restricted domains in a group and their asymptotic behaviors*, Comput. Math. Appl. 60 (2010), 2653–2665.

82. J. Chung, S.-Y. Chung and D. Kim, *Generalized Pompeiu equation in distributions*, Appl. Math. Lett. 19 (2006), 485–490.

83. J. Chung, D. Kim and J. M. Rassias, *Hyers-Ulam stability on a generalized quadratic functional equation in distributions and hyperfunctions*, J. Math. Phys. 50 (2009), Article 113519, 14 pages. (doi:10.1063/1.3263147)

84. J. K. Chung, S.-M. Jung and P. K. Sahoo, *On a functional equation on groups*, J. Korean Math. Soc. 38 (2001), no. 1, 37–47.

85. S.-Y. Chung and Y.-S. Lee, *Stability of a Jensen type functional equation*, Banach J. Math. Anal. 1 (2007), no. 1, 91–100.

86. J. G. van der Corput, *Goniometrische functies gekarakteriseerd door een functionaal betrekking*, Euclides 17 (1940), 55–75.

87. S. Czerwik, *On the stability of the quadratic mapping in normed spaces*, Abh. Math. Sem. Hamburg 62 (1992), 59–64.

88. S. Czerwik, *On the stability of the homogeneous mapping*, C. R. Math. Rep. Acad. Sci. Canada 14 (1992), 268–272.

89. S. Czerwik, *The stability of the quadratic functional equation*, In 'Stability of Mappings of Hyers-Ulam Type' (edited by Th. M. Rassias and J. Tabor), Hadronic Press, Palm Harbor, Florida, 1994, pp. 81–91.

90. S. Czerwik, Functional Equations and Inequalities in Several Variables, World Scientific, Hackensacks, New Jersey, 2002.

91. S. Czerwik, *Some results on the stability of the homogeneous functions*, Bull. Korean Math. Soc. 42 (2005), no. 1, 29–37.

92. T. M. K. Davison, *The complete solution of Hosszú's functional equation over a field*, Aequationes Math. 11 (1974), 114–115.

93. M. M. Day, Normed Linear Spaces, Ergebnisse der Mathematik und ihrer Grenzgebiete, Springer, Berlin, 1958.

94. G. Dolinar, *Generalized stability of isometries*, J. Math. Anal. Appl. 242 (2000), 39–56.

95. B. R. Ebanks, Pl. Kannappan and P. K. Sahoo, *A common generalizations of functional equations characterizing normed and quasi-inner-product spaces*, Canad. Math. Bull. 35 (1992), 321–327.

96. B. R. Ebanks, Pl. Kannappan and P. K. Sahoo, *Cauchy differences that depend on the product of arguments*, Glas. Mat. Ser. III 27 (1992), 251–261.

97. E. Elhoucien, Y. Manar and Th. M. Rassias, *Hyers-Ulam stability of the quadratic and Jensen functional equations on unbounded domains*, J. Math. Sci. Adv. Appl. 4 (2010), no. 2, 287–301.

98. V. A. Faiziev and Th. M. Rassias, *The space of (ψ, γ)-pseudocharacters on semigroups*, Nonlinear Funct. Anal. Appl. 5 (2000), no. 1, 107–126.

99. V. A. Faiziev, Th. M. Rassias and P. K. Sahoo, *The space of (ψ, γ)-additive mappings on semigroups*, Trans. Amer. Math. Soc. 354 (2002), no. 11, 4455–4472.

100. V. A. Faiziev and P. K. Sahoo, *On the stability of Drygas functional equation on groups*, Banach J. Math. Anal. 1 (2007), no. 1, 43–55.

101. V. A. Faiziev and P. K. Sahoo, *On the stability of the quadratic equation on groups*, Bull. Belg. Math. Soc. Simon Stevin 15 (2008), no. 1, 135–151.

102. I. Fenyö, *On an inequality of P. W. Cholewa*, In 'General Inequalities 5' (edited by W. Walter), Birkhäuser, Basel, 1987, pp. 277–280.

103. J. W. Fickett, *Approximate isometries on bounded sets with an application to measure theory*, Studia Math. 72 (1981), 37–46.

104. G. L. Forti, *An existence and stability theorem for a class of functional equations*, Stochastica 4 (1980), 23–30.

105. G. L. Forti, *The stability of homomorphisms and amenability, with applications to functional equations*, Abh. Math. Sem. Univ. Hamburg 57 (1987), 215–226.

106. G. L. Forti, *Hyers-Ulam stability of functional equations in several variables*, Aequationes Math. 50 (1995), 143–190.

107. G. L. Forti, *Comments on the core of the direct method for proving Hyers-Ulam stability of functional equations*, J. Math. Anal. Appl. 295 (2004), no. 1, 127–133.

108. G. L. Forti, *Elementary remarks on Ulam-Hyers stability of linear functional equations*, J. Math. Anal. Appl. 328 (2007), no. 1, 109–118.

109. W. Förg-Rob and J. Schwaiger, *On the stability of a system of functional equations characterizing generalized hyperbolic and trigonometric functions*, Aequationes Math. 45 (1993), 285–296.

110. W. Förg-Rob and J. Schwaiger, *On the stability of some functional equations for generalized hyperbolic functions and for the generalized cosine equation*, Results Math. 26 (1994), 274–280.

111. Z. Gajda, *On stability of the Cauchy equation on semigroups*, Aequationes Math. 36 (1988), 76–79.

112. Z. Gajda, *On stability of additive mappings*, Int. J. Math. Math. Sci. 14 (1991), 431–434.

113. P. Găvruta, *A generalization of the Hyers-Ulam-Rassias stability of approximately additive mappings*, J. Math. Anal. Appl. 184 (1994), 431–436.

114. P. Găvruta, *On the stability of some functional equations*, In 'Stability of Mappings of Hyers-Ulam Type' (edited by Th. M. Rassias and J. Tabor), Hadronic Press, Palm Harbor, Florida, 1994, pp. 93–98.

115. P. Găvruta, *On the Hyers-Ulam-Rassias stability of mappings*, In 'Recent Progress in Inequalities' (edited by G. V. Milovanović), Kluwer, Dordrecht, 1998, pp. 465–470.

116. P. Găvruta, *An answer to a question of Th. M. Rassias and J. Tabor on mixed stability of mappings*, Bul. Ştiinţ. Univ. Politeh. Timiş. Ser. Mat. Fiz. 42 (1997), 1–6.

117. P. Găvruta, *Hyers-Ulam stability of Hosszú's equation*, In 'Functional Equations and Inequalities, Volume 518' (edited by Th. M. Rassias), Kluwer, Dordrecht, 2000, pp. 105–110.

118. P. Găvruta, M. Hossu, D. Popescu and C. Căprău, *On the stability of mappings*, Bull. Appl. Math. Tech. Univ. Budapest 73 (1994), 169–176.

119. P. Găvruta, M. Hossu, D. Popescu and C. Căprău, *On the stability of mappings and an answer to a problem of Th. M. Rassias*, Ann. Math. Blaise Pascal 2 (1995), 55–60.

120. P. Găvruta, S.-M. Jung and K.-S. Lee, *Remarks on the Pexider equations modulo a subgroup*, Far East J. Math. Sci. 19 (2005), no. 2, 215–222.

121. R. Ger, *Stability of addition formulae for trigonometric mappings*, Zeszyty Naukowe Politechniki Slaskiej, Ser. Mat. Fiz. 64 (1990), 75–84.

122. R. Ger, *Superstability is not natural*, Rocznik Nauk. Dydak. WSP w Krakowie, Prace Mat. 159 (1993), 109–123.

123. R. Ger and P. Šemrl, *The stability of the exponential equation*, Proc. Amer. Math. Soc. 124 (1996), 779–787.

124. J. Gevirtz, *Stability of isometries on Banach spaces*, Proc. Amer. Math. Soc. 89 (1983), 633–636.

125. N. Ghobadipour, A. Ebadian, Th. M. Rassias and M. Eshaghi Gordji, *A perturbation of double derivations on Banach algebras*, Commun. Math. Anal. 11 (2011), no. 1, 51–60.

126. E. Glowacki and Z. Kominek, *On stability of the Pexider equation on semigroups*, In 'Stability of Mappings of Hyers-Ulam Type' (edited by Th. M. Rassias and J. Tabor), Hadronic Press, Palm Harbor, Florida, 1994, pp. 111–116.

127. F. P. Greenleaf, Invariant Means on Topological Groups, Van Nostrand Math. Studies, Vol. 16, New York, 1969.

128. P. M. Gruber, *Stability of isometries*, Trans. Amer. Math. Soc. 245 (1978), 263–277.

129. H. Haruki and Th. M. Rassias, *New generalizations of Jensen's functional equation*, Proc. Amer. Math. Soc. 123 (1995), 495–503.

130. H. Haruki and Th. M. Rassias, *A new functional equation of Pexider type related to the complex exponential function*, Trans. Amer. Math. Soc. 347 (1995), 3111–3119.

131. K. J. Heuvers, *A characterization of Cauchy kernels*, Aequationes Math. 40 (1990), 281–306.

132. K. J. Heuvers, *Another logarithmic functional equation*, Aequationes Math. 58 (1999), 260–264.

133. K. J. Heuvers and Pl. Kannappan, *A third logarithmic functional equation and Pexider generalizations*, Aequationes Math. 70 (2005), 117–121.

134. M. Hosszú, *On the functional equation* $F(x + y, z) + F(x, y) = F(x, y + z) + F(y, z)$, Period. Math. Hungar. 1 (1971), 213–216.

135. D. H. Hyers, *On the stability of the linear functional equation*, Proc. Natl. Acad. Sci. USA 27 (1941), 222–224.

136. D. H. Hyers, G. Isac and Th. M. Rassias, *On the asymptoticity aspect of Hyers-Ulam stability of mappings*, Proc. Amer. Math. Soc. 126 (1998), 425–430.

137. D. H. Hyers, G. Isac and Th. M. Rassias, Stability of Functional Equations in Several Variables, Birkhäuser, Boston, 1998.

138. D. H. Hyers and Th. M. Rassias, *Approximate homomorphisms*, Aequationes Math. 44 (1992), 125–153.

139. D. H. Hyers and S. M. Ulam, *On approximate isometries*, Bull. Amer. Math. Soc. 51 (1945), 288–292.

140. D. H. Hyers and S. M. Ulam, *Approximate isometries of the space of continuous functions*, Ann. of Math. 48 (1947), 285–289.

141. G. Isac, *Opérateurs asymptotiquement linéaires sur des espaces localement convexes*, Colloq. Math. 46 (1982), 67–72.

142. G. Isac and Th. M. Rassias, *On the Hyers-Ulam stability of* ψ-*additive mappings*, J. Approx. Theory 72 (1993), 131–137.

143. G. Isac and Th. M. Rassias, *Functional inequalities for approximately additive mappings*, In 'Stability of Mappings of Hyers-Ulam Type' (edited by Th. M. Rassias and J. Tabor), Hadronic Press, Palm Harbor, Florida, 1994, pp. 117–125.

144. G. Isac and Th. M. Rassias, *Stability of* ψ-*additive mappings: Applications to nonlinear analysis*, Int. J. Math. Math. Sci. 19 (1996), 219–228.

145. K. Jarosz, Perturbations of Banach Algebras, Lecture Notes in Mathematics 1120, Springer, Berlin, 1985.

146. B. E. Johnson, *Approximately multiplicative functionals*, J. London Math. Soc. (2) 34 (1986), 489–510.

147. B. E. Johnson, *Approximately multiplicative maps between Banach algebras*, J. London Math. Soc. (2) 37 (1988), 294–316.

148. K.-W. Jun, G.-H. Kim and Y.-W. Lee, *Stability of generalized gamma and beta functional equations*, Aequationes Math. 60 (2000), 15–24.

149. K.-W. Jun and H.-M. Kim, *Remarks on the stability of additive functional equation*, Bull. Korean Math. Soc. 38 (2001), no. 4, 679–687.

150. K.-W. Jun and H.-M. Kim, *On the stability of Apollonius' equation*, Bull. Belg. Math. Soc. Simon Stevin 11 (2004), no. 4, 615–624.

151. K.-W. Jun and H.-M. Kim, *Stability problem of Ulam for generalized forms of Cauchy functional equation*, J. Math. Anal. Appl. 312 (2005), no. 2, 535–547.

152. K.-W. Jun and H.-M. Kim, *Ulam stability problem for a mixed type of cubic and additive functional equation*, Bull. Belg. Math. Soc. Simon Stevin 13 (2006), no. 2, 271–285.

153. K.-W. Jun and Y.-H. Lee, *On the Hyers-Ulam-Rassias stability of a pexiderized quadratic inequality*, Math. Inequal. Appl. 4 (2001), 93–118.

154. K.-W. Jun and D.-W. Park, *Stability of isometries between finite dimensional Hilbert spaces*, J. Korean Math. Soc. 32 (1995), no. 1, 103–108.

155. K.-W. Jun, D.-S. Shin and B.-D. Kim, *On Hyers-Ulam-Rassias stability of the Pexider equation*, J. Math. Anal. Appl. 239 (1999), 20–29.

156. S.-M. Jung, *On the Hyers-Ulam-Rassias stability of approximately additive mappings*, J. Math. Anal. Appl. 204 (1996), 221–226.

157. S.-M. Jung, *On solution and stability of functional equation* $f(x + y)^2 = af(x)f(y) + bf(x)^2 + cf(y)^2$, Bull. Korean Math. Soc. 34 (1997), 561–571.

158. S.-M. Jung, *On the superstability of the functional equation* $f(x^y) = yf(x)$, Abh. Math. Sem. Univ. Hamburg 67 (1997), 315–322.

159. S.-M. Jung, *Hyers-Ulam-Rassias stability of functional equations*, Dyn. Syst. Appl. 6 (1997), 541–566.

160. S.-M. Jung, *On the modified Hyers-Ulam-Rassias stability of the functional equation for gamma function*, Mathematica (Cluj) 39 (62) (1997), 235–239.

161. S.-M. Jung, *On a modified Hyers-Ulam stability of homogeneous equation*, Int. J. Math. Math. Sci. 21 (1998), 475–478.

162. S.-M. Jung, *On modified Hyers-Ulam-Rassias stability of a generalized Cauchy functional equation*, Nonlinear Stud. 5 (1998), 59–67.

163. S.-M. Jung, *Hyers-Ulam-Rassias stability of Jensen's equation and its application*, Proc. Amer. Math. Soc. 126 (1998), 3137–3143.

164. S.-M. Jung, *On the stability of the functional equation for exponential function*, Manuscript.

165. S.-M. Jung, *On the stability of the functional equation* $f(xy) = f(x)^y$, Mathematica (Cluj) 40 (63) (1998), no. 1, 89–94.

166. S.-M. Jung, *Superstability of homogeneous functional equation*, Kyungpook Math. J. 38 (1998), 251–257.

167. S.-M. Jung, *On the Hyers-Ulam stability of the functional equations that have the quadratic property*, J. Math. Anal. Appl. 222 (1998), 126–137.

168. S.-M. Jung, *On the stability of gamma functional equation*, Results Math. 33 (1998), 306–309.

169. S.-M. Jung, *On the Hyers-Ulam-Rassias stability of a quadratic functional equation*, J. Math. Anal. Appl. 232 (1999), 384–393.

170. S.-M. Jung, *On the stability of the Lobačevskiĭ equation*, Funct. Differ. Equ. 6 (1999), 111–123.

171. S.-M. Jung, *On the superstability of the functional equation* $f(x^y) = f(x)^y$, In 'Functional Equations and Inequalities' (edited by Th. M. Rassias), Kluwer, Dordrecht, 2000, pp. 119–124.

172. S.-M. Jung, *Quadratic functional equations of Pexider type*, Int. J. Math. Math. Sci. 24 (2000), 351–359.

173. S.-M. Jung, *Stability of the quadratic equation of Pexider type*, Abh. Math. Sem. Univ. Hamburg 70 (2000), 175–190.

174. S.-M. Jung, Hyers-Ulam-Rassias Stability of Functional Equations in Mathematical Analysis, Hadronic Press, Palm Harbor, Florida, 2001.

175. S.-M. Jung, *On the stability of a multiplicative functional equation*, J. Math. Anal. Appl. 254 (2001), 247–261.

176. S.-M. Jung, *Hyers-Ulam-Rassias stability of isometries on restricted domains*, Nonlinear Stud. 8 (2001), no. 1, 125–134.

177. S.-M. Jung, *Stability of the orthogonality equation on bounded domains*, Nonlinear Anal. 47 (2001), 2655–2666.

178. S.-M. Jung, *Asymptotic properties of isometries*, J. Math. Anal. Appl. 276 (2002), 642–653.

179. S.-M. Jung, *Hyers-Ulam stability of Butler-Rassias functional equation*, J. Inequal. Appl. 2005 (2005), no. 1, 41–47.

180. S.-M. Jung, *On an asymptotic behavior of exponential functional equation*, Acta Math. Sin. (Engl. Ser.) 22 (2006), no. 2, 583–586.

181. S.-M. Jung, *A fixed point approach to the stability of a Volterra integral equation*, Fixed Point Theory Appl. 2007 (2007), Article ID 57064, 9 pages. (doi:10.1155/2007/57064)

182. S.-M. Jung, *A fixed point approach to the stability of isometries*, J. Math. Anal. Appl. 329 (2007), no. 2, 879–890.

183. S.-M. Jung, *Stability of the orthogonality equation on a bounded domain*, In 'Inequalities and Applications' (edited by Th. M. Rassias and D. Andrica), Cluj Univ. Press, Cluj-Napoca, 2008, pp. 125–145.

184. S.-M. Jung, *Hyers-Ulam stability of Fibonacci functional equation*, Bull. Iranian Math. Soc. 35 (2009), no. 2, 217–227.

185. S.-M. Jung, *Functional equation $f(x) = pf(x-1) - qf(x-2)$ and its Hyers-Ulam stability*, J. Inequal. Appl. 2009 (2009), Article ID 181678, 10 pages. (doi:10.1155/2009/181678)

186. S.-M. Jung and B. Chung, *Remarks on Hyers-Ulam stability of Butler-Rassias functional equation*, Dyn. Contin. Discrete Impuls. Syst. Ser. A Math. Anal. 13 (2006), 193–197.

187. S.-M. Jung and B. Kim, *On the stability of the quadratic functional equation on bounded domains*, Abh. Math. Sem. Univ. Hamburg 69 (1999), 293–308.

188. S.-M. Jung and B. Kim, *Stability of isometries on restricted domains*, J. Korean Math. Soc. 37 (2000), no. 1, 125–137.

189. S.-M. Jung and T.-S. Kim, *A fixed point approach to the stability of the cubic functional equation*, Bol. Soc. Mat. Mexicana (3) 12 (2006), no. 1, 51–57.

190. S.-M. Jung, T.-S. Kim and K.-S. Lee, *A fixed point approach to the stability of quadratic functional equation*, Bull. Korean Math. Soc. 43 (2006), 531–541.

191. S.-M. Jung and Y.-H. Kye, *Stability of a generalized Hosszú's functional equation*, In 'Inequalities and Applications' (edited by Th. M. Rassias and D. Andrica), Cluj Univ. Press, Cluj-Napoca, 2008, pp. 147–153.

192. S.-M. Jung and Z.-H. Lee, *A fixed point approach to the stability of quadratic functional equation with involution*, Fixed Point Theory Appl. 2008 (2008), Article ID 732086, 11 pages. (doi:10.1155/2008/732086)

193. S.-M. Jung and S. Min, *A fixed point approach to the stability of the functional equation $f(x + y) = F[f(x), f(y)]$*, Fixed Point Theory Appl. 2009 (2009), Article ID 912046, 8 pages. (doi:10.1155/2009/912046)

194. S.-M. Jung, M. S. Moslehian and P. K. Sahoo, *Stability of a generalized Jensen equation on restricted domains*, J. Math. Inequal. 4 (2010), no. 2, 191–206.

195. S.-M. Jung and J. M. Rassias, *Stability of general Newton functional equations for logarithmic spirals*, Adv. Difference Equ. 2008 (2008), Article ID 143053, 5 pages. (doi:10.1155/2008/143053)

196. S.-M. Jung and J. M. Rassias, *A fixed point approach to the stability of a functional equation of the spiral of Theodorus*, Fixed Point Theory Appl. 2008 (2008), Article ID 945010, 7 pages. (doi:10.1155/2008/945010)

197. S.-M. Jung and Th. M. Rassias, *A fixed point approach to the stability of a logarithmic functional equation*, In 'Nonlinear Analysis and Variational Problems' (edited by P. M. Pardalos, Th. M. Rassias and A. A. Khan), Springer, New York, 2010, pp. 99–109.

198. S.-M. Jung and P. K. Sahoo, *Hyers-Ulam-Rassias stability of an equation of Davison*, J. Math. Anal. Appl. 238 (1999), 297–304.

199. S.-M. Jung and P. K. Sahoo, *On the stability of a mean value type functional equation*, Demonstratio Math. 33 (2000), 793–796.

200. S.-M. Jung and P. K. Sahoo, *Hyers-Ulam stability of the quadratic equation of Pexider type*, J. Korean Math. Soc. 38 (2001), 645–656.

201. S.-M. Jung and P. K. Sahoo, *A functional equation characterizing cubic polynomials and its stability*, Int. J. Math. Math. Sci. 27 (2001), no. 5, 301–308.

202. S.-M. Jung and P. K. Sahoo, *Stability of a functional equation for square root spirals*, Appl. Math. Lett. 15 (2002), 435–438.

203. S.-M. Jung and P. K. Sahoo, *On the stability of the Pompeiu functional equation*, Math. Inequal. Appl. 5 (2002), no. 2, 257–261.

204. S.-M. Jung and P. K. Sahoo, *Hyers-Ulam stability of a generalized Hosszu functional equation*, Glas. Mat. Ser. III 37 (57) (2002), 283–292.

205. S.-M. Jung and P. K. Sahoo, *Stability of a functional equation of Drygas*, Aequationes Math. 64 (2002), 263–273.

206. S.-M. Jung and P. K. Sahoo, *Stability of a functional equation that arises in the theory of conditionally specified distributions*, Kyungpook Math. J. 44 (2004), no. 1, 111–116.

207. S.-M. Jung and P. K. Sahoo, *Superstability of the generalized orthogonality equation on restricted domains*, Proc. Indian Acad. Sci. Math. Sci. 114 (2004), no. 3, 253–267.

208. H. H. Kairies, *Die Gammafunktion als stetige Lösung eines Systems von Gauss-Funktionalgleichungen*, Results Math. 26 (1994), 306–315.

209. Pl. Kannappan, *On cosine and sine functional equations*, Ann. Polon. Math. 20 (1968), 245–249.

210. Pl. Kannappan, *On sine functional equation*, Stud. Sci. Math. Hun. 4 (1969), 331–333.

211. Pl. Kannappan, *Quadratic functional equation and inner product spaces*, Results Math. 27 (1995), 368–372.

212. Pl. Kannappan and G.-H. Kim, *The stability of the generalized cosine functional equations*, Ann. Acad. Paedagog. Crac. Stud. Math. 1 (2001), 49–58.

213. Pl. Kannappan and P. K. Sahoo, *Cauchy difference – a generalization of Hosszú functional equation*, Proc. Natl. Acad. Sci. India 63 (1993), 541–550.

214. H. Khodaei and Th. M. Rassias, *Approximately generalized additive functions in several variables*, Int. J. Nonlinear Anal. Appl., to appear.

215. G.-H. Kim, *On the stability of the quadratic mapping in normed spaces*, Int. J. Math. Math. Sci. 25 (2001), 217–229.

216. G.-H. Kim, *The stability of the generalized sine functional equations, II*, J. Inequal. Pure Appl. Math. 7 (2006), 1–10.

217. G.-H. Kim, *A generalization of Hyers-Ulam-Rassias stability of the G-functional equation*, Math. Inequal. Appl. 10 (2007), 351–358.

218. G.-H. Kim and S. S. Dragomir, *The stability of the generalized d'Alembert and Jensen functional equations, II*, Int. J. Math. Math. Sci. 2006 (2006), 1–12.

219. G.-H. Kim and Y.-W. Lee, *On the stability of a beta type functional equation*, J. Appl. Math. Computing 14 (2004), 429–445.

220. H.-M. Kim, *Stability for generalized Jensen functional equations and isomorphisms between C^*-algebras*, Bull. Belg. Math. Soc. Simon Stevin 14 (2007), no. 1, 1–14.

221. H.-M. Kim, K.-W. Jun and J. M. Rassias, *Extended stability problem for alternative Cauchy-Jensen mappings*, J. Inequal. Pure Appl. Math. 8 (2007), Issue 4, Article 120, 17 pages.

222. H.-M. Kim, S.-Y. Kang and I.-S. Chang, *On the stability for cubic functional equation of mixed type*, Dyn. Syst. Appl. 17 (2008), 583–594.

223. H.-M. Kim, J. M. Rassias and Y.-S. Cho, *Stability problem of Ulam for Euler-Lagrange quadratic mappings*, J. Inequal. Appl. 2007 (2007), Article ID 10725, 15 pages. (doi:10.1155/2007/10725)

224. Z. Kominek, *On a local stability of the Jensen functional equation*, Demonstratio Math. 22 (1989), 499–507.

225. T. Koshy, Fibonacci and Lucas Numbers with Applications, John Wiley & Sons, New York, 2001.

226. M. A. Krasnoselskii and P. P. Zabreiko, Geometrical Methods of Nonlinear Analysis, Springer, Berlin, 1984.

227. M. Kuczma, Functional Equations in a Single Variable, PWN – Polish Scientific Publishers, Warszawa, 1968.

228. M. Kuczma, An Introduction to the Theory of Functional Equations and Inequalities, PWN – Uniwersytet Śląski, Warszawa, 1985.

229. M. Kuczma, B. Choczewski and R. Ger, Iterative Functional Equations, Encyclopedia of Mathematics and its Applications, Cambridge Univ. Press, Cambridge, 1990.

230. S. Kurepa, *On the quadratic functional*, Publ. Inst. Math. Acad. Serbe Sci. Beograd 13 (1959), 57–72.

231. K. Lajkó, *Applications of extensions of additive functions*, Aequationes Math. 11 (1974), 68–76.

232. Y.-H. Lee and K.-W. Jun, *A generalization of the Hyers-Ulam-Rassias stability of the Pexider equation*, J. Math. Anal. Appl. 246 (2000), 627–638.

233. Y.-H. Lee and K.-W. Jun, *On the stability of approximately additive mappings*, Proc. Amer. Math. Soc. 128 (2000), 1361–1369.

234. Y.-W. Lee and B.-M. Choi, *The stability of Cauchy's gamma-beta functional equation*, J. Math. Anal. Appl. 299 (2004), 305–313.

235. Y.-W. Lee and B.-M. Choi, *Stability of a beta-type functional equation with a restricted domain*, Commun. Korean Math. Soc. 19 (2004), 701–713.

236. Y.-W. Lee and G.-H. Kim, *Approximate gamma-beta type functions*, Nonlinear Anal. 71 (2009), e1567–e1574.

237. J. Lindenstrauss and A. Szankowski, *Non linear perturbations of isometries*, Astérisque 131 (1985), 357–371.

238. L. Losonczi, *On the stability of Hosszú's functional equation*, Results Math. 29 (1996), 305–310.

239. G. Maksa, *Problems 18*, In 'Report on the 34th ISFE,' Aequationes Math. 53 (1997), 194–194.

240. B. Margolis and J. B. Diaz, *A fixed point theorem of the alternative for contractions on a generalized complete metric space*, Bull. Amer. Math. Soc. 74 (1968), 305–309.

241. K. Maurin, Methods of Hilbert Spaces, Monografie Matematyczne Vol. 45, Polish Scientific Publishers (PWN), Warszawa, 1972.

242. M. Mininni, *Coincidence degree and solvability of some nonlinear functional equations in normed spaces: a spectral approach*, Nonlinear. Anal. Theory Mech. Appl. 1 (1977), 105–122.

243. M. Mirzavaziri and M. S. Moslehian, *A fixed point approach to stability of a quadratic equation*, Bull. Braz. Math. Soc. 37 (2006), no. 3, 361–376.

244. F. Moradlou, H. Vaezi and C. Park, *Fixed points and stability of an additive functional equation of n-Apollonius type in C^*-algebras*, Abstr. Appl. Anal. 2008 (2008), Article ID 672618, 13 pages. (doi:10.1155/2008/672618)

245. M. S. Moslehian, *Orthogonal stability of the Pexiderized quadratic equation*, J. Difference Equ. Appl. 11 (2005), no. 11, 999–1004.

246. M. S. Moslehian, *On the stability of the orthogonal Pexiderized Cauchy equation*, J. Math. Anal. Appl. 318 (2006), no.1, 211–223.

247. M. S. Moslehian, *Almost derivations on C^*-ternary rings*, Bull. Belg. Math. Soc. Simon Stevin 14 (2007), no. 1, 135–142.

248. M. S. Moslehian, *Ternary derivations, stability and physical aspects*, Acta Appl. Math. 100 (2008), no. 2, 187–199.

249. M. S. Moslehian and Th. M. Rassias, *Orthogonal stability of additive type equations*, Aequationes Math. 73 (2007), no. 3, 249–259.

250. M. S. Moslehian and Th. M. Rassias, *Stability of functional equations in non-Archimedean spaces*, Appl. Anal. Discrete Math. 1 (2007), no. 2, 325–334.

251. M. S. Moslehian and Th. M. Rassias, *Generalized Hyers-Ulam stability of mappings on normed Lie triple systems*, Math. Inequal. Appl. 11 (2008), no. 2, 371–380.

252. M. S. Moslehian and G. Sadeghi, *Stability of two types of cubic functional equations in non-Archimedean spaces*, Real Anal. Exchange 33 (2007), no. 2, 375–384.

253. M. S. Moslehian and G. Sadeghi, *Stability of linear mappings in quasi-Banach modules*, Math. Inequal. Appl. 11 (2008), no. 3, 549–557.

254. A. Najati and Th. M. Rassias, *Stability of homomorphisms and (θ, φ)-derivations*, Appl. Anal. Discrete Math. 3 (2009), no. 2, 264–281.

255. A. Najati and Th. M. Rassias, *Stability of a mixed functional equation in several variables on Banach modules*, Nonlinear Anal. 72 (2010), 1755–1767.

256. A. Najati and Th. M. Rassias, *Stability of the Pexiderized Cauchy and Jensen's equations on restricted domains*, Commun. Math. Anal. 8 (2010), no. 2, 125–135.

257. A. Najdecki, *On stability of a functional equation connected with the Reynolds operator*, J. Inequal. Appl. 2007 (2007), Article ID 79816, 3 pages. (doi:10.1155/2007/79816)

258. C. T. Ng, *Jensen's functional equation on groups*, Aequationes Math. 39 (1990), 85–90.
259. M. Omladič and P. Šemrl, *On nonlinear perturbations of isometries*, Math. Ann. 303 (1995), 617–628.
260. Z. Páles, *Remark 27*, In 'Report on the 34th ISFE,' Aequationes Math. 53 (1997), pp. 200-201.
261. C. Park, *On the stability of the linear mapping in Banach modules*, J. Math. Anal. Appl. 275 (2002), 711–720.
262. C. Park, *On an approximate automorphism on a C^*-algebra*, Proc. Amer. Math. Soc. 132 (2004), 1739–1745.
263. C. Park, *Hyers-Ulam-Rassias stability of a generalized Euler-Lagrange type additive mapping and isomorphisms between C^*-algebras*, Bull. Belg. Math. Soc. Simon Stevin 13 (2006), no. 4, 619–631.
264. C. Park, *Fixed points and Hyers-Ulam-Rassias stability of Cauchy-Jensen functional equations in Banach algebras*, Fixed Point Theory Appl. 2007 (2007), Article ID 50175, 15 pages. (doi:10.1155/2007/50175)
265. C. Park, *Hyers-Ulam-Rassias stability of homomorphisms in quasi-Banach algebras*, Banach J. Math. Anal. 1 (2007), no. 1, 23–32.
266. C. Park, *Fixed points and stability of an AQCQ-functional equation in non-Archimedean normed spaces*, Abstr. Appl. Anal. 2010 (2010), Article ID 849543, 15 pages. (doi:10.1155/2010/849543)
267. C. Park and J.-H. Kim, *The stability of a quadratic functional equation with the fixed point alternative*, Abstr. Appl. Anal. 2009 (2009), Article ID 907167, 11 pages. (doi:10.1155/2009/907167)
268. C. Park and J. M. Rassias, *Stability of the Jensen-type functional equation in C^*-algebras: a fixed point approach*, Abstr. Appl. Anal. 2009 (2009), Article ID 360432, 17 pages. (doi:10.1155/2009/360432)
269. C. Park and Th. M. Rassias, *Hyers-Ulam stability of a generalized Apollonius type quadratic mapping*, J. Math. Anal. Appl. 322 (2006), no. 1, 371–381.
270. C. Park and Th. M. Rassias, *Fixed points and generalized Hyers-Ulam stability of quadratic functional equations*, J. Math. Inequal. 1 (2007), no. 4, 515–528.
271. C. Park and Th. M. Rassias, *On the stability of orthogonal functional equations*, Tamsui Oxf. J. Math. Sci. 24 (2008), no. 4, 355–365.
272. C. Park and Th. M. Rassias, *Fixed points and stability of the Cauchy functional equation*, Aust. J. Math. Anal. Appl. 6 (2009), Issue 1, Article 14, 9 pages.
273. C. Park and Th. M. Rassias, *Fixed points and stability of functional equations*, In 'Nonlinear Analysis and Variational Problems' (edited by P. M. Pardalos, Th. M. Rassias and A. A. Khan), Springer, New York, 2010, pp. 125–134.
274. D.-W. Park and Y.-H. Lee, *The Hyers-Ulam-Rassias stability of the pexiderized equations*, Nonlinear Anal. 63 (2005), e2503–e2513.
275. W.-G. Park and J.-H. Bae, *On a bi-quadratic functional equation and its stability*, Nonlinear Anal. 62 (2005), no. 4, 643–654.
276. W.-G. Park and J.-H. Bae, *A multidimensional functional equation having quadratic forms as solutions*, J. Inequal. Appl. 2007 (2007), Article ID 24716, 8 pages. (doi:10.1155/2007/24716)
277. W.-G. Park and J.-H. Bae, *A functional equation originating from elliptic curves*, Abstr. Appl. Anal. 2008 (2008), Article ID 135237, 10 pages. (doi:10.1155/2008/135237)
278. J. C. Parnami and H. L. Vasudeva, *On Jensen's functional equation*, Aequationes Math. 43 (1992), 211–218.
279. V. Radu, *The fixed point alternative and the stability of functional equations*, Fixed Point Theory 4 (2003), no. 1, 91–96.
280. J. M. Rassias, *On approximation of approximately linear mappings by linear mappings*, J. Funct. Anal. 46 (1982), no. 1, 126–130.
281. J. M. Rassias, *On a new approximation of approximately linear mappings by linear mappings*, Discuss. Math. 7 (1985), 193–196.
282. J. M. Rassias, *Solution of a problem of Ulam*, J. Approx. Theory 57 (1989), no. 3, 268–273.

283. M. Th. Rassias, *Solution of a functional equation problem of Steven Butler*, Octogon Math. Mag. 12 (2004), 152–153.

284. M. Th. Rassias, Problem-Solving and Selected Topics in Number Theory, Springer, Berlin, 2011.

285. Th. M. Rassias, *On the stability of the linear mapping in Banach spaces*, Proc. Amer. Math. Soc. 72 (1978), 297–300.

286. Th. M. Rassias, *The stability of linear mappings and some problems on isometries*, In 'Mathematical Analysis and Its Applications' (edited by S. M. Mazhar, A. Hamoui and N. S. Faour), Pergamon Press, New York, 1985.

287. Th. M. Rassias, *On the stability of mappings*, Rend. Sem. Mat. Fis. Milano 58 (1988), 91–99.

288. Th. M. Rassias, *Seven problems in mathematical analysis*, In 'Topics in Mathematical Analysis' (edited by Th. M. Rassias), World Scientific, Singapore, 1989.

289. Th. M. Rassias, *The stability of mappings and related topics*, In 'Report on the 27th ISFE,' Aequationes Math. 39 (1990), pp. 292–293.

290. Th. M. Rassias, *On a modified Hyers-Ulam sequence*, J. Math. Anal. Appl. 158 (1991), 106–113.

291. Th. M. Rassias, *Problem 18*, In 'Report on the 31st ISFE,' Aequationes Math. 47 (1994), pp. 312–313.

292. Th. M. Rassias, *Remark and problem 19*, In 'Report on the 31st ISFE,' Aequationes Math. 47 (1994), pp. 313–314.

293. Th. M. Rassias, *On a problem of S. M. Ulam and the asymptotic stability of the Cauchy functional equation with applications*, Int. Ser. Numer. Math. 123 (1997), 297–309.

294. Th. M. Rassias, Inner Product Spaces and Applications, Addison Wesley Longman, Pitman Research Notes in Mathematics Series, No. 376, 1997.

295. Th. M. Rassias, Approximation Theory and Applications, Hadronic Press, Palm Harbor, Florida, 1998.

296. Th. M. Rassias, New Approaches in Nonlinear Analysis, Hadronic Press, Palm Harbor, Florida, 1999.

297. Th. M. Rassias, *Properties of isometric mappings*, J. Math. Anal. Appl. 235 (1999), 108–121.

298. Th. M. Rassias, *On the stability of the quadratic functional equation*, Stud. Univ. Babeş-Bolyai Math. 45 (2000), no. 2, 77–114.

299. Th. M. Rassias, *On Hille's functional equations, Robinson's functional equation and related functional equations*, Panamer. Math. J. 10 (2000), no. 3, 83–98.

300. Th. M. Rassias, Functional Equations and Inequalities, Kluwer, Dordrecht, 2000.

301. Th. M. Rassias, Survey on Classical Inequalities, Kluwer, Dordrecht, 2000.

302. Th. M. Rassias, *On the stability of functional equations and a problem of Ulam*, Acta Appl. Math. 62 (2000), no. 1, 23–130.

303. Th. M. Rassias, *The problem of S. M. Ulam for approximately multiplicative mappings*, J. Math. Anal. Appl. 246 (2000), no. 2, 352–378.

304. Th. M. Rassias, *On the stability of functional equations in Banach spaces*, J. Math. Anal. Appl. 251 (2000), no. 1, 264–284.

305. Th. M. Rassias, *Isometries and approximate isometries*, Int. J. Math. Math. Sci. 25 (2001), no. 2, 73–91.

306. Th. M. Rassias, *On the stability of functional equations originated by a problem of Ulam*, Mathematica 44 (67) (2002), no. 1, 39–75.

307. Th. M. Rassias, *On the stability of approximate isometries*, Tamsui Oxf. J. Math. Sci. 18 (2002), no. 1, 45–56.

308. Th. M. Rassias, Functional Equations, Inequalities and Applications, Kluwer, Dordrecht, 2003.

309. Th. M. Rassias, *On the stability of minimum points*, Mathematica 45 (68) (2003), no.1, 93–104.

310. Th. M. Rassias and P. Šemrl, *On the behavior of mappings which do not satisfy Hyers-Ulam stability*, Proc. Amer. Math. Soc. 114 (1992), 989–993.

311. Th. M. Rassias and P. Šemrl, *On the Hyers-Ulam stability of linear mappings*, J. Math. Anal. Appl. 173 (1993), 325–338.

312. Th. M. Rassias and J. Tabor, *What is left of Hyers-Ulam stability?*, J. Nat. Geom. 1 (1992), 65–69.

313. Th. M. Rassias and J. Tabor, *On approximately additive mappings in Banach spaces*, In 'Stability of Mappings of Hyers-Ulam Type' (edited by Th. M. Rassias and J. Tabor), Hadronic Press, Palm Harbor, Florida, 1994, pp. 127–134.

314. Th. M. Rassias and S. Xiang, *On approximate isometries in Banach spaces*, Nonlinear Funct. Anal. Appl. 6 (2001), no. 2, 291–300.

315. Th. M. Rassias and S. Xiang, *On the stability of approximate isometries*, Tamsui Oxf. J. Math. Sci. 18 (2002), no. 1, 45–56.

316. J. Rätz, *On approximately additive mappings*, In 'General Inequalities 2' (edited by E. F. Beckenbach), Birkhäuser, Basel, 1980, pp. 233–251.

317. J. Rätz, *On Wigner's theorem: Remarks, complements, comments, and corollaries*, Aequationes Math. 52 (1996), 1–9.

318. A. Redouani, E. Elhoucien and Th. M. Rassias, *The superstability of d'Alembert's functional equation on step 2 nilpotent groups*, Aequationes Math. 74 (2007), no. 3, 226–241.

319. W. Rudin, Functional Analysis, McGraw-Hill, New York, 1991.

320. P. K. Sahoo, *Hyers-Ulam stability of an Abel functional equation*, In 'Stability of Functional Equations of Ulam-Hyers-Rassias Type' (edited by S. Czerwik), Hadronic Press, Palm Harbor, Florida, 2003, pp. 143–149.

321. P. K. Sahoo and Pl. Kannappan, Introduction to Functional Equations, CRC Press, Boca Raton, Florida, 2011.

322. P. K. Sahoo and T. Riedel, Mean Value Theorems and Functional Equations, World Scientific, Singapore, 1998.

323. P. K. Sahoo and L. Székelyhidi, *A functional equation on $\mathbb{Z}_n \oplus \mathbb{Z}_n$*, Acta Math. Hungar. 94 (2002), 93–98.

324. P. K. Sahoo and L. Székelyhidi, *On the general solution of a functional equation on $\mathbb{Z} \oplus \mathbb{Z}$*, Arch. Math. (Basel) 81 (2003), no. 2, 233–239.

325. J. Schwaiger, *Remark 10*, In 'Report on the 30th ISFE,' Aequationes Math. 46 (1993), 289–289.

326. P. Šemrl, *The stability of approximately additive functions*, In 'Stability of Mappings of Hyers-Ulam Type' (edited by Th. M. Rassias and J. Tabor), Hadronic Press, Palm Harbor, Florida, 1994, pp. 135–140.

327. P. Šemrl, *The functional equation of multiplicative derivation is superstable on standard operator algebras*, Integr. Equat. Oper. Theory 18 (1994), 118–122.

328. P. Šemrl, *Hyers-Ulam stability of isometries on Banach spaces*, Aequationes Math. 58 (1999), 157–162.

329. C. S. Sharma and D. F. Almeida, *The first mathematical proof of Wigner's theorem*, J. Nat. Geom. 2 (1992), 113–124.

330. F. Skof, *Sull'approssimazione delle applicazioni localmente δ-additive*, Atti Accad. Sci. Torino 117 (1983), 377–389.

331. F. Skof, *Proprietá locali e approssimazione di operatori*, Rend. Sem. Mat. Fis. Milano 53 (1983), 113–129.

332. F. Skof, *Approssimazione di funzioni δ-quadratiche su dominio ristretto*, Atti Accad. Sci. Torino Cl. Sci. Fis. Mat. Natur. 118 (1984), 58–70.

333. F. Skof, *On two conditional forms of the equation $\|f(x + y)\| = \|f(x) + f(y)\|$*, Aequationes Math. 45 (1993), 167–178.

334. F. Skof, *On the stability on functional equations on a restricted domain and a related topic*, In 'Stability of Mappings of Hyers-Ulam Type' (edited by Th. M. Rassias and J. Tabor), Hadronic Press, Palm Harbor, Florida, 1994, pp. 141–151.

335. F. Skof, *On asymptotically isometric operators in normed spaces*, Istit. Lombardo Accad. Sci. Lett. Rend. A 131 (1997), 117–129.

336. F. Skof and S. Terracini, *Sulla stabilità dell'equazione funzionale quadratica su un dominio ristretto*, Atti Accad. Sci. Torino Cl. Sci. Fis. Mat. Natur. 121 (1987), 153–167.

337. V. Y. Stetsenko, *New two-sided estimates for the spectral radius of a linear positive operator*, Dokl. Acad. Nauk Tadzhik SSSR 33 (1991), 807–811.

338. R. Swain, *Approximate isometries in bounded spaces*, Proc. Amer. Math. Soc. 2 (1951), 727–729.

339. L. Székelyhidi, *The stability of linear functional equations*, C. R. Math. Rep. Acad. Sci. Canada 3 (1981), 63–67.

340. L. Székelyhidi, *On a stability theorem*, C. R. Math. Rep. Acad. Sci. Canada 3 (1981), 253–255.

341. L. Székelyhidi, *On a theorem of Baker, Lawrence and Zorzitto*, Proc. Amer. Math. Soc. 84 (1982), 95–96.

342. L. Székelyhidi, *Remark 17*, In 'Report on the 22nd ISFE,' Aequationes Math. 29 (1985), pp. 95–96.

343. L. Székelyhidi, *Note on Hyers's theorem*, C. R. Math. Rep. Acad. Sci. Canada 8 (1986), 127–129.

344. L. Székelyhidi, *Fréchet equation and Hyers's theorem on noncommutative semigroups*, Ann. Polon. Math. 48 (1988), 183–189.

345. L. Székelyhidi, *The stability of the sine and cosine functional equations*, Proc. Amer. Math. Soc. 110 (1990), 109–115.

346. J. Tabor, *On approximate by linear mappings*, Manuscript.

347. J. Tabor, *On approximately linear mappings*, In 'Stability of Mappings of Hyers-Ulam Type' (edited by Th. M. Rassias and J. Tabor), Hadronic Press, Palm Harbor, Florida, 1994, pp. 157–163.

348. J. Tabor, *Hosszú's functional equation on the unit interval is not stable*, Publ. Math. Debrecen 49 (1996), 335–340.

349. J. Tabor, *Remark 20*, In 'Report on the 34th ISFE,' Aequationes Math. 53 (1997), pp. 194–196.

350. J. Tabor, *Superstability of the Cauchy, Jensen and isometry equations*, Results Math. 35 (1999), 355–379.

351. J. Tabor and J. Tabor, *Homogeneity is superstable*, Publ. Math. Debrecen 45 (1994), 123–130.

352. L. Tan and S. Xiang, *On the Aleksandrov-Rassias problem and the Hyers-Ulam-Rassias stability problem*, Banach J. Math. Anal. 1 (2007), no. 1, 11–22.

353. T. Trif, *Hyers-Ulam-Rassias stability of a linear functional equation with constant coefficients*, Nonlinear Funct. Anal. Appl. 5 (2006), 881–889.

354. S. M. Ulam, A Collection of Mathematical Problems, Interscience, New York, 1960.

355. J. Väisälä, *Isometric approximation property of unbounded sets*, Results Math. 43 (2003), 359–372.

356. A. Vogt, *Maps which preserve equality of distance*, Stud. Math. 45 (1973), 43–48.

357. P. Volkmann, *Zur Stabilität der Cauchyschen und der Hosszúschen Funktionalgleichung*, Seminar 5 (1998), 1–5.

358. E. P. Wigner, Gruppentheorie und ihre Anwendungen auf die Quantenmechanik der Atomspektren, Fried. Vieweg, 1931.

359. D. Yang, *Remarks on the stability of Drygas' equation and Pexider-quadratic equation*, Aequationes Math. 68 (2004), 108–116.

360. D. Yang, *The stability of the quadratic functional equation on amenable groups*, J. Math. Anal. Appl. 291 (2004), 666–672.

361. D. Yang, *The quadratic functional equation on groups*, Publ. Math. Debrecen 66 (2005), 327–348.

362. M. Youssef, E. Elhoucien and Th. M. Rassias, *On the Hyers-Ulam stability of the quadratic and Jensen functional equations on a restricted domain*, Nonlinear Funct. Anal. Appl., to appear.

363. M. Youssef, E. Elhoucien and Th. M. Rassias, *Hyers-Ulam stability of the Jensen functional equation in quasi-Banach spaces*, Nonlinear Funct. Anal. Appl., to appear.

364. E. Zeidler, Nonlinear Functional Analysis and Its Applications I (Fixed Point Theorems), Springer, New York, 1985.

Index